VIII International Conference on Solar-Terrestrial Relations and Physics of Earthquake Precursors 2017

E3S Web of Conferences Volume 20 (2017)

Paratunka, Russia
25 - 29 September 2017

Editors:

Boris Shevtsov
Irina Myagkova
Vladimir Kozlov

ISBN: 978-1-5108-5056-9

Printed from e-media with permission by:

Curran Associates, Inc.
57 Morehouse Lane
Red Hook, NY 12571

Some format issues inherent in the e-media version may also appear in this print version.

Printed by Curran Associates, Inc. (2017)

For additional information, please contact EDP Sciences – Web of Conferences at the address below.

EDP Sciences – Web of Conferences
17, Avenue du Hoggar
Parc d'Activité de Courtabœuf
BP 112
F-91944 Les Ulis Cedex A
France

Phone: +33 (0) 1 69 18 75 75
Fax: +33 (0) 1 69 28 84 91

contact-edps@webofconferences.org

Additional copies of this publication are available from:

Curran Associates, Inc.
57 Morehouse Lane
Red Hook, NY 12571 USA
Phone: 845-758-0400
Fax: 845-758-2633
Email: curran@proceedings.com
Web: www.proceedings.com

TABLE OF CONTENTS

edp sciences

E3S Web of Conferences

Statement of Peer review

In submitting conference proceedings to E3S Web of Conferences, I certify to the Publisher that I adhere to the **Policy on Publishing Integrity** of the journal in order to safeguard good scientific practice in publishing.

1. All articles have been subjected to peer review administered by the proceedings editors.
2. Reviews have been conducted by expert referees, who have been requested to provide unbiased and constructive comments aimed, whenever possible, at improving the work.
3. Proceedings editors have taken all reasonable steps to ensure the quality of the materials they publish and their decision to accept or reject a paper for publication has been based only on the merits of the work and the relevance to the journal.

Title, date and place of the conference

VIII International Conference "Solar-Terrestrial Relations and Physics of Earthquake Precursors"
September 25-29, 2017
Paratunka, Kamchatka region, Russia

Proceedings editor(s):

Prof. Boris Shevtsov, Dr. Irina Myagkova, Dr. Vladimir Kozlov

Date and editor's signature

29.09.2017

17, avenue du Hoggar - PA de Courtabœuf – BP 112 - 91944 Les Ulis Cedex A (France)
Tél. : 33 (0)1 69 18 75 75 - Fax : 33(0)1 69 07 45 17 – www.edpsciences.org

Influence of heating experiments on parameters of Schumann resonances

Irina Agranat[1,*], *Vladimir* Sivokon[1]

[1] Institute of Cosmophysical Research and Radio Wave Propagation FEB RAS, 684034, Kamchatka region, Elizovskiy district, Paratunka, Mirnaya str., 7, Russia

Abstract. Recently the significant increase in intensity of researches in the field of the fissile impact on geophysical processes in various environments is noted. Special attention is paid to a research of impact on an ionosphere of a potent short-wave radio emission of heating stands. Today experiments on change of an ionosphere are made generally at stands HAARP, EISCAT in Tromse (Norway). Within the Russian campaign (Tomsk) EISCAT/heating (AARI_HFOX) made from October 19 to October 30, 2016 experiments on impact on an ionosphere F-layer by the radiation potent HF. For assessment of impact of these experiments on geophysical processes mathematical methods carried out the analysis of change of the parameters of the Schumann resonances received on the basis of data from the station of constant observation of the Schumann resonances in the city of Tomsk, the Tomsk State University (Russia).

1 Introduction

Heating stands unlike traditional methods of diagnostics of an ionosphere give the chance to influence targeted natural ionospheric processes, and, perhaps, to operate these processes. Impact on an ionosphere by powerful Shortwave radiation leads to the fact that the ionosphere (F layer) is heated — in the field of a wave electrons of ionospheric plasma disperse, and disperse nonuniformly, so, the conduction changes and the periodic structure of ionospheric currents is created.

Besides the scientific results received in works with heating stands such as discovery of low-frequency radiation of ionospheric current systems (Getmantsev's effect), research of plasma instabilities, optical luminescence of the indignant area, is available also a possibility of applied application of these installations.

HAARP on Alaska and "EISCAT" heating stands in Norway technically are the most potent. These stands have the high effective power of radiation, a wide frequency range of "heating", a possibility of radiation in various modes, including the continuous, pulse and amplitude-modulated with a modulation frequency from extremely low-frequency to very low-frequency range, a possibility of scanning of a beam.

Schuman's resonance is called the phenomenon of formation of standing electromagnetic waves of low and ultralow frequencies between the Earth's surface and an

* Corresponding author: agranat@ikir.ru

ionosphere.

The Earth and its ionosphere are a huge spherical resonator which cavity is filled with the low conductivity environment. If the electromagnetic wave which arose in this environment after rounding of the globe coincides again with characteristic phase (enters a resonance), then it can exist long time.

Average monthly experimentally observed values of the most legibly expressed resonant frequencies make about 8, 14, 20, 26, 32 Hz. At higher frequencies resonances become almost imperceptible. For the main – the lower, most intensive spectrum line variations of a resonant frequency within 7-11 Hz are possible, but mostly within a day the dispersion of resonant frequencies usually lies in limits ± (0.1-0.2) Hz.

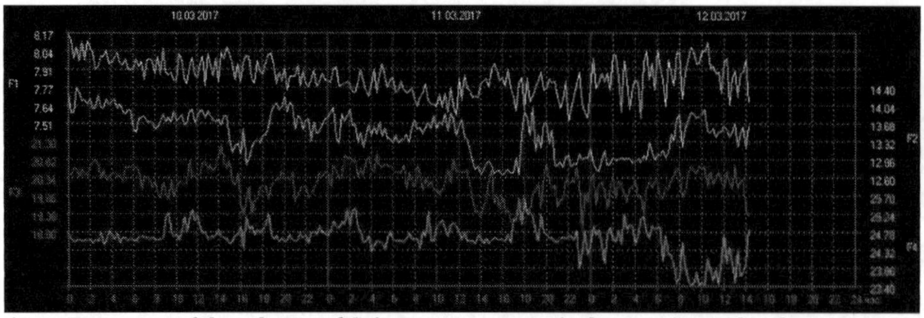

Fig. 1. Fragments of the relation of Schumann resonance`s frequencies on local time according to complex monitoring in Tomsk, local time is expressed in the Tomsk Summer Standard Time (TSST). TLDV =UTC+ 7 hours

The central frequencies of resonances are defined by wave guide parameters Earth - an ionosphere, namely height of the lower bound of D-layer and a profile of its conduction which, in turn, are defined by energy of the ionizing radiation of the Sun and the mesosphere. The size of cooperative energy of these radiations on the day party exceeds radiant energy on the night party, depending on position of the Sun on an ecliptic, by 200 - 700 times that involves the considerable difference of heights and profiles of conduction of D-layer.

Energy factor (q), as well as frequency, is the parameter of the frequency characteristic of the resonator Earth - ionosphere and also depends mainly on a profile of conduction of a D-layer.

Amplitude of a resonance is defined, first of all, by energy of a rating from the world storm centers which activity depends on their local time of day. On the other hand, energy in a point of reception is defined by its arrangement of rather these centers, namely: the azimuths influencing a ratio of amplitudes of horizontal components of the field; the remoteness defining attenuation owing to losses and owing to change of section of a wave guide; a phase in the point of reception defining change of amplitude according to a phase of a standing wave, and also parameters of local storm activity.

For these reasons amplitude of resonances cyclically changes with the daily period, but its value and the nature of change can differ for various points of observation at preservation of separate common regularities

Single-pass variations arise owing to single-pass impact on an ionosphere of the potent external factors changing parameters of its ionization. The splashes in solar space radiation which are result of solar flashes and splashes in galactic space radiation concern to those.

On the basis of the aforesaid it is possible to draw a conclusion that studying influence of experiments on heating of an ionosphere on parameters of Schumann resonances will help to draw a conclusion on the nature of impact of such experiments on a geophysical situation, and also on activity of people.

2 Method of investigations

In October, 2016 we were given an opportunity to analyse and study this influence or to draw a conclusion on its absence. A number of experiments on studying of the phenomena in a F-layer of an ionosphere caused by an ionosphere rating potent HF - radiation was conducted to the period from October 19 to October 30 within the Russian campaign EISCAT/heating (AARI_HFOX).

Knowing the period and time of impact on an ionosphere, we can estimate change of parameters of the Shumann resonances at the corresponding time terms to draw a conclusion about influence of experiments at heating stands on a geophysical situation.

As the frequency and good quality of the Schumann resonances, are parameters of the frequency characteristic of the resonator Earth — an ionosphere, it is logical to assume that these parameters are not subject to influence of experiments on heating while amplitude of a resonance is defined by energy of a rating from the world storm centers, therefore is more subject to influence from the external phenomena.

For an assessment of a geophysical situation during an experiment data of complex monitoring in Tomsk, (data source — http://sosrff.tsu.ru/) are taken.

The data submitted on the website http://sosrff.tsu.ru/ are digitized and kept in the form of a sequence of data from 14 till October 29, this period includes a row of control days and days of carrying out an experiment.

For the analysis of influence the following technique, for each of components was chosen:

1) mean daily value of each parameter D is calculated

2) the expectation for daily values of each parameter M is calculated

3) M/D relation – the relative size of the interesting parameter is calculated.

The step 3 allows to receive the normalized values of parameter.

4) by imposing of schedules of original values of parameter and received on a step 3 relations, the visual assessment of nature of change of input data in the dates corresponding to days of carrying out experiments of EISCAT becomes.

3 Results of investigation and discussion

During the investigation it was detected the increasing of the absolute and relative amplitude of the component a1, a2, and a3 during the period is present from October 20 to October 29, that speaks about relation of value of these components of Schuman resonances on heating of an ionosphere.

The assumption made theoretically that the frequency and the energy factor are probably not subject to influence from experiments on an ionosphere rating potent HF - radiation, is also confirmed, results for one of components of frequency and energy factor of the Schuman resonances are given in figures.

DOI: 10.1051/e3sconf/20172001001

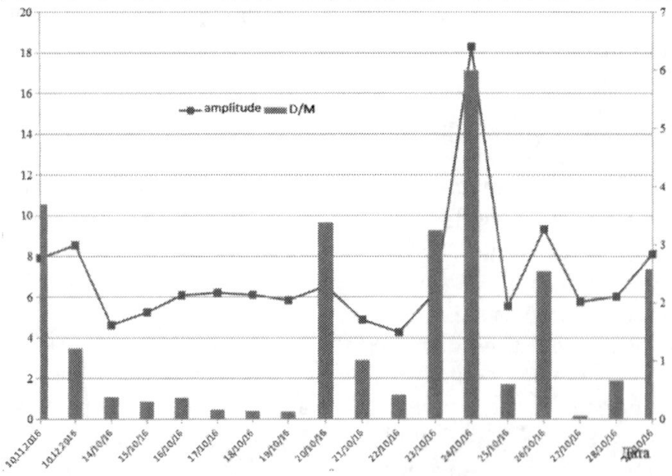

Fig.2. Result of the analysis for a component a1 of amplitude of the Schumann resonances.

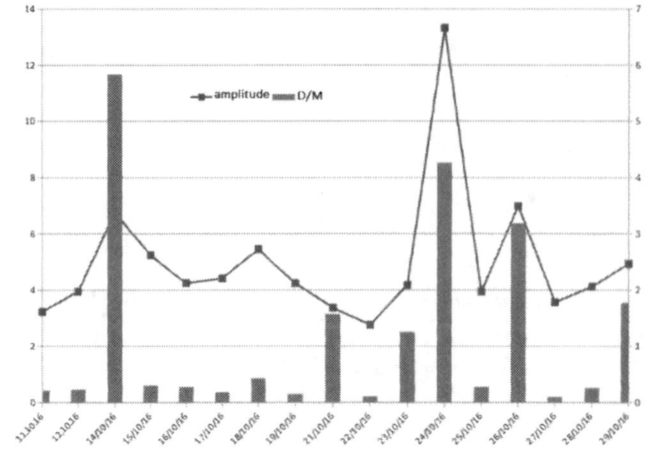

Fig. 3. Result of the analysis for a component a2 of amplitude of the Schumann resonances.

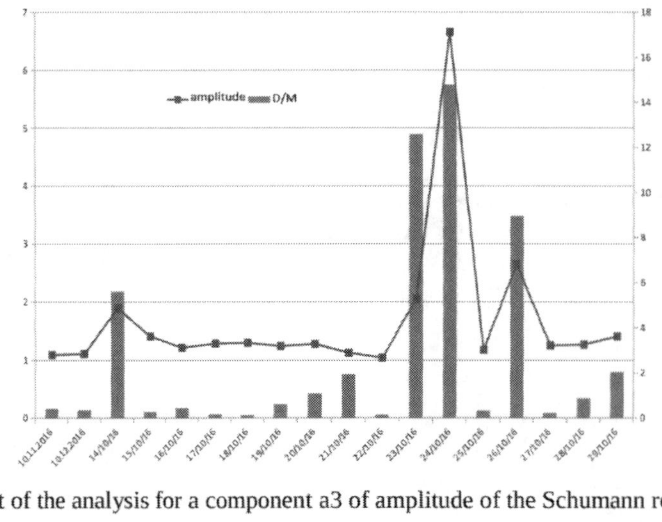

Fig. 4. Result of the analysis for a component a3 of amplitude of the Schumann resonances.

 DOI: 10.1051/e3sconf/20172001001

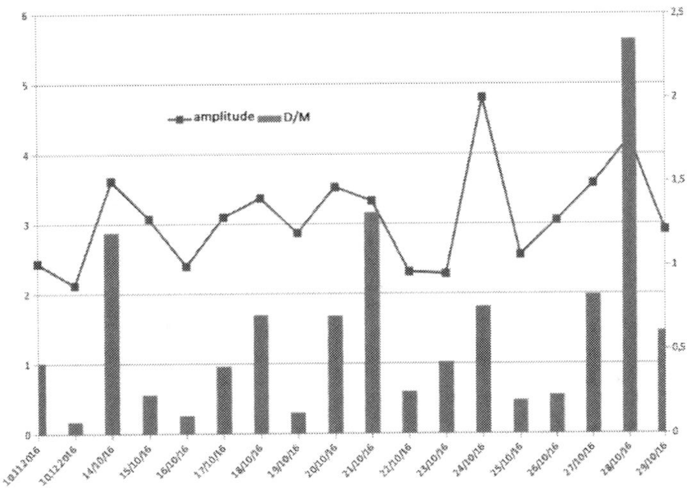

Fig. 5. Result of the analysis for a component a4 of amplitude of the Schumann resonances.

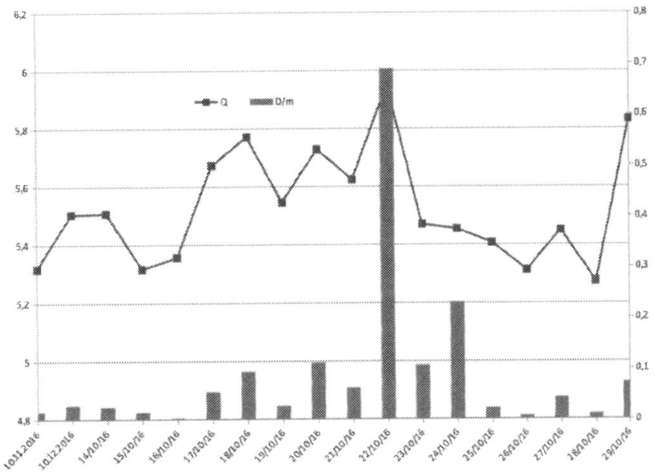

Fig. 6. Result of the analysis for a component q1 of energy factor of the Schumann resonances.

DOI: 10.1051/e3sconf/20172001001

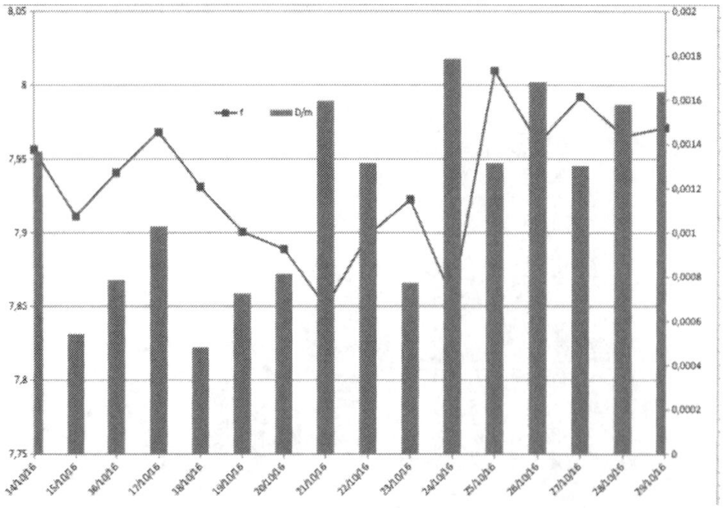

Fig. 7. Result of the analysis for a component f1 of frequency of the Schumann resonances.

4 Conclusions

In a number of the countries researches for the benefit of assessment of a possibility of simulated influence on natural geophysical processes are actively conducted now. Works are in the considered area at a stage of fundamental and search and conceptual researches.

The existing methods of the fissile impact on a geophysical situation, except for nuclear detonations, have much a smaller power engineering, than geophysical indignations of natural origin. For example, indignations of a magnetic field of Earth during magnetic storms in tens of times exceed what managed to be created by means of heating stands now. That is why the main idea by which foreign experts when developing new technologies of creation of geophysical weapon are guided is use of the resonance and trigger phenomena. It means activization of natural natural processes with a release of their big internal energy by means of much less potent impact on a geophysical situation.

Recently the significant increase in intensity of researches in the field of the fissile impact on geophysical processes in various environments is noted. The majority of these researches has the closed character. However at the modern level of development of technologies creation of the geophysical weapon capable to make global impact on the population and infrastructure of the states is represented improbable and also to estimate technical capability of development of perspective tools (weapon), the bound to simulated influence on an ionosphere (a magnetic field of Earth).

At the same time it is impossible to exclude that by analogy with the forecast of space weather during the conducted geophysical researches efficient methods of prediction of other anomalies and cataclysms can be developed. It is apparent that preliminary information on the forthcoming adverse geophysical situation can be used in the military purposes that speaks about relevance of a problem of geophysical ensuring activity of armed forces.

References

1. Gurevich A.V. Uspehi fizicheskih nauk, **177**, 11, 1145-1177 p. (2007)
2. Sivokon V.P., Kalugin I.A., Kobylkin V.S., Popov A.V. Vestnik KamchatGTU, **37,** 14-18 p.
3. Grach S.M., Komrakov G.P., Schvartz M.M., Yurishev M.A. Izvestiya vyschih uchebnyh zavedeniy. Radiofizika, **41,** 8, 966-977 p
4. Blagoveshchenskaya N. F., Borisova T. D., Kornienko V. A., Robinzon T.R., Yoman T.K., Frolov V.L., Ritveld M.T. Solnechno-zemnaya fizika, **2,** 12, 206–209 p. (2008)
5. K.Devis. *Radiovolny v ionosphere*, 504 p. (1973)
6. Sivokon V.P., Druzhin G.I. Geomagnetism i aeronomiya, **46,** 4, 521-524 p. (2006)
7. Blagoveshchenskaya N. F., Borisova T. D., Kornienko V. A., Moskvin I. V., Rietveld M. T., Frolov V. L., Uryadov V. P., Kagan L. M., Yampolski Yu. M., Galushko V. L., Koloskov A. V., Kasheev S. B., Zalizovski A. V., Vertogradov G. G., Vertogradov V. G., Kelley M. C. 24, 2333–2345 p. (2006)
8. Borisova T. D., Blagoveshchenskaya N. F., Moskvin I. V., Rietveld M. T., Kosch M. J., Thid´e B. Geophys., 20, 1479–1486 p. (2002)

Role of excited atoms in Lidar measurements of the middle and upper atmosphere

Vasily Bychkov[1,⋆], *Andrey* Perezhogin[1,⋆⋆] and *Il'ya* Seredkin[1]

[1] *Institute of Cosmophysical Research and Radio Wave Propagation FEB RAS (IKIR FEB RAS), 684034, Mirnaya str., 7, Paratunka, Elizovskiy district, Kamchatka, Russia*

Abstract. The paper presents the results of a lidar study of the atmosphere in the altitude region of 25-400 km, obtained at the lidar station of Kamchatka in 2008-2016. The development of methods for lidar measurements and signal processing is discussed. The method that takes into account the aftereffects of photoelectric multipliers and the technique to measure the background signal is described. These tools allow the authors to include the mesosphere and the upper atmosphere into the altitude region investigated by the lidar method. The results of the investigation of the backscattering lidar signal from the altitude region of 100-400 km caused by the resonance scattering on exited components of upper atmosphere are discussed.

Keywords: atmosphere, lidar, stratosphere, mesosphere, scattering, sounding

1 Introduction

Increased backscattering at the wavelength of 532 nm in the region of the upper atmosphere and its correlation with the critical frequency of ionospheric F2 layer on some days was detected during the investigation of the aftereffect of photomultiplier (PMT) Hamamatsu H8259-01 applied in the lidar in Kamchatka. Resonance behavior of a backscattering signal was discovered in the region of 100-400 km. The backscattering signal in the region of 150-300 km was investigated in the papers [1–3]. According to the data of ionosphere and lidar observations at the wavelength of 532 nm it was shown that during soft (0.1-10 keV) electron precipitations into the atmosphere, the total lidar signal from the region of heights of 100-400 km may correlate with plasma content in the region of maximum of nighttime ionosphere F2 layer. Analysis of the geophysical state, which accompanied the phenomenon, allowed to make the conclusion that the possible physical mechanism explaining these correlations is the resonance scattering at exited ions of nitrogen atom forming during soft (0.1-10 keV) electron precipitations into the ionosphere.

The content of nitrogen ions at ionospheric heights is units of percentage by day and negligibly small by night. The papers [4, 5] present the results of registration of the resonance scattering at exited oxygen ions, which is the main charged component in the ionospheric F2 layer. The wavelength of the sounding emission of 561.107 nm was chosen for the study of resonant scattering by atomic oxygen ions, as the most appropriate for the equipment.

⋆e-mail: vasily.v.bychkov@gmail.com
⋆⋆e-mail: d72156@gmail.com

2 Equipment

In the lidar observations of 2015-2016 a dual-frequency lidar with a laser Brilliant-B was used to generate 532.08 nm radiation and a dye laser TDL-90 with an pumping laser YG982E to generate radiation with the wavelength of more than 532.08 nm. For example, we describe the scheme for the wavelength of 561.107 nm. Dye Rhodamine 590 was used. The both lasers had the frequency of 10 Hz. Pulse energy at the wavelength of 532.08 nm was $\sim$ 400 mJ and about 150 mJ at the wavelength of 561.107 nm. The receiving telescope has a parabolic mirror with a diameter of 60 sm. The angle of view of the receiver was chosen equal 0.2 mrad.

The scheme of possible dipole transitions falling within the radiation band of TDL-90 and Brilliant - B is shown in Table 1, where OII is O^{+} ion, NII is N^{+} ion, NIII is N^{++} ion. It is taken into account that based on the ratings, the radiation spectrum widths at the half maximum are 0.04 nm for the laser Brilliant-B and 0.005 nm for laser TDL.

Table 1. Dipole transitions of excited ions of oxygen and nitrogen atoms falling within the radiation bands of lasers TDL-90 with Rh-590 dye and Brilliant-B [6].

	Compo-nent	Wavelength Air (nm)	A_{ki} (s^{-1})	Lower Level	Term	J	Upper Level	Term	J
1	OII	561.1072	2,14e+06	$2s^2 2p^2(^1S)3s$	^2S	$^1/_2$	$2s^2 2p^2(^3P)4p$	$^2P^\circ$	$^1/_2$
2	NIII	532.0870	5.68e+07	$2s2p(^3P^\circ)3p$	^2D	5/2	$2s2p(^3P^\circ)3d$	$^2F^\circ$	7 /2
3	NII	532.0958	2.52e+07	$2s2p^2(^4P)3p$	$^5P^\circ$	1	$2s2p^2(^4P)3d$	^5P	2

The signal received by the telescope is sent to a light guide as a parallel beam through the lenses of a receiving unit. A spectrum dividing plate is set in the light guide at the angle of 45o degrees to the direction of beam propagation. Radiation which has more than 532 nm wavelength is omitted by the plate. Radiation with the wavelength of 532 nm and less is reflected at the angle of 90° into another light guide. Then, the separated radiation fluxes are sent to two PMT Hamamatsu H8259-01, connected to photon counters Hamamatsu M8784. The duration of the counter gate is 10 μs which corresponds to the layer of 1.5 km taking into the account the beam two-way propagation.

Control of ionosphere state was carried out according to the data of "Parus" ionosonde which replaced the Automatic Ionospheric Station (AIS) in August 2015.

3 Measurement method and signal processing

In the investigation of aerosol distribution in the middle atmosphere, the scattering ratio was taken as the main parameter. It was determined as $R = (\beta_a+\beta_m)/\beta_m = 1+\beta_a/\beta_m$, here βa, βm are the coefficients of back aerosol and molecular scattering. The dynamic range of back scattering signals is very large and cannot be provided by a one PMT. The near zone (from 0 to 20 km) signals were excluded by electronically PMT locking. Nevertheless, R(H) profile recovered only to the heights of 50-60 km. Then the scattering ratio rapidly increases caused by aftereffect pulses. The results of investigation of photomultiplier tube (PMT) aftereffect were published in the paper [7]. It was shown that in the region of 60-90 km, the aftereffect pulses may be considered by detection of an exponent in a signal from the region of $\sim$100-150 km by least square method and its subtraction from the basic signal. Application of this method allows us to recover the profile of the relation of scattering $R = (\beta_a+\beta_m)/\beta_m$ from the heights of $\sim$50-60 km to 75-90 km. Thus, the whole region of the middle atmosphere (25-80 km) becomes available for investigation by lidar sounding method.

We should mention the technique to measure the background signal which was suggested within the course of investigation of PMT aftereffect. Analysis of a night summed signal showed that the signal at altitudes of 150-600 km behaves differently on different days. Usually on the graph, it is a straight line parallel to the height axis. However, on some days the signals decreases after 100 km and remains inclined to the height axes up to 500-600 km. Examples of such signals, obtained on March 28 and September 14, 2008, are shown in Fig. 1. The technical settings of the lidar were the same on the both days. The time of signal accumulation was about the same (~7.5 hours). The signal value differences at the height of 600 km are explained by full Moon during the observations on September 14 and its absence on March 28, 2008. The difference in signal behavior in the region of 100-500 km cannot be explained by the aftereffect. Moon illumination on September 14 created additional background. The signal total level increased and PMT aftereffects had to manifest brighter.

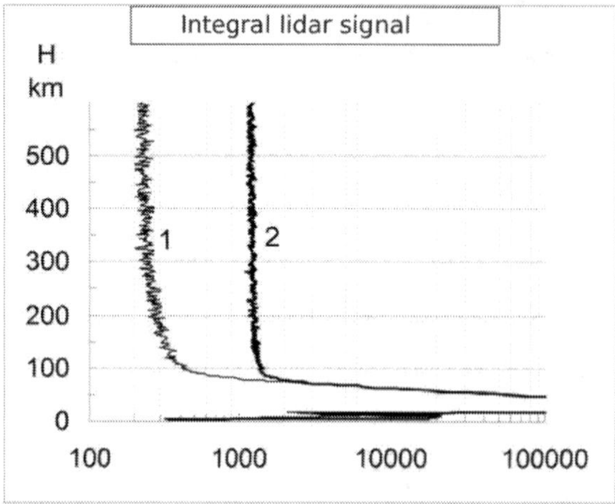

Figure 1. Total lidar signal obtained on March 28, 2008 (1) and September 14, 2008 (2).

It was concluded that background measurements within the interval of 2-4 ms (150-600 km) after a laser pulse sending may contain some additional signals. For this reason the background signal was measured as the mean value measured from the 20th to 24th ms with a step of 10 μs after each laser pulse sending. The background signal measured in such a way does not contain aftereffect pulses and is provided by good data collection.

In the course of investigation of PMT aftereffect during the sounding at the wavelength of 532 nm, an unusual burst of the total lidar signal from the region of 200-400 km was detected. Further studies have shown that the increase in the lidar signal from the heights of the upper atmosphere is explained by resonance scattering on excited atomic nitrogen atoms caused by super thermal electron precipitation (0.1-10 keV) into the ionosphere [1–3]. Fig. 2 illustrates the results of ionospheric and lidar sounding which were discussed in these papers. To show the importance of proper treatment of the background signal during the measurements in the upper atmosphere, the figures were somewhat modified.

All the curves of Fig. 2a, 2c were plotted according to the lidar data with 15-minute accumulation time (9000 laser pulses). They are a total signal from the layers of 100 km thick. The curves in Fig. 2a, 2c are plotted based on the lidar data applying background values measured at 20-24 ms after each laser pulse sending. When plotting the curves for Fig. 3a, 3c, we used the background value as an average over 100 strobes (150 km) in the region of 350-500 km.

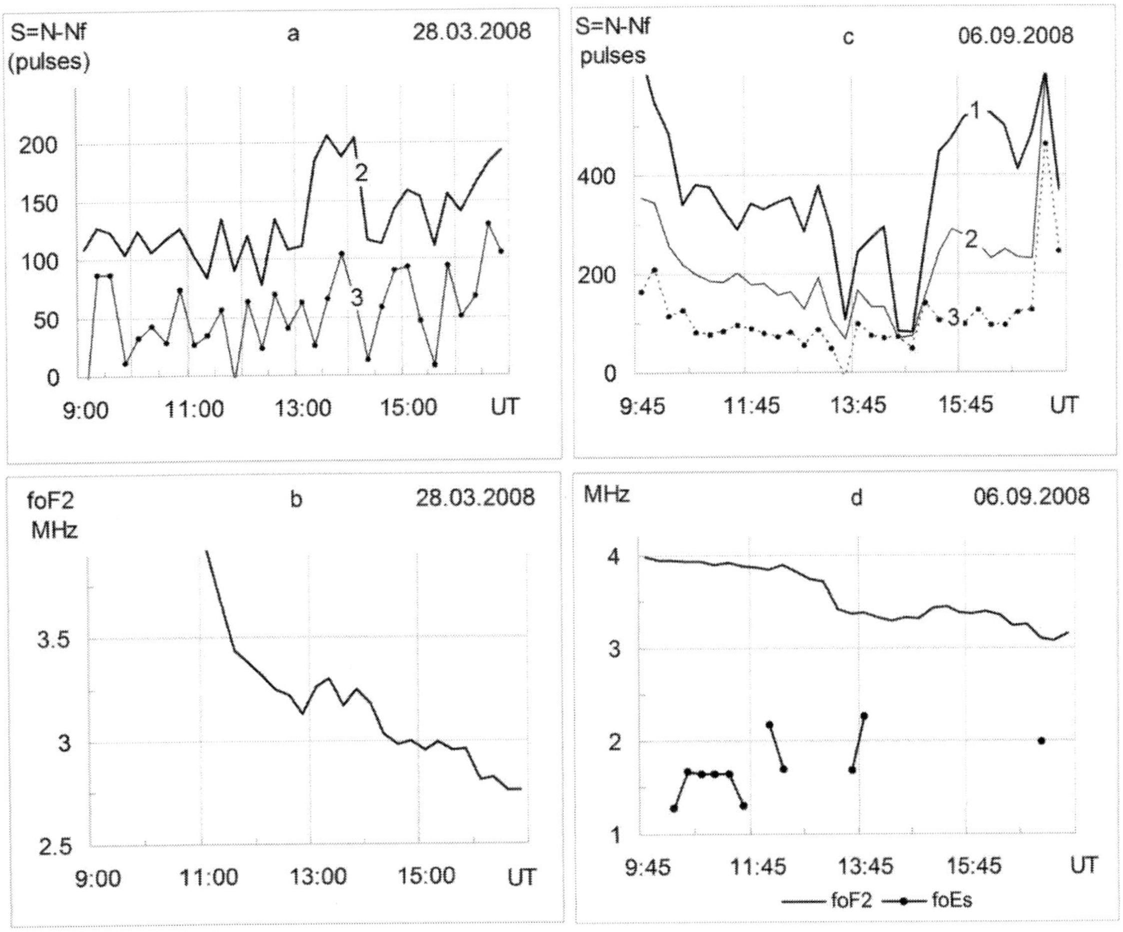

Figure 2. Total lidar "signal-background" (a, c) from the regions of 100-200 km (curve 1) and 200-300 km (curves 2, 3) and foF2 graph (b, d) during lidar observations on March 28 and September 6, 2008.

The curves 2a, 2c have synchronous changes with foF2 that is explained by the resonance scattering at excited ions of atomic nitrogen. The dependence of curve 3a and 3c forms on the choice of background averaging interval in the region of 300-600 km and signal decrease for such a choice of the background values allow us to make the conclusion that the background signal measured before the beginning of 5 ms after laser pulse sending cannot be used to investigate the upper atmosphere. In the middle atmosphere, estimating the background signal by its average value over the upper strobe can lead to unpredictably underestimated results. For example, to values of the scattering ratio R smaller than 1.

4 Some results in the upper atmosphere

The papers [4, 5] present Nh-profiles of atomic oxygen ion excited states obtained in the result of two-frequency sounding at the wavelength of 532.08 and 561.107 nm. The frequency of 561.107 nm corresponds to dipole transition between excited states of atomic oxygen long-lived ion. Fig. 3 illustrates the Nh-profiles of atomic nitrogen excited states obtained from the same initial data as for

Fig. 2. As long as the resonance scattering signal at atomic nitrogen ions is several times less than the scattering signal at oxygen ions, the profiles include only the time interval which corresponds to the maximum value of resonance scattering signal shown in Fig. 2. The time of signal accumulation is 75 min and is marked in the both figures.

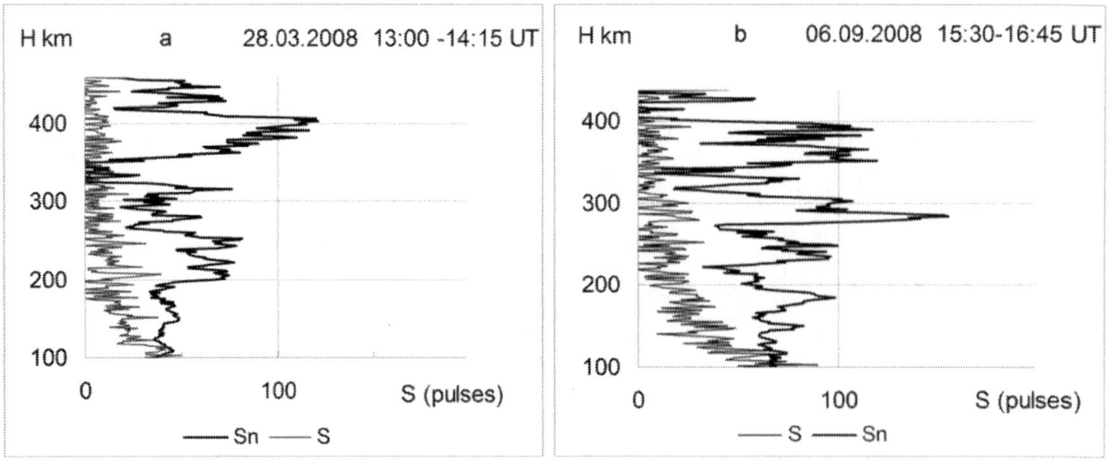

Figure 3. Total lidar signal S and normalized signal Sn on March 28 and September 6, 2008.

Signal S is the backscattering total signal accumulated over the 75 minutes except the measured background. Sn signal is the S signal normalized to a geometric factor ($H^2/100^2$) relatively its value at the height of 100 km. Sn data are smoothed by the moving average method with the window of 10.5 km. In fact, these are curves of the recorded content of nitrogen atom ion excited states. Figure 3a contains two separate areas of increased scattering, which may be the basis for estimating the characteristic energy of the spectra of precipitated electrons. A sharp decrease in the signal Fig. 3a at altitudes of 200 and ~ 350 km is characteristic of the rate of ionization caused by monoenergetic electron fluxes. Electron precipitations may occur within very short time periods. Decrease of accumulation time should result in the detection of such defined areas that requires recorded signal general level increase.

5 Conclusions for mesospheric observations

Resonance scattering on excited components of the upper atmosphere should also be manifested in the mesosphere during relativistic electron precipitation. The paper [Bychkov et al, 2012] illustrates the data on the correlation between the ratio of scattering $R = (\beta_a+\beta_m)/\beta_m$ in layers of ~5 km thick and ionospheric parameter f_{min}. Explanations of this correlation have not been proposed. The total coefficient of scattering was determined according to the results of lidar observations at the wavelength of 532 nm.

Analysis of ionospheric and lidar data on January 18 and 23, February 18 and 19, 2008 discovered correlations between the values of ionospheric parameter f_{min} and the total ratio of scattering R in layers of ~ 5 km thick in the mesospheric region. Correlation coefficient values were 0.68, 0.77, 0.5 and 0.5, respectively. When calculating the molecular scattering coefficient, we used temperature data measured on Aura satellite.

On all the mentioned days according to Demeter satellite flying westward and eastward to Kamchatka during the lidar observations, significant fluxes of relativistic electrons with the energies in

 DOI: 10.1051/e3sconf/20172001002

range 90-526 keV were recorded when the satellite crosses the latitude of the lidar station of Kamchatka (53N). Ionization rate in the mesosphere region caused by the precipitation of these electrons was calculated from the spectrum of electrons measured on Demeter satellite on January 18, 2008. The obtained ionization rate exceeds the values generally accepted in modeling of D layer of ionosphere by more than one order.

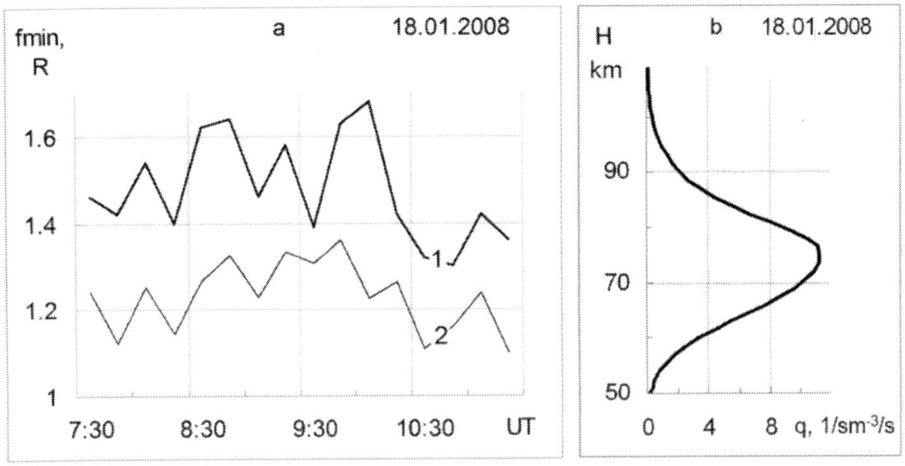

Figure 4. Values of f_{min} (4a, 1) and of lidar signal (4a, 2) from the region of 64.5-70.5 km on January 18, 2008 (a), and ionization rate caused by precipitated electrons measured on Demeter at 10:10 UT (b) [8].

As an example, Fig. 4 illustrates the data from the paper [8]. Fig. 4 shows f_{min} values and the average relation of scattering coefficients R in the region of 64.5-70.5 km. Correlation coefficient between the curves is 0.68. Fig. 4b represents the ionization rate caused by relativistic electrons calculated according to the data from Demeter satellite. The calculations were made according to the spectrum of precipitated electrons measured on the satellite Demeter by the method described in the paper [9].

In the same paper [8] the authors estimate the possibility of water vapor condensation in the mesosphere. Water vapor pressure over ice necessary for the condensation to begin was calculated by Magnus' formula. To estimate water content in the mesosphere, the measurement data on Aura satellite were used. It was obtained that in winter time during average temperature conditions in the mesosphere, the water content is two orders less than that necessary for condensation and for aerosol formation. Aerosol formations were observed in the mesosphere on January 18, 2008 and on other days in the region of heights of 60-80 km.

The results of the papers [1–3] allow us to suggest the mechanism of formation of additional lidar signals in the mesosphere. Changes of f_{min} are associated with the change of free electron content in the upper layers of ionospheric D region. Increase of f_{min} means the increase of absorption of incident radio wave low frequencies. In the absence of collisions of free electrons with neutral particles, an electromagnetic wave oscillates a free electron. A moving electron radiates at the same frequency and absorption does not occur. A radio wave loses its energy during collisions of a vibrating electron with a neutral particle. Absorption coefficient per a path unit $K \sim N/\nu$, where N is free electron content, ν is the frequency of collisions of electrons with neutrals. In the Earth atmosphere, absorption of radio waves with the frequency of units of MHz mainly occurs in the region of about 70 km where absorption coefficient is maximal. The content of free electrons decreases at the height of less than 70 km. At higher altitudes the frequency of collisions of electrons with neutrals drops with content of neutrals decreasing. Thus, increase of f_{min} indicates the growth of electron content which means ionization rate increase in nighttime conditions.

The main component of the atmosphere at these heights is molecular nitrogen. Ionization by energetic electrons is accompanied by photochemical reactions, for example, by the processes of dissociation of molecular nitrogen, nitrogen oxide and many other reactions causing the formation of excited atoms and ions of atomic nitrogen. The conditions for participation of these ions in the process of resonance scattering at the heights of 70 km do not differ from the conditions of the upper atmosphere though the frequency of ion-neutral collisions at the heights of ~ 70 km increases. The frequency of electron-neutral collisions may be calculated as:

$$\nu[N] = 0.81 \cdot 10^{-10}(T/M)^{1/2} \cdot N(c^{-1})$$

where T is temperature, M – molecular weight, N – particle concentration in the atmosphere [10] . Substituting N equal to 10^{15} sm^{-3} for the height of 70 km, we obtain $\nu \sim 5 \cdot 10^5 s^{-1}$, i.e. 500 kHz. Taking into account that the frequency of ion-neutral collisions is at least two orders less than that of electrons, it becomes clear that for excited ions with radiative lifetime of less than 10^{-6} s, the increase of collisions at the heights of 70 km does not play significant role.

Thus, the lidar signal increase from mesospheric heights during relativistic electron precipitations may be explained by the appearance of excited ions of atomic nitrogen and the conditions for resonance scattering at the wavelength of 532 nm. This suggestion explains the correlation of the lidar signal in the region of heights of 70 km with ionospheric parameter f_{min} observed in 2008.

6 Conclusions

Analysis of the obtained results allows us to make the following conclusion:

- it is shown that a correct account of the background signal during lidar studies of the upper and middle atmosphere can be of fundamental importance;

- when analyzing the results of mesospheric sounding, we should take into the account the frequencies corresponding to dipole transitions between excited states of atmospheric main components in laser radiation spectrum. Precipitations of relativistic electrons may cause additional lidar signal and formation of imaginary aerosol formations in the mesosphere;

- lidar investigations of the upper atmosphere may provide the data on the spectra of precipitated electrons with the energies of 0.1-10 keV;

- the main features of resonance scattering manifestation may be the inclination to the height axis of lidar total nighttime signal at the heights of more than 100-150 km and the correlation of total signals from different regions of the upper atmosphere between each other. Ionospheric data may serve as aditional source of information.

The paper was supported by FEB RAS Grant No. 12-I-OFN-16, RFBR Grant No. 16-05-00901a, RSCF No. 14-11-00194.

References

[1] V. Bychkov, Y.A. Nepomnyashchii, A. Perezhogin, B. Shevtsov, N. Polekh, Atmospheric and Oceanic Optics **27**, 297 (2014)

[2] V. Bychkov, Y.F. Nepomnyashchii, A. Perezhogin, B. Shevtsov, Atmospheric and Oceanic Optics **28**, 303 (2015)

[3] V.V. Bychkov, Y.A. Nepomnyashchiy, A.S. Perezhogin, B.M. Shevtsov, *Resonance scattering at excited atoms and ions of the upper atmosphere as a possible mechanism for ionosphere investigations*, in *XXI International Symposium Atmospheric and Ocean Optics. Atmospheric Physics* (International Society for Optics and Photonics, 2015), pp. 968073–968073

[4] V.V. Bychkov, A.S. Perezhogin, I.N. Seredkin, B.M. Shevtsov, *Lidar investigations of the scattering of the upper and middle atmosphere*, in *XXII International Symposium Atmospheric and Ocean Optics. Atmospheric Physics* (International Society for Optics and Photonics, 2016), pp. 100355R–100355R

[5] V.V. Bychkov, A.S. Perezhogin, I.N. Seredkin, B.M. Shevtsov, V.N. Marichev, Atmospheric and Oceanic Optics **30**, 209 (2017)

[6] *Nist atomic spectra database lines form*, `https://physics.nist.gov/PhysRefData/ASD/lines_form.html`

[7] V. Bychkov, A. Perezhogin, B. Shevtsov, V. Marichev, P. Novikov, A. Cheremisin, Optika Atmos. Okeana **24**, 107 (2011)

[8] V. Bychkov, A. Perezhogin, B. Shevtsov, V. Marichev, G. Matvienko, A. Belov, A. Cheremisin, Atmospheric and Oceanic Optics **25**, 228 (2012)

[9] M. Demin, M.: Fizmatlit **2**, 92 (2008)

[10] N. Shefov, A. Semenov, V.Y. Khomich, Geos, Moscow (2006)

Natural factor impact on atmospheric electric field variations in Kamchatka

Pavel Firstov[1,2,*], *Nina* Cherneva[1], and *Rinat* Akbashev[2]

[1] Institute of Cosmophysical Research and Radio Wave Propagation FEB RAS
[2] Kamchatka Branch of Geophysical Service RAS

Abstract. The paper briefly describes a site network which registers atmospheric electric field strength (AEF V'). The scheme of natural processes affecting the formation of the local atmospheric electric field is considered. AEF V' disturbances on ground flux meters are described. They were recorded when two eruptive clouds from Shiveluch volcano eruption were passing by. Key words: atmospheric electric field, potential gradient, electrostatic flux meter, volcanic clouds

1 Introduction

Electric field of the near ground atmospheric layer is an important object for the investigation in the context of development of numerical characteristics of the process referring to the Global Electric Circuit (GEC) and in the development of the approaches to determine and to describe the nature and mechanisms of electric effects at the crust-atmosphere boundary. The principles of GEC formation are described in detail in many papers from which a review [1] deserves special attention. This paper shows the latest results which give better understanding of the physical processes occurring in the GEC.

Of special interest is the investigation of Atmospheric Electric Field (AEF) variations in the conditions of ever-present disturbances not only of global but regional and local scales as well. It is accepted that the height of the «equalizer level», that is an isopotential surface along which the potential equalizes for quite a short time, is ~60 km. Consequently, variations in AEF V' potential gradient on the surface caused by potential distribution inhomogenuities on the «equalizer level» should also have a size close to the regional scale. Smaller AEF V' scales falling outside the regional limits refer to local scales.

In spite of many investigations of AEF behavior, the nature and the mechanisms of local electric fields require a detailed study in each concrete area that is due to the complexity of the problem and the multifactorial local processes forming the AEF. In Kamchatka peninsular, investigation of AEF is of special interest because of the specific natural conditions of this region at the background of industrially clean atmosphere of Kamchatka.

Intensity of geodynamic processes in Kuril-Kamchatka subduction zone (earthquakes and volcano eruptions), fast changeability of meteorological values caused by high cyclonic activity in the north-western part of the Pacific ocean determine the diversity of local processes affecting the AEF. Thus, the complex of meteorological, geophysical and space

*Corresponding author: firstov@emsd.ru

factors, simultaneously acting and significantly changing in time, complicate the AEF behavior in Kamchatka.

2 Instrumentation and observation method

Investigation of atmospheric electric field in Kamchatka has been carried out from the beginning of 90-ies of the latest century when an electrostatic flux meter «Pole-2» with analogue registration [2] was installed at «Paratunka» (PRT) observatory of IKIR FEB RAS (Fig. 2a) to measure AEF *V'*. Then, a system for AEF *V'* data storage and processing in the digital form was developed [3, 4]. Moreover, a lightning direction finder is in operation at «Paratunka» observatory [5]. It is included into to the global WWLLN network and the data from it allow us to determine the source coordinates of volcanic lightning discharges accompanying explosive eruptions in Kamchatka [6]. The information on the planet lightning activity including that on lightning discharges from a definite threshold accompanying eruptions is available on the site of WWLLN network http://webflash.ess.washington.edu/.

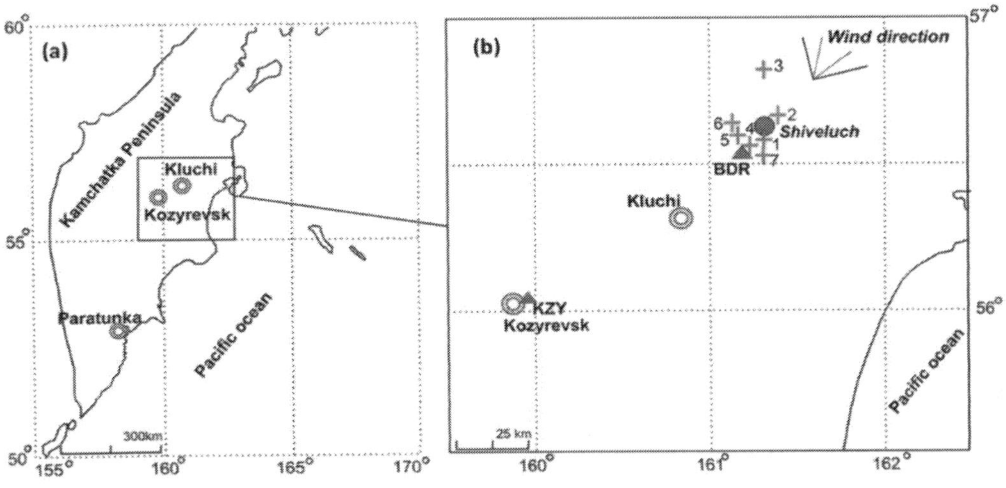

Fig. 1. Scheme of arrangement of AEF strength registration sites on Kamchatka peninsula (a), arrangement of observation sites in the region of the Northern Group of volcanoes and location of discharges from volcanic lightning during Shiveluch volcano eruption on November 16, 2014 (b).

AEF V' registration in the central part of Kamchatka peninsula is of special interest as long as it is possible to compare the behaviors of AEF V' unitary and annual variations of the continental and coastal regions of Kamchatka. The Northern Group of volcanoes is also located here including four active volcanoes of Kamchatka (Kluchevskoy, Bezymyanny, Ploskiy Tolbachik, Shiveluch volcanoes) (Fib. 1b). Frequent explosive eruptions of these volcanoes increase the possibility of registration of AEF V' under volcanic plumes. Thus, an electrostatic flux meter EF-4 was installed at «Kozyrevsk» (KZY) site of Kamchatka Branch of Geophysical Service RAS (KB GS RAS) in 2013 (Fig. 2b). It is described in the paper [7]. Moreover, a meteorological station Vasiala wxt520 is in operation at KZY which allows us to measure weather parameters that is of great importance for AEF V' data interpretation. Some time later, EF-4 was installed at Kluchi (KLY) seismic site.

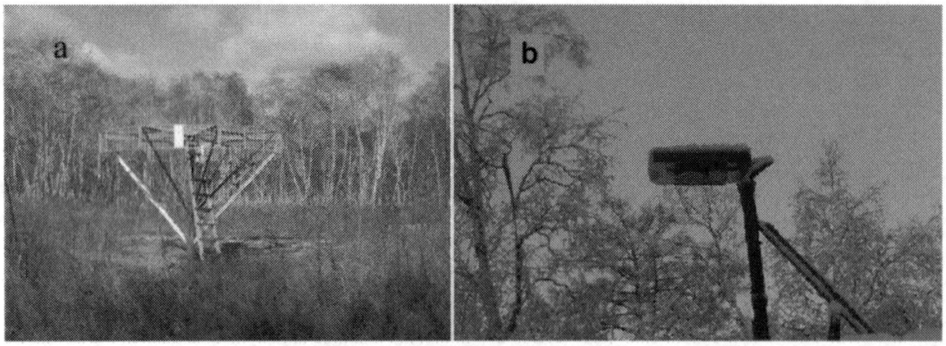

Fig. 2. General view of the equipotential surface on which «Pole-2» flux meter is mounted at PRT observatory (a), flux meter EF-4 installed in KZY with obturator downwards (b).

Monitoring of the volcanic activity of the Northern Group is carried out by Kamchatka Branch of the Federal Research Center «Geophysical Service of RAS» (KB FRC GS RAS) based on the investigation of some volcano seismicity. Radiotelemetric seismic station (RTSS) network is in operation in the region of the Northern Group of volcanoes. The closest to the volcano «Baydarnaya» (BDR) and «Semkorok» RTSS are located 10 and 15 km from the volcano crater (Fib. 2b) and register the time of an explosion beginning with the accuracy up to a second.

Meteorological parameters (air pressure, temperature and humidity) are observed at «Petropavlovsk» and «Kluchi» meteorological observatories of Kamchatka Division on Hydrometeorology and Environment Monitoring located in Petropavlovsk-Kamchatksiy and Kluchi. Height balloon atmospheric sounding is also carried out twice a day and the obtained data are published on the site http://www.esrl.noaa.gov/raobs/intl/intl2000.wmo.

3 Main natural factors affecting local electric field

Based on the analysis of long-term observations of AEF V' variations applying meteorological data, that of cosmic radiation intensity, geomagnetic field variations, radon exhalations and VLF electromagnetic radiation, the papers [8, 9] suggest a phenomenological model of natural and near space process impact on AEF state (Fig. 3).

One of the local scales is the height of radon mixing layer upper boundary which varies from hundreds of meters in winter on a windless night to 6-8 kilometers on a hot and windy summer day. We should also take into account that radon penetration into the atmosphere may be blocked by soil upper layer freezing and snow cover [10]. Some universal scale for this area can, evidently, be determined by geological-tectonic structure of the region. Conductivity of the near ground layer is primary determined by light ion concentration which depends not only on the ionization by radon but by many factors which form local AEFs sometimes masking the unitary variations determined by GEC.

The most significant factors affecting the local AEF are the following [8, 9]:

- radon outflow into the atmosphere,
- cosmic ray flux variations,
- light and heavy ion balance change at the times of dawn and dusk,
- ionopheric electric current potential effect of AEF potential.

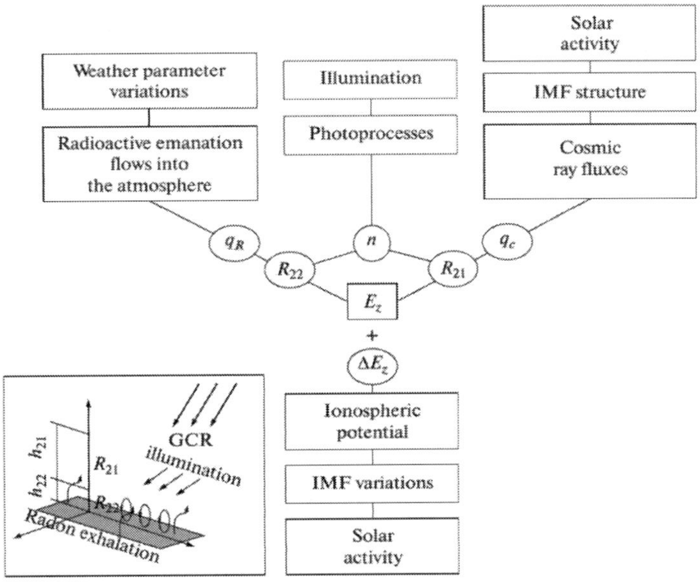

Fig. 3. Scheme of the processes of atmospheric electric field formation in the presence of factors affecting its value in the near ground layer. H_{21} and H_{22} ionization regions in the atmosphere by cosmic ray and radon exhalation with resistance R_{22} and R_{21} in these regions.

AEF strength is non-uniformly distributed along the height. For the total difference of the earth-ionosphere potential of about 300 kV, ~ 270 kV fall in the «lower» 20-kilometer layer and 30 kV fall within the rest 80 km. That means that the resistance of the «lower» layer significantly determines the current in the whole earth-ionosphere column.

We denote the resistance of the column upper part, depending only on external factors, by R_1, and the resistance of the column lower part by $R_2 = R_{21} + R_{22}$, where: R_{21} is the resistance of h_{21} layer that value is *determined by cosmic rays*; R_{22} is the resistance of the variable thickness layer h_{22} in which cosmic ray ionization is *enlarged by radon ionization* (Fig. 3). The thickness of h_{22} depends on the air turbulent mixing and potential drop in «radon» layer can be written as $U_{22}=U\cdot R_{22}/(R_1+R_{21}+R_{22})$.

The main factor determining the local AEF is the atmospheric near ground layer resistance which primary depends on the ionizing effect of cosmic rays and radon. Moreover, the diurnal variations of radon outflow from the soil are determined by atmospheric pressure variations and the seasonal ones depend on the soil upper layer permeability determined by soil temperature.

Meteorological factors play a key role in the formation of AEF annual variation. Higher values of field intensity in winter are associated with the fact that radon outflow in winter months drops due to the soil freezing and thick snow cover in Kamchatka [10]. This process may also be affected by significant decrease of sunlight (Paratunka 53°N) that causes the increase of heavy ion quantity due to photoattachment. Heavy ions have low mobility and the atmospheric near ground layer conductivity decreases in winter time that cause AEF E_Z increase.

Air conductivity is determined by light ion concentration, charge and mobility. «Average» light ion mobility in the air at the sea level equals $\mu = 1.5$ sm^2/(V·s). Heavy ions, dust particles and molecular clusters with attached molecular ions and electrons, have low mobility and do not almost participate in electric current rise. Thus, at dawn, E_Z slowly grows to 10% within three hours due to the increased turbulence [9].

4 Electrization of eruptive clouds forming during Shiveluch volcano eruptions

Natural factors, unique for the Russian Federation, which form electric structures of the local scale in Kamchatka peninsula, are explosive eruptions of volcanoes. Eruptive clouds forming during this process are electric structures which are somewhat close to meteorological clouds. Sheveluch volcano has actively erupted within the latest five years. It is the most northern one among the active volcanoes of Kamchatka (56°47' N, 157°56' E) with extrusive dome of 2500 m above the sea level (Fig. 1b). Within the latest decade, its eruptions, determined by slow magma squeezing out and extrusive dome formation, are periodically accompanied by strong ash explosions. At that, in the cases of wind southern and south-western directions, eruptive cloud trajectories pass the sites KZY and KZV.

It is generally know that volcanic clouds (plumes) bear an electric charge which is recorded by ground-based devices located near the cloud propagation traces. Eruptive cloud electrization takes place by two reasons, by magma breakage (damage) and gravitative differentiation of ash particles in eruptive column and plume [11, 12]. In this case, bipolar charges are formed and volcanic lightning of different intensity occur.

An explosive eruption with eruptive cloud height of 13000 m above the sea level occurred on Sheveluch volcano on November 16, 2014 at 10:17. At KZY (Kozyrevsk site) located 110 km to the south-west, a positive anomaly with the duration of 1.5 hour was registered ~ 2 after the event on a AEF V' record (Fig. 4a). Positive disturbance coincided in time with eruptive plume passage at the distance of 25 km from KZY at the heights of 8 and 12 km, that was fixed by the data obtained on the basis of wind stratification of the atmosphere and satellite images [6].

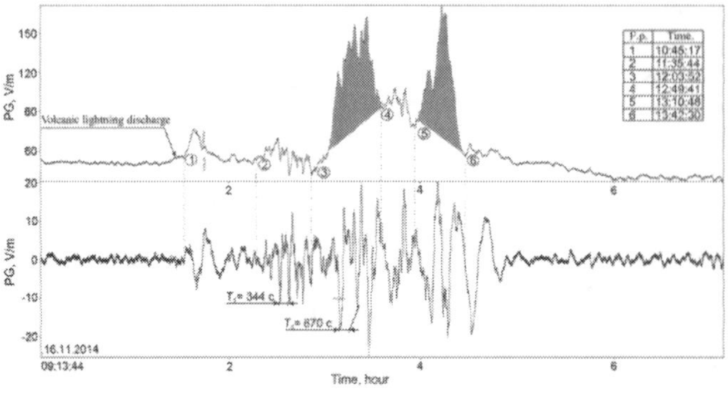

Fig. 4. A fragment of AFE strength record at KZY site with characteristic reference points (a); AEF V' curve after high frequency filtering with frp = 0.0008333 Hz (20 min) (b). Numbers in circles indicate the reference points in the signal record from a passing eruptive cloud.

Fair weather conditions (FWC) were observed before and after the eruption. The AEF V' background value was ~ 50 V/m (Fig. 4a). Seven discharges were registered after the eruption within 20 min. Discharge locations are shown in Fig. 4a. The first weak variations of AEF V' were observed from 10:45 and almost in two hours after the eruption, clear beginnings of two anomalies are distinguished (at 12:04 and at 13:10) with the total duration of 1.5 hour (Fig. 4a) when AEF V' value reached its maximum value of 170 B/m. FWC give the ground to consider the AEF V' variations as a consequence of a passing eruptive plume.

DOI: 10.1051/e3sconf/20172001003

We can note that «higher-frequency» oscillations with the amplitude of ~10 V/m are observed before the clearly defined AEF V' anomalies within 78 minutes. The anomalies are also complicated by «higher-frequency» oscillations. The signal with the duration of more than tree hours, in which five segments could be defined, was filtered by a high-frequency filter with frp = 0.0008333 Hz (20 min). The segments distinguished by kinematic and dynamic parameters are well defined on the filtered signal (Fig. 4b).

The beginning of the signal almost coincides with the time of the latest discharge from the volcanic lightning which occurred at the distance of 90 km from KZY (Fig. 1b). Oscillations with 12-15-minute periods are registered in AEF V' «high-frequency» range variations both on the first and the last segments. For the second and third segments, oscillations with shorter periods of 5-6 minutes are characteristic. Evidently, charged particles interact in the result of convective and turbulent processes in an eruptive cloud and regions with different AEF V' value are formed.

As it was shown in the paper [13], volcanic clouds have a tripole structure. The upper part of a cloud compound of vapor and aerosols has a positive charge. The middle part, represented by fine ash, has a negative charge. And the lower part, enriched in large-sized ash in the result of gravitation, also has a positive charge. At significant distances from the volcano, when eruptive clouds are mainly composed of aerosol and ash of micrometric dimensions, they have a positive charge that is observed in our case.

A great success for the observers of eruptive cloud charge evolution was the Siveluch volcano eruption on June 14, 2017 when a lucky combination of wind direction and FWC was observed that allowed us to record AEF V' variations during an eruptive cloud passage over KLY and KZV sites.

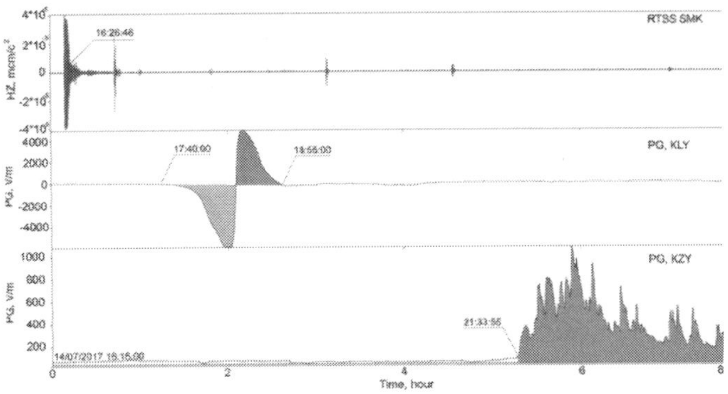

Fig. 5. Record fragment of ground displacement rate at RTSS SMK (a) and AEF strength at KLY and KZY (b, c) during the passage of the eruptive cloud from Shiveluch volcano eruption on June 14, 2017.

In contrast to the previous eruption, much greater quantity of ash was thrown into the atmosphere during this eruption that is indicated by AEF V' large values both at KLY and KZY. The significant amount of ash precipitated at KLY at the distance of 45 km from the volcano (~300 g/m2 according to N.A. Zharinov's oral report) agrees with the large scale of bipolar signal amplitude of $6 < V' < 5$ kV/m (Fig. 5b). Based on the signal form, we can assume that the eruptive cloud discharge is a horizontal dipole the axis of which is oriented along the propagation line [14]. The time of signal arrival, whereas its duration is an hour and fifteen minutes at KLY site, agrees with the wind stratification of the atmosphere obtained during atmosphere height sounding at Kluchi.

The one-polar signal from the eruptive cloud plume at KZY lasted for more than three hours with $V'_{MAX} = 900$ V/m (Fig. 5c).

5 Conclusions

Natural factors, unique for the Russian Federation, which form electric structures of the local scale in Kamchatka peninsula, are explosive eruptions of volcanoes. Eruptive clouds forming during this process are electric structures which are somewhat close to meteorological clouds. Sheveluch volcano has actively erupted within the latest five years. It is the most northern one among the active volcanoes of Kamchatka (56°47' N, 157°56' E) with extrusive dome of 2500 m above the sea level (Fig. 1b). Within the latest decade, its eruptions, determined by slow magma squeezing out and extrusive dome formation, are periodically accompanied by strong ash explosions. At that, in the cases of wind southern and south-western directions, eruptive cloud trajectories pass the sites KZY and KZV.

References

1. E.A. Mareev, Uspekhi Fizicheskikh Nauk, **180**(5), 527- 533 (2010)
2. O.P. Rulenko, A.V. Ivanov, A.V. Shumeyko, Doklady Akademii Nauk, **326**(6), 980-982 (1992)
3. A.V. Buzevich, S.E. Smirnov, V.I. Filimonov, P.P. Firstov, *Proc. of the International Conference «Solar-terrestrial bonds and electromagnetic precursors of earthquakes»* (Petropavlovsk-Kamchatskiy: IKIR DVO RAN), 36-38 (1998)
4. S.E. Smirnov, *Proc. of the school-seminar «Meteorological basis of magnetic observations of Siberia and Far East»* (Petropavlovsk-Kamchatskiy, IKIR FEB RAS), 7-10 (2003)
5. G.I. Druzhin, D.V. Tarasenko, V.M. Pukhov, A.V. Zlygostev, *Proc. of the II International Conference «Solar-terrestrial bonds and electromagnetic precursors of earthquakes»* (Petropavlovsk-Kamchatskiy: IKIR DVO RAN), 32-33 (2001)
6. P.P. Firstov, R.R. Akbashev, R.R Holzworth, N.V. Cherneva, B.M. Shevtsov, Izvestiya Rossiyskoy Akademii Nauk. Fizika Atmosfery i Okeana – Izvestiya. Atmospheric and Oceanic Optics, **53**(1) 29–37 (2017)
7. V.A. Efimov, D.M. Oreshkin, P.P. Firstov, R.R. Akbashev, Seysmicheskie pribory – Seismic instruments, **49**(4) 35-46 (2013)
8. E.A. Ponomarev, N.V. Cherneva, P.P. Firstov, *Proc. of the 7th Intern. Conf. "Problems of Geocosmos"* (St. Petersburg, Russia, 26-30 May 2008), 211-217 (2008)
9. E.A. Ponomarev, N.V. Cherneva, P.P. Firstov, Geomagnetizm i Aeronomiya – Geomagnetism and Aeronomy, **3**(51) 405-411 (2011)
10. P.P. Firstov, N.V. Cherneva, E.A. Ponomarev, A.V. Buzevich, Bulletin of KRAESC (Earth sciences, KGU), **1**(7) 102-109 (2006)
11. M. R. James, S.J. Lane, J.S. Gilbert, J. Geol. Soc. London, **155**, 587-590 (1998)
12. T.A. Mather, R.G. Harrison, Serv Geophys, **37**, 387-432, (2006)
13. T. Miura, T. Koyaguchi, Y. Tanaka, Bull Volcanol, **64**, 75–93 (2002)
14. N.V. Cherneva, E.A. Ponomarev, P.P. Firstov, A.V. Buzevich, Bulletin of KRAESC (Earth sciences, KGU), **2**(10), 60-64 (2007)

Characteristics of Schumann Resonance Parameters at Kuju Station

Akihiro Ikeda[1,*], *Teiji* Uozumi[2], *Akimasa* Yoshikawa[3], *Akiko* Fujimoto[2], *Shuji* Abe[2], *Hiromasa* Nozawa[1], and *Manabu* Shinohara[1]

[1]National Institute of Technology Kagoshima College, Kirishima, Japan
[2]International Center for Space Weather Science and Education, Kyushu University, Fukuoka, Japan
[3]Department of Earth and Planetary Sciences, Kyushu University, Fukuoka, Japan

Abstract The ground magnetic field variation in the extremely low frequency (ELF) range has been measured by an induction magnetometer at Kuju, Japan (KUJ; M.Lat. = 23.4 degrees, M. Lon. = 201.0 degrees) since 2003. The first mode of the Schumann resonance (SR) around 8 Hz can be seen at KUJ. The SR in H (horizontal northward component) shows maximum peaks around 08 UT and 15 UT. In the case of D (horizontal eastward component), the SR shows its maximum peak around 08 UT. These peaks are coincident with the enhancement of lightning activity in Africa and Asia. Thus, we found the influence of the lightning activity on the observed SR at KUJ.

1 Introduction

The Schumann resonance (SR), whichwas detected by Blaser and Wagner [1960, 1962] for the first time, is the global resonance of electromagnetic waves generated by the global lightning activity.The resonance is formed by the Earth-ionosphere cavity and the specific resonance frequencies, which are about 8, 14, 21, and 26 Hz, appear in the ground magnetic field variation.

The diurnal variations of SR parameters reflect the properties of both the global lightning activity and the state of the Earth-ionosphere cavity. The variation in the amplitude of SR in a day show peaks which correspond with the three major thunderstorm centers, namely, African, Asian, and American [e.g., Sátori et al., 1996;Zhou et al., 2013].

In addition, SR parameters connect with the earth's climate [e.g., Williams, 1992; Price, 2000]. Williams [1992] suggested that the variation of SR frequency indicates the variations in the tropical temperature. Thus, the variation of SR can be used to monitor the earth's climate.

In order to use the SR parameters for monitoring theearth's climate, we need a better understanding of the SR detected at our observatory Kuju, Japan (KUJ; M.Lat. = 23.4 degrees, M. Lon. = 201.0 degrees).We present the effects of the global lightning activity to the fundamental mode of the SR at KUJ.

*Corresponding author: a-ikeda@kagoshima-ct.ac.jp

2 Data Set

The ground magnetic field variation in the extremely low frequency (ELF) range has been measured by an induction magnetometer at KUJ since 2003. The observation is a part of activities by International Center for Space Weather Science and Education, Kyushu University. The components of ground magnetic field used in this study are horizontal northward component (H) and horizontal eastward component (D). The sampling rate of the ground magnetic field data is 50 Hz. In this study, we examined the magnetic field data obtained from 2003 to 2012.

To identify SR parameters, which are frequency and power of the fundamental mode of the SR, we calculated PSD (power spectral density) every 10 seconds segment by using FFT (First Fourier Transformation). Since, the fundamental mode of SR is about 7.8 Hz, we find a peak frequency ranging 6.0 Hz to 9.0 Hz for running average of the PSD data. Then

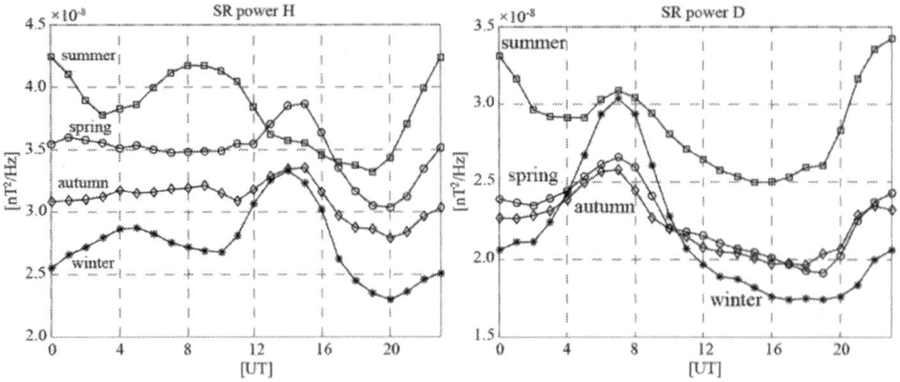

Fig. 2. The mean daily variation of SR power for spring (March, April, and May), summer (June, July, and August), autumn (September, October, and November), and winter (December, January, and February).

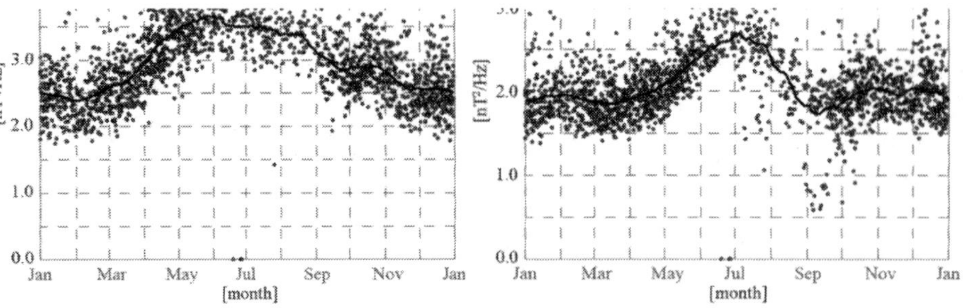

Fig. 1. The daily value of the SR power for H and D as a function of the day number. The lines indicate 30-day-centered moving average of the data.

is buried. The peak at 08 UT corresponds with the peak time of the lightning in Asia. Thus, the SR at Kuju indicates that the lightning activity from Africa affects through the year and the lightning activity from Asian center becomes stronger in summer.In the case of D, the most prominent peak at 07 UT seems to be associated with the lightning activity from Asian center. While the peaks at 23 UT are noticeable both in H and D, they bear no reference to the three major regions of lightning activity.

As stated above, we found the response of the SR power at Kuju to the global lightning activity. The variation of the SR power seems to be affected by the thunderstorm regions which become active in relation to the observation site. However, the relationship between the peak at 23 UT and the global lightning activity is unclear.Setman and Fraser [1991] suggested that the local height of the ionospheric D region for a given observation site also affects the SR amplitude. In addition, Melnikov et al. [2004] indicated the terminator effect on the SR, and concluded that the effect is caused by the lateral ionospheric gradient across the terminator region. Thus,we need to take into account the state of the local ionosphere for the better understanding.

Acknowledgement

The ELF-range magnetic field data is taken by International Center for Space Weather Science and Education, Kyushu University. The authors wish to thank Dr. Rehman, A. U. and Mr. S. Abematsu for their supportive comments on an earlier draft of this paper.

References

1. Balser, M., Wagner C. A., Observation of earth-ionosphere cavity resonances, Nature, 188, 638-641, 1960.
2. Balser, M., Wagner, C. A., On frequency variations of the earth-ionosphere cavity modes, *J. Geophys. Res. 67*, 4081-4083, 1962.
3. Sátori, G., and B. Zieger, Spectral characteristics of Schumann resonances observed in Central Europe, *J. Geophys. Res. 101*, D23, 29,663-29,669, 1996.
4. Zhou, H., H. Yu, B. Cao, and X. Qiao, Diurnal and seasonal variations in the Schumann resonance parameters observed at Chinese observatories, *Journal of Atmospheric and Solar-Terrestrial Physics, 98*, 86-96, 2013.
5. Williams, E. R., The Schumann resonance: a global tropical thermometer, *Science, 256* (5060), 1184-1187, 1992.
6. Price, C., Evidence for a link between global lightning activity and upper tropospheric water vapor, *Nature, 406* (6793), 290-293, 2000.
7. Whipple, F. J. W., and E. J. Scrase, Point discharge in the electric field of the earth, *Met. Off. Geophys. Mem., London, 68*, 20, 1936.
8. Setman, D.D., and B. J. Fraser, Simultaneous observations of Schumann resonances in California and Australia: Evidence for intensity modulation by the local height of the D region, *J. Geophys. Res. 96*, 15,973-15,984, 1991.
9. Melnikov, A., C. Price, G. Sátori, and M. Füllekrug, Influence of solar terminator passages on Schumann resonance parameters, *Journal of Atmospheric and Solar-Terrestrial Physics, 66*, 1187-1194, 2004.

Seasonal variations of mesopause temperature and the amplitude of the VLF signals of the Novosibirsk radio station during 2009-2016

Alexey Korsakov[1,*], *Vladimir* Kozlov[1,2], *Anastasia* Ammosova[1], *Petr* Ammosov[1], *Galina* Gavrilyeva[1], *Igor* Koltovskoi[1], and *Yegor* Pavlov[2]

[1]Yu.G. Shafer Institute of Cosmophysical Research and Aeronomy SB RAS, 677980 Lenin Ave, 31, Yakutsk, Republic of Sakha (Yakutia), Russian Federation
[2]Physical Technical Institute of the North-Eastern Federal University, 677000 Belinsky St., 58, Yakutsk, Republic of Sakha (Yakutia), Russian Federation

Abstract. Dynamics of seasonal variations of the amplitude of the VLF radio signal received in Yakutsk from the navigation station near Novosibirsk and the P-branches of the OH band (6-2) radiation intensity in the wavelength range 835 - 853 nm are considered. The radiation variations give information about mesopause region measured at the Maimaga station (130 km from Yakutsk). The observation from 2009 to 2016 covers period with minimum and maximum solar activity. The mesopause temperature and the VLF signal increase with increasing solar flux F10.7 in winter. The mesopause temperature seasonal variations and the VLF signal strength for the Novosibirsk-Yakutsk path are regularly inverted from year to year. By decade data averaging the VLF radio signal strength dependence on the temperature of the atmosphere at the OH excitation height can be expressed by a linear function. The coefficient of determination: R^2 = 0.59, the anticorrelation coefficient: r_{10} = - 0.77. The variations of the VLF radio noise and the radio station signal for the eight-year interval are similar to solar activity (F10.7 index). The signal level of the radio station and radio noise registered in the winter is more sensitive to variations of F10.7 index in 24th solar cycle activity.

1 Introduction

The variations of parameters of electromagnetic signals of radio navigation stations are registered in the very low frequency range (VLF: 3-30 kHz) for sounding the lower ionosphere. The radiation intensity is also recorded in the wavelength range from 835 to 853 nm, where the p-branches of the OH (6-2) emission band provide information of the mesopause region condition.

The excited hydroxyl molecule makes $2 \cdot 10^4$ collisions per second before the radiation which is enough to thermalization with the environment [1]. Thus the OH rotational temperature calculated from night sky spectra is equal to the neutral atmosphere

* Corresponding author: korsakovaa@ikfia.ysn.ru

temperature at the mesopause region altitudes. The region altitude, temperature and thickness along with the lower ionosphere conductivity vary on different space-time scales. Ionizing radiation for the lower ionosphere (D and E regions) is the most energetic part of the solar X-ray spectrum, the intense solar line of hydrogen Lyman-α (121.5 nm) and extreme ultraviolet radiation [2]. The mesopause region temperature is characterized by large seasonal variations reaching 30 K at latitudes ~ 40° N [3]. For latitudes 60° N the seasonal variations increase to ~ 60 K [4, 5]. The mesopause region altitude in winter is ~ 100 km. After the transition of the atmosphere general circulation to summer regime the mesopause region shifts lower up to an altitude of 86 km [6]. It was found that for the latitude region 40 - 60° N the summer temperature of the hydroxyl molecular band response on the solar activity variation was equal to 2.5 ± 1 K/100SFU and the winter temperature response estimated as 6.5 ± 2K/100SFU [7, 8]. SFU is the solar flux unit radio emission at a wavelength of 10.7 cm, equal to 10^{-22} W/(m^2·Hz). The amplitude of the annual component varies from 18.2 K for the years of minimum solar activity to 64.7 K for the years of the maximum and the amplitude of the semiannual component increases from 6.8 K to 26.6 K [9].

The VLF radio waves sensitivity to various geophysical phenomena depends on the geographical location of the transmitter and receiver, the signal propagation direction (relative to the Earth's magnetic field), the propagation path length and the signal frequency [10]. It is shown an anticorrelation relationship between the mesopause temperature and the intensity of narrow-band VLF radio signals reflected from the lower ionosphere at the middle and tropical latitudes [11]. The variability of the mesopause temperature and the VLF signals amplitude by ~ 72% is due to seasonal changes in solar radiation, and ~ 28% have other sources. There are quasi-semiannual mesopause temperature and nighttime ionosphere parameters variations in dynamics of the received VLF radio stations signals [12].

2. Registration methods

The OH(6-2) rotational temperature were obtained by the infrared digital spectrograph. The device is installed at the Maimaga optical station (63° N, 129.5° E) which located at 120 km north of the Yakutsk. The observations were carried out in cloudless nights, with the sun at least 9° below the horizon. The emission spectrum of the night sky is observed in the range from 836 nm to 853 nm, where the P-branch of OH(6-2) band emission is located. The model spectrum whose deviation from the actual one is less than the registration noise is considered to correspond most closely to the reality; and the rotational temperature determined based on this spectrum corresponds to the temperature at the mesopause height. During such sampling, rotational temperature values with systematic errors exceeding random ones are excluded from further processing. The estimates indicate that random errors in temperature measurements vary from 2 to 5 K depending on the signal-to-noise ratio [13].

VLF signals from the stations of the radio technical long-range navigation system (RSDN-20) are continuously recorded in Yakutsk from 2009 to the present. Transmitters are located near Krasnodar, Novosibirsk and Khabarovsk. In the intervals between radio

 DOI: 10.1051/e3sconf/20172001005

pulses of navigation stations at the same frequencies (11.904, 12.649 and 14.881 kHz) radio noise is recorded. The method makes it possible to record amplitude and phase variations of the signal and radio noise in the 372 Hz band [14, 15].

The radio noise measuring in VLF range is taken into account that the natural radio noise consists of fluctuation and impulse components. The main contribution to pulsed radio noise is made by radio pulses of nearby lightning discharges (atmospherics). Fluctuation radio noise is a continuous random sequence superimposed on each other by pulses of distant lightning discharges, which have experienced large attenuation during propagation. According to the recommendations of the International Telecommunication Union, measurements of radio noise at frequencies below 30 MHz (http://www.itu.int/publ/R-REC/en) it should be carried out in a small frequency band or at several frequencies. We registered the radio noise level at three close frequencies of 11.904 kHz, 12.649 kHz and 14.881 kHz in 372 Hz band in radio stations off air time intervals. It is assumed that the intensity of a radio noise in a narrow frequency band can be characterized by the median value of the spectral component amplitude. The contribution of the impulse component decreases with the median averaging of the values at each stage. For the first stage, the median averaging is performed in selected interval (0.4 sec) during the reception of the corresponding radio station pulse in each packet of RSDN-20. The duration of the packet, and hence the frequency of each radio station pulse on each of the three frequencies is 3.6 seconds. At the second stage, a median averaging of radio station amplitude values is performed for each of the three frequencies in the interval of 3 minutes (50 packets of RSDN-20). Similarly, for radio noise, a median averaging is performed in the "empty" dispatch of the packet (on condition in the absence radio stations signals).

3. Experimental data

The ground base observations by infrared digital spectrograph are carried out in the territory of central Yakutia from the middle of August to the middle of May. The seasonal changes of temperature can be estimated as a sum of an annual, semiannual and terannual harmonics with amplitudes of 28.6 K, 10.6 K and 3 K respectively [9]. The rotational temperature OH(6-2) seasonal variations and the variations in the VLF signal field strength of the radio station Novosibirsk (14.881 kHz) for nighttime conditions of Novosibirsk-Yakutsk VLF propagation path (15 - 19 UT) for the period 2009 - 2016 are shown in Figure 1.

DOI: 10.1051/e3sconf/20172001005

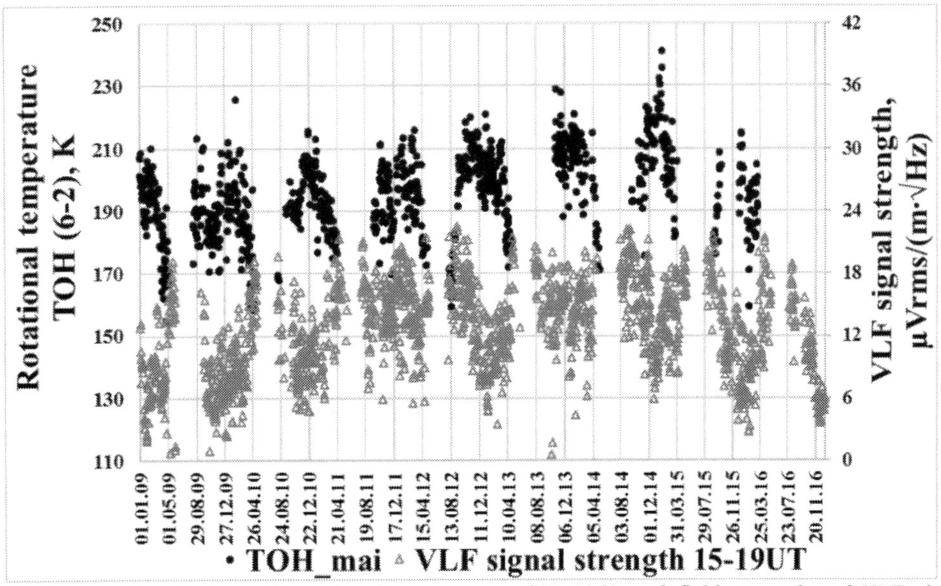

Fig. 1. Seasonal variations of rotational temperature OH (6-2) and field strengths of VLF signal Novosibirsk (14.881 kHz) for the period 15 - 19 UT 2009 - 2016.

For a convenience of comparison, the signal strength values outside the periods of the rotational temperature variation observation are excluded. The VLF field strength variations were distorted due to damage ground bus of the receiver from November 2011 to April 2012. Excluding this interval, the temperature seasonal variations of the mesopause region and the field strength received by the VLF signal for the Novosibirsk-Yakutsk path are regularly inverted from year to year, as in researches for similar VLF paths [11].

Based on the available data of the rotational temperature and electromagnetic field strength received the VLF signal of the radio station Novosibirsk for 7 years from 2009 to 2016 the median averaging seasonal night variations were prepared. For this case the cross-correlation coefficient is $r = -0.52$. The data are shown in Figure 2. As can be seen from the figure, both the rotational temperature and the field strength of the VLF signal variations have significant fluctuations.

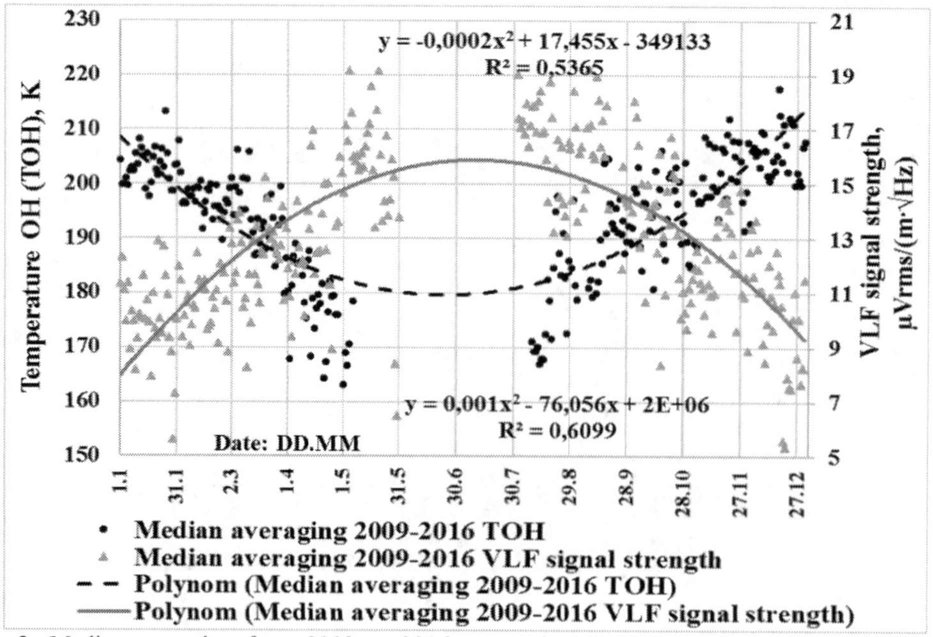

Fig. 2. Median averaging from 2009 to 2016 seasonal night time variations of the rotational temperature and Novosibirsk station VLF radio signal strength.

A decade (10 day) data averaging of the OH rotational temperature and the VLF signal strength was performed to reduce the fluctuations and to determine the relationship. The anticorrelation coefficient is r_{10} = - 0.77. The VLF signal strength dependence on the atmospheric temperature at the OH excitation height presented in Figure 3. The dependence can be expressed by a linear equation (1):

$$A_{Novosibirsk} = -0.16 \cdot TOH + 44.42, \quad (1)$$

where the field strength of the VLF signal received in Yakutsk from the radio station Novosibirsk $A_{Novosibirsk}$ is measured in μVrms/(m √Hz), and the OH rotational temperature *TOH* is measured in degrees K. The coefficient of determination for the linear dependence is R^2 = 0.59, which corresponds to the anticorrelation coefficient r_{10} = - 0.77.

 DOI: 10.1051/e3sconf/20172001005

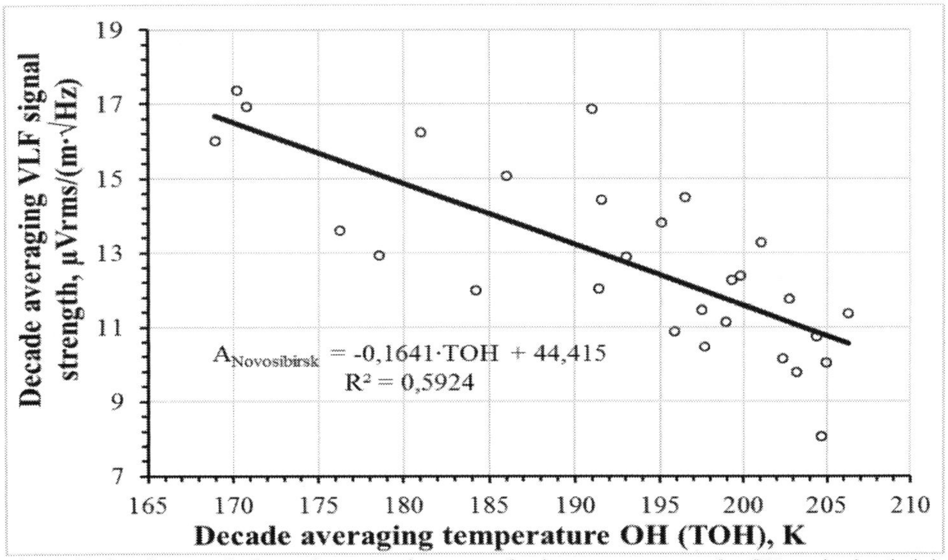

Fig. 3. VLF signal strength dependence on the atmospheric temperature at the OH excitation height.

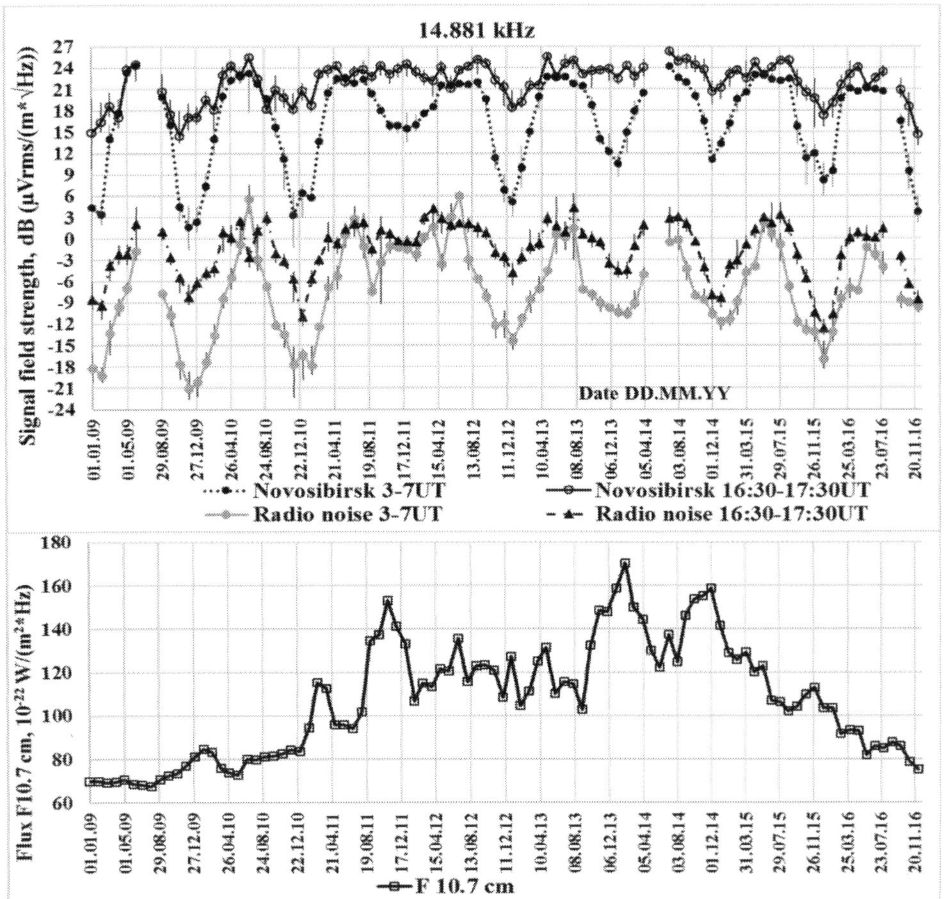

Fig. 4. Variations of the Novosibirsk radio station signal field strength and radio noise (14.881 kHz), the solar radio flux F 10.7 cm 2009-2016.

Seasonal variations of the signal field strength of the Novosibirsk radio station and radio noise (14.881 kHz), in daytime (3-7 UT) and nighttime (16:30-17:30 UT) Novosibirsk-Yakutsk propagation path conditions (2.64 Mm) and also data on solar activity - the solar radio flux at 10.7 cm wavelength (F10.7 index) (US Dept. Of Commerce, NOAA, Space Weather Prediction Center (SWPC). URL: http://www.swpc.noaa.gov/products/solar-cycle-progression) for the period 2009 to 2016 are shown in Figure 4. Monthly median averaging was performed. The quartiles of 25% and 75% are presented as fluctuations. The signal field strength seasonal variations are most pronounced in daytime. The variations associated with the increase D region ionization efficiency from December to June: the solar zenith angle decreases over the path, so the gradient of the ionosphere electron concentration increases and VLF attenuation decreases. The asymmetry is observed in the amplitude seasonal variations which manifested the fact that the average median values of these parameters during the autumn equinox (September) are closer to the summer solstice. The recorded parameters of VLF radio signals during the vernal equinox are closer to the parameters of the winter solstice. This asymmetry agrees with the seasonal asymmetry of the D region electron concentration profiles [16].

The strength of the radio signal field increases by 3 ± 1 dB and 4 ± 1.5 dB for registered in daytime (3 - 7 UT) and night time (16:30 - 17:30 UT) conditions respectively with increasing solar activity (2014) in summer period. Variations in the signal level (from the minimum to the maximum solar activity 2009-2014) were 9 ± 2 dB for both day and night propagation conditions in winter. In 2015 - 2016 years the signal level decrease to the values of 2009-2010. The behaviour of the signal strength interannual variation of Novosibirsk radio station on an eight-year interval both for summer and for winter periods is similar to the behaviour of solar activity variations (F10.7 index). The field strength variations of the radio noise (14.881 kHz) are 7 ± 2 dB and 3.3 ± 1.5 dB when registered in daytime (3 - 7 UT) and night time (16:30 - 17:30 UT) conditions respectively in summer. Variations in the radio noise level (from the minimum to the maximum of solar activity 2009-2014) amounted to 10.4 ± 2 dB in daytime and 6.5 ± 1.5 dB in night time conditions of registration in winter. There is a decrease radio noise level in the winter 2015 – 2016 approaching to the values 2009 – 2010.

4. Discussion

It should be noted that the signal level of the radio station and radio noise registered in winter, relative to the summer season, is more sensitive to F10.7 index variations from the minimum to the maximum of the 24th solar cycle activity. The seasonal variations of lower ionosphere parameters were considered earlier [17]. The effective recombination coefficient is lower in winter. The differences are due to seasonal variations in the mesosphere region meteorological parameters. Hence there is the high sensitivity of the received VLF signal parameters in winter season relative to summer, which is also noted for sudden ionospheric disturbance effects [18].

For the solar cycle activity maximum, the VLF signals level of radio stations noontime recorded (low and medium latitudes for the most part of the selected extended radio paths crossing the equator), approximately by 0.3 ± 0.1 dB/Mm more than in the solar minimum period [19]. The variations correspond to our analysis data. The greater attenuation in the solar cycle activity minimum can be partially due to a decrease in the Lyman-α flux, the main ionization source of the atmosphere at altitudes 65 – 80 km, while an intensity of cosmic particles flux increased. The cosmic particles flux causes an increase the lower altitude electron concentration, where the VLF radio attenuation is greater due to even greater concentration of neutral particles.

5. Conclusions

Seasonal variations in the mesopause region temperature and the field strength of VLF signal received for the Novosibirsk – Yakutsk propagation path are in anticorrelation relationship regularly from year to year.

The VLF radio signal strength dependence on the temperature of the atmosphere at the OH excitation height can be expressed by a linear function. The linear function slope is -0.16 and the constant term (line offset) is 44.42. The coefficient of determination for the linear dependence is R^2 = 0.59, which corresponds to the anticorrelation coefficient r_{10} = -0.77.

The behavior of the interannual variations of the field strength of the VLF radio noise and the Novosibirsk radio station signal for the eight-year interval are similar to the behavior of solar activity (F10.7 index) both for summer and for winter. The signal level of the radio station and radio noise registered in the winter season, relative to the summer season, is more sensitive to variations of F10.7 index from minimum to maximum of the 24th solar cycle activity.

The study was supported by RFBR, research projects No. 15-45-05005 r_vostok_a, 16-35-00121-mol_a, 16-35-00204-mol_a, 15-05-05320-a, 17-05-00855 A.
The study was supported by the program of complex scientific research in the Republic of Sakha (Yakutia) 2016-2020.

References

1. N.N. Shefov, A.I. Semenov, V.Yu. Khomich, *Izluchenie verkhnei atmosfery – indikator ee struktury i dinamiki* (GEOS, Moscow, 2006), 741 (in Russian)
2. M.I. Panasyuk, et al. *Fizicheskie usloviya v kosmicheskom prostranstve. Model' kosmosa: v 2 t.* (KDU, Moscow, 2007), **1**, 871
3. D.C. Senft, G.C. Papen, C.S. Gardner, J.R. Yu, D.A. Krueger, C.Y. She, Geophys. Res. Lett., **21**, 821-824 (1994)
4. P.J. Espy, J. Stegman, Phys. Chem. Earth., **27**, 543–553 (2002)
5. A.M. Ammosova, P.P. Ammosov, Proc. SPIE, **8696**, 86960S-86964S (2012)
6. C.Y. She, , U. von Zahn, J. Geophys. Res., **103**, 5855 – 5863 (1998)
7. A.I. Semenov, N.N. Shefov, V.I. Perminov, V.Yu. Khomich, H.M. Fadel, Geomagnetism and Aeronomy, **45:2**, 236-240 (2005)
8. G.S. Golitsyn, A.I. Semenov, N.N. Shefov, V.Yu. Khomich, Phys. Chem. Earth., **31:(1-3)**, 10-15 (2006)

DOI: 10.1051/e3sconf/20172001005

9. A.M. Ammosova, G.A. Gavrilyeva, P.P. Ammosov, I.I. Koltovskoi, Proc. SPIE, **10035**, 1003560S1-4 (2016)
10. A.B. Orlov, G.V. Azarnin, Problemi difraktsii i rasprostraneniya voln, **10**, 3-107 (1970) (in Russian)
11. I. Silber, C. Price, C.J. Rodger, C. Haldoupis, J. Geophys. Res.-Atmos., **118**, 4244-4255 (2013)
12. I. Silber, C. Price, C.J. Rodger, Atmos. Chem. Phys., **16**, 3279-3288 (2016)
13. P.P. Ammosov, G.A. Gavrilyeva, A.M. Ammosova, I.I. Koltovskoi, Advances in Space Research,. **54**, 2518 – 2524 (2014)
14. R.R. Karimov, V.I. Kozlov, A.A. Korsakov, V.A. Mullayarov, V.P. Melchinov, Current problems in remote sensing of the Earth from space, **9:4**, 57-62 (2012)
15. V.I. Kozlov, A.A. Korsakov, L.D. Tarabukina, N.S. Duiukova, Bulletin of South Ural State University. Series of "Mathematics. Mechanics. Physics", **9:1**, 57-64 (2017)
16. Ya.L. Al'pert, *Rasprostranenie electromagnitnikh voln i ionosphera* (Nauka, Moscow, 1972), 564 (in Russian)
17. A.D. Danilov, A.G. Simonov, Ionosfernye issledovaniya, **34**, 54-72 (1981) (in Russian)
18. A.D. Danilov, A.B. Orlov, L.P. Morozova, V.P. Kishchuk, Geomagnetism and Aeronomy, **23:2**, 311-313 (1983)
19. N.R. Thomson, M.A. Clilverd, Journal of Atmospheric and Solar-Terrestrial Physics, **62:7**, 601-608 (2000)

Effect of the electromagnetic environment in pipes in Yakutia

Vladimir I. Kozlov[1,2,*], *Larisa M.* Abzaletdinova[1] , *Lena D.* Tarabukina[1,2] , *Alexey A.* Korsakov[1] and *Yuri M.* Grigoryev[2]

[1]Yu.G. Shafer Institute of Cosmophysical Research and Aeronomy SB RAS, 677000, Yakutsk, Russia
[2]M.K. Ammosov North-Eastern Federal University, 677000, Yakutsk, Russia

Abstract. The results of measurements of telluric currents flowing through pipelines laid in the cryolithozone in Yakutia are presented during geomagnetic storms and lightning discharges occurring in the 10 km zone around the pipeline. The currents reach tens of amperes. Construction of a lightning density distribution map. The effect of a direct current flowing in a 150 m long pipeline with a diameter of 110 mm laid in permafrost on the change in the pipe-to-ground potential is measured. Based on the analysis of variations in the horizontal component of the magnetic activity in the current 24th cycle of solar activity, it is established that the frequency of occurrence of a certain level of the rate of change in geomagnetic activity can be expressed by a power law with an order of magnitude of about two at low values. The exponent decreases to unity as the geomagnetic activity increases. The probability of occurrence of time intervals with given threshold values of the rate of change of geomagnetic activity is determined.

1 Introduction

Geomagnetic field fluctuations and charged (thunderstorm) clouds create electric fields and, accordingly, currents in pipelines, called geomagnetically induced currents (GIT). Pipelines in Yakutia, for example, the pipeline Eastern Siberia-Pacific Ocean (ESPO), lay at a depth of 2 m. In the transitional sections with different conductivity, there is a difference in electrical potentials between the pipeline and the earth, and electrochemical corrosion increases. The entry point of wandering currents will serve as a cathode, and the point of flow of currents will serve as an anode. Local corrosion of underground pipelines often manifests itself in the form of ulcers penetrating deep into the metal. Studies on the determination of GIT give values up to tens of amperes [1] and hundreds [2]. The induced current in the main objects can cause significant fluctuations in the potentials of the pipeline-ground (PPG). Changes in PPG can reach up to 1 V [3]. Because of the change in the potential "pipeline-ground" associated with telluric currents, the following problems arise. Positive values of PPG lead to corrosion. Fluctuations of PPG more than 0.85 V

* e-mail: vkozlov@ikfia.ysn.ru

 DOI: 10.1051/e3sconf/20172001006

relative to the natural potential of the metal-ground (-0.66 V) can lead to increased corrosion, even taking into account the potential of protection 0.25 V. Fluctuations of PPG up to 1.2 V can cause the production of hydrogen at the boundary between the metal and the coating, which will lead to a violation of the bond between the coating and the metal. Possible damage to electronic equipment connected to the pipeline [4].

2 Methods of investigation

An experimental study of the influence of the magnitude of the direct current flowing through the pipeline on the change in the potential difference between the pipeline and the ground was carried out on a part of the buried pipeline with the following parameters: length 150 m, diameter 110 mm, thickness 11 mm. This section of the gas pipeline between Yakutsk and Tabaga crosses the bulk gravel road, creating a heterogeneity in the ground around the pipeline. It was obtained (Fig. 1) that the potential difference "pipeline-ground" (ΔUp-gr) depending on the current in the pipeline (Ip) is expressed as ΔUp-gr = 0.46 * Ip-0.005, at R^2 = 0.9998. In our measurements, in the absence of current in the pipeline segment under test, the potential is equal to Up-gr = -0.33 V. For the operating pipeline with a quiet geomagnetic situation, the measured potential Up-gr = -0.26-0.33 V. When the current passes through Pipeline there was a displacement of the potential in the positive direction, which causes increased corrosion in the pipeline.

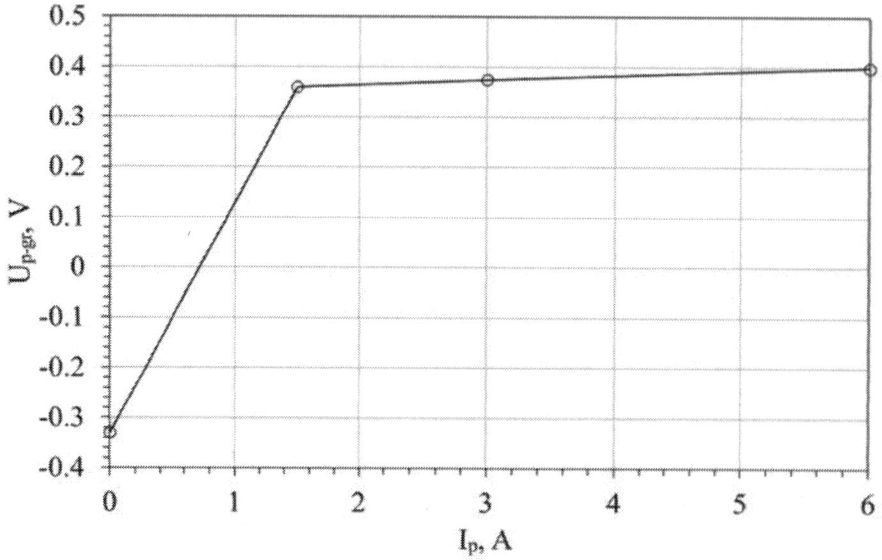

Figure 1. The experimental values of the potential difference ΔUp-gr "pipeline-ground" as a function of the magnitude of the constant electric field I p of the pipeline.

It was found in [5] that the potential "pipeline-ground" in the parts of the Punga-Vuktyl gas pipeline, where accidents often occurred, varied from -4.5 to +2.5 V. Test measurements of the pipeline-ground potential showed: if the current is zero, the potential will also be zero, and if the current rises to 5 A, the potential will linearly increase to 1.6 V. If the current goes in the opposite direction, the potential will change to -0.5 V. The potential difference (ΔUp -gr) "pipeline-ground" depending on the current in the pipeline is expressed as ΔUp-gr = 0,23 * Ip. The data correspond to our results, except for the saturation of the potential and the steepness of the dependence, which is determined by the state of the local soil. In our case, the pipeline surrounds the frozen ground.

It was shown in [6] that for geological auroral electrodes, as the causes of large horizontal components of magnetic field variations (H) in auroral regions, the geographical northern component (X) is clearly larger than the eastern (Y), but dX / dt and dY / dt are almost equal. This is important when assessing the risks of occurrence of geomagnetically induced currents (GIT) in the main objects, since the horizontal geoelectric field can have large values in any direction, and not only parallel to the electrode. An important parameter for the investigation of the appearance of GIT is the probability of occurrence of large values of the time derivative of the horizontal component of the geomagnetic field dB / dt. The value of dB / dT) used to estimate GIT values has a high positive correlation with the number of sunspots (0.86) and aa-index (0.8998). 92.665% of the obtained geomagnetic activity can be explained by the joint action of the indices of the number of sunspots and aa-index [7].

3 Results and discussion

Measurements of electric currents induced in the branches of the Tas-Yuryakh-Mirny oil pipeline with a diameter of 0.5 m during ground-based lightning discharges in the 1-10 km strip from the pipeline were measured. The same measurements were taken at the branch of a gas pipeline 110 mm in diameter and about 20 km long to the village of Tabaga from the main gas pipeline "Srednevilyuiskoye gas field - Mastakh-Berge-Yakutsk".

To observe the electromagnetic field oscillations emitted by induced currents in the pipeline, three horizontal magnetic loop antennas were used, located at different distances from the pipeline. To determine the location of the thunderstorm near the observation site, a one-point positioning system for lightning was installed at the IKFIA located at the polygon near Yakutsk and the Wolrd Wide Lightning Location Network (WWLLN) system [8], one of whose stations is located in Yakutsk [9]. It is obtained that the thunderstorm front moves with an average velocity of 44 km h^{-1} and has an average area of 5000-10000 km^2 [10]. It follows that in the subpolar latitudes, lightning strikes of one thunderstorm occur within a radius of 10 km around the pipeline for 1-2 hours, even during a thunderstorm moving across the pipeline.

From the current flowing through the Tas-Yuryakh-Mirny pipeline during lightning discharges, the induced signal in the loop antenna located at a distance of 3.5 m from the pipe was 63 times greater than the current signal induced in the loop antenna located at a distance 150 m. The distance to the lightning discharge was estimated from the time difference between the time of the lightning flash and the sound of thunder. A thunderstorm crossed the pipeline at a distance of 5-10 km from the observation point according to the data of the thunderstorm seekers.

Similar observations were made on a gas pipeline 30 km from Yakutsk when the thunderstorm was 3-10 km from the place of observation and crossed the branch of the pipeline to the village Tabaga. In these measurements, it was found that during a near-thunderstorm, the magnitude of the signal of the magnetic component of the electromagnetic field measured in the immediate vicinity of the pipeline was 190 times greater than the magnitude of the signal induced in the loop antenna located 400 m from the pipe. Calculated currents in the pipeline induced by ground-based lightning strike at a distance of 3-7 km from the observation site reached 10 A.

The average annual number of lightning discharges in Yakutia in 2009-2016 is shown in Fig. 2. The density of discharges is corrected for the lightning discharge detection efficiency of the WWLLN system, which is 11-15% and about 30% for lightning discharges with a current of more than 100 kA [11]. Density was calculated for the summer seasons (June 1 - August 31) in 2009-2016, when thunderstorms are active in the territory under consideration.Measurements of the currents in the pipeline during geomagnetic storms were carried out on a branch to. Tabaga from the main gas pipeline "Srednevilyuiskoye gas field - Mastah-Berge-Yakutsk" and branches between distribution stations No. 2 of Yakutsk and Pavlovsk, a pipeline 0.53 m in diameter from the gas distribution station No. 2 of Yakutsk comes under the ground to the village of Hatassy and further. A double gas pipeline is laid under the riverbed of the Lena River, on the opposite side of the river there is a gas distribution station in the village of Pavlovsk.

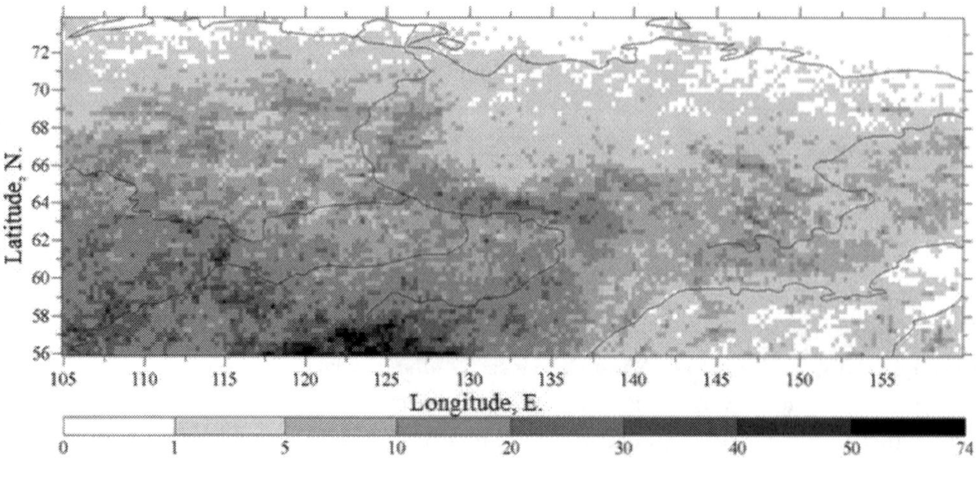

Figure 2. The average annual number of lightning discharges in Yakutia in 2009-2016.

To measure the induced currents flowing in the gas pipeline during geomagnetic disturbances, two magnetometers were used. One sensor was installed next to the pipeline (1 m distance), the second sensor (used for comparison) was installed at a distance of 600 m from the pipeline. The induced current was determined from the difference in the magnitudes of the magnetic disturbances measured at two sensors, proximal and far from the pipeline, taking into account the calibration. Test measurements to assess the sensitivity of the magnetic sensor were carried out using a current passing through the wire, which was located at a distance of 1 m from the sensor. During a calm geomagnetic situation, the magnetometers placed on the pipeline and at a distance of 600 m from the pipeline showed a similar noise pattern. This indicates that there are no induced quasi-permanent currents in the pipeline during small perturbations of the geomagnetic field.

Current measurements in the pipeline were made during three powerful geomagnetic storms. During a magnetic storm on January 21, 2005, the current in the pipeline 0.11 in diameter reached values of up to 3 A. During the storm on May 28, 2011, the current in the pipeline with a diameter of 0.53 m reached values of up to 0.5 A. During the storm August 27, 2013, the current in the pipeline with a diameter of 0.53 m reached values of up to 0.3 A. An investigation of the measured current flowing through the pipeline during strong geomagnetic storms showed that the current reaches tens of amperes and lasts from several hours to several days.

The method of experimental investigation of the effect of geomagnetic disturbances on extended trunk objects can be reduced to the identification of time periods in which the value of the rate of increase of the magnetic field (dB / dt) above the threshold levels:> 5 nT / min (active field), corrosion increases sharply in the absence of additional protection ; > 20 nT / min (storm), corrosion increases dramatically even with protection; > 30 nT / min (extreme storm) effects on energy systems [12].

It was shown in [13] that a closer functional connection is observed between the HIT and the magnitude of the magnetic field increase rate (dB / dt) than with the horizontal component of the geomagnetic field H. A good correlation of the dX / dt variations with the induced currents in the transmission line directed along the X is shown in [14].

It is proposed in [7] to use the published data of the International Real-time Magnetic Observatory Network as the magnitude of the magnetic field increase rate (dB / dt). Therefore, the study of variations in geomagnetic activity is carried out using the data of the rate of increase in the strength of the horizontal component (dB / dt) of the Yakutsk Observatory of the interregional data collection node for servicing the Russian-Ukrainian segment of the world network of the INTERMAGNET geomagnetic observatories. In the base INTERMAGNET data (Rate of Chanch dB / dt) of the Yakutsk Observatory (Yakutsk, YAK, 61.96 / 129.66) are presented with a minute resolution (http://www.intermagnet.org/data-donnee/dataplot-eng.php?type=xyz).

Consider the probability of exceeding a certain level dB / dt by the example of a very strong magnetic storm August 05, 2011. From 17 to 20 hours, the disturbance level reached G4, and from 20 to 23 hours - G3. August 06 from 02 to 05 hours - G1. from 23 to 02 hours - G2, from 02 to 05 hours - G1. And then until 08:00 on August 07, 2011 the storm intensity gradually decreased. For the entire day August 05, 2011 time intervals with the rate of change of the magnetic field > 5 nT were 7.15%, > 20 nT 1.18%, > 30 nT 0.42% b < -5 nT 6.6%, < -20 nT 1.32%, < -30 nT 0.56%. The frequency of occurrence of a given level dB / dt is determined by the expression n = 444.84 (dB / dt) -1.717 where $R^2 = 0.85$. Or the probability of exceeding the level can be expressed as: lg (P) = -0.057 (dB / dt) - 0.1946, with $R^2 = 0.99$.

When considering the dependence of the current in the pipeline on the rate of magnetic field variations, the example of a geomagnetic storm on May 28, 2011 found that the current in a pipeline, depending on the rate of magnetic field variations, dB / dt [nT / min], can be expressed as (Fig. 3): I = 0.024 * (dB / dt) + 0.07. That is, at 5 nT / min, the GIT

 DOI: 10.1051/e3sconf/20172001006

values appearing in the pipeline reach 1 A. And this, in accordance with our measurements, leads to positive values of PPG, and hence to increased corrosion.

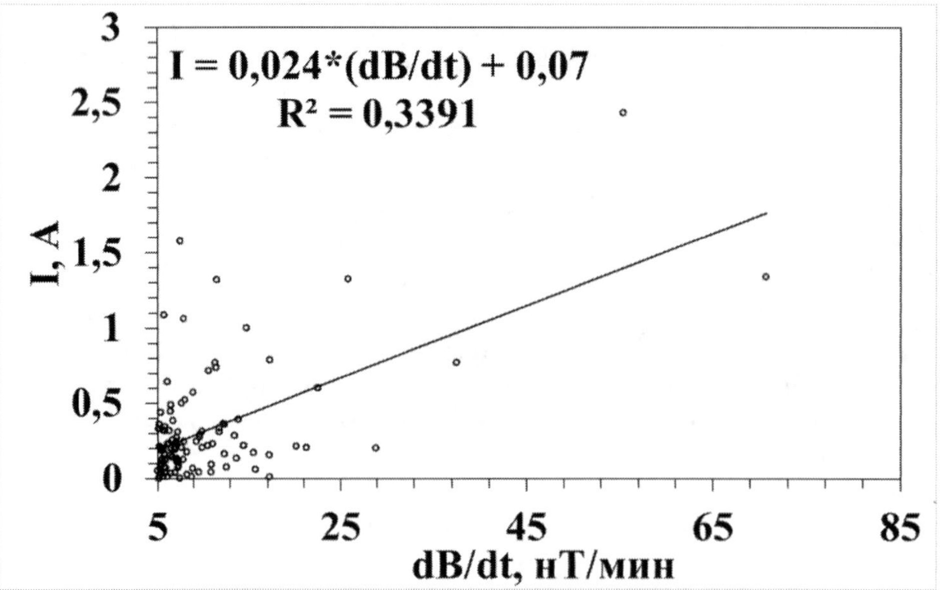

Figure 3. Current in the pipeline caused by variations of the magnetic field May 28, 2011.

At the magnetic station of Yakutsk in 2008-2013, the geomagnetic situation was characterized by the presence of isolated magnetic storms exceeding 305 times that of dB / dt > 5 nT / min, and -dB / dt> 20 nT / min-40 times. Exceeding the levels of the rate of increase of the magnetic field intensity in 2014-2016 were observed: > 5 nT / min 118 storms, > 10 nT / min 86 storms, > 20 nT / min 41 storms and > 30 nT / min 25 storms. The maximum amount of magnetic disturbance is observed in spring and autumn.

Differential distribution of the occurrence of perturbations for storms 2014-2016. Is a power law and can be expressed by the formula $y = 33{,}6x^{-1{,}1}$, with the value of the determinant $R^2 = 0{,}72$. The maximum observed disturbance is 163 nT / min.

The characteristic individual differential distribution of the occurrence of the variation in the rate of change of the magnetic field dB / dt, nT / min under the magnetic storm G4 on June 22, 2015, is expressed by a power law. In the range of values from 1 to 10-20 nT / min, the distribution has a steeper slope, expressed by the formula $y = 781\ x^{-1.7}$, at $R^2 = 0.92$, compared to the interval from 10 to 100 nT / min, where the distribution has more Sloping slope, expressed by the formula $y = 76\ x^{-0.94}$, with $R^2 = 0.73$.

The 14th magnetic storms for 2015 and January 2016 (the maximum geomagnetic activity in the 24th cycle of solar activity) were chosen to show the maximum values of the change in dB / dt for 2014-2016. The value of the exponent of a single power law for events with a maximum variation in the rate of change of the magnetic field dB / dt from 6 nT / min to 100 nT / min varies from -3.1 to -2.5, and for events of the maximum range from 100 nT / min Up to 163 nT / min the value of the indicator varies from -2.2 to -1.2. That is, as

DOI: 10.1051/e3sconf/20172001006

the value of dB / dt increases, the value of the indicator decreases in absolute value, and therefore the distribution is being adjusted.

For magnetic storms, the probability of exceeding the threshold levels of the rate of increase of the magnetic disturbance was obtained at the maximum of the geomagnetic activity of the 24th solar cycle: 37% for > 5 nT / min, 8.3% for > 20 nT / min, 4.6% for > 30 nT / Min.

The distribution of the directionality of the variations dB / dt has a chaotic character without pronounced directivity, in contrast to the variation of the horizontal component of the magnetic field H, which has a pronounced directionality.

In the 24th cycle, the maximum geomagnetic activity was observed in 2015 and in 2016 on the declining branch of solar activity. According to [15], the 24th solar cycle is the beginning of an epoch of "lowered" solar activity. It is assumed [16] that the epoch of "reduced" solar activity will last 5 solar cycles. It is expected that these solar cycles will be of low and medium activity. It is assumed in [15] that in this era solar flares can not significantly exceed the powerful flares already observed during instrumental observations. This makes it possible to assume that the conclusions drawn from our analysis of the magnitudes of the variations of magnetic activity from 2008 to 2016 can be applied to a future forty-year interval.

4 Conclusions

During strong geomagnetic storms and during a thunderstorm at a distance of 10 km around the pipeline, geomagnetic induced currents appear in it, reaching tens or even hundreds of amperes. As a result, the potential of the soil-pipeline increases and the corrosion of the pipeline is possibly increased.

The research is partially funded by RFBR No. 15-45-05005 r_vostok_a and a comprehensive research program in the Republic of Sakha (Yakutia).

References

[1] S. V. Avakyan, N.A. Voronin, Issledovanie Zemli iz kosmosa, **3**, 14-29 (2011) (in Russian)
[2] J. M. Torta, S. Marsal, M. Quintana, Earth Planets and Space, **66:1**, 87 (2014)
[3] D. H. Boteler, Advances in Space Research, **26:1**, 15-20 (2000)
[4] Yu. I. Garber, Stroitel'stvo truboprovodov, **9**, 28-30 (1992) (in Russian)
[5] A.V. Furkin, M.V. Tret'yakova, R.V. Aginey, Trudy Komi nauchnogo tsentra Ural'skogo otdeleniya Rossiyskoy akademii nauk, **1:17**, 80-83 (2014) (in Russian)

DOI: 10.1051/e3sconf/20172001006

[6] A. Viljanen, Geophysical Research Letters, **24:6**, 631–634 (1997)
[7] E.O. Falayi, A.B. Rabiu, Frontiers in Science, **2:1**, 1-5 (2012)
[8] R. L. Dowden, J. B. Brundell, C. J. Rodger, J. Atmos. Solar-Terr. Phys., **64:7**, 817-879 (2002)
[9] L. D. Tarabukina, V. I. Kozlov, R. R. Karimov, V. A. Mullayarov, Proc. SPIE **9680**, 96805S (2015)
[10] V. I. Kozlov, V. A. Mullayarov, A. E. Vasilyev, Russian Meteorology and Hydrology, **3**, 27-31 (2003)
[11] S. F. Abarca, K. L. Corbosiero, T. J. Galarneau Jr., J. Geophys. Res. **115**, D18206 (2010)
[12] J. Koen, C.T. Gaunt, Geomagnetically induced currents at mid-latitudes. URSI, Maastricht, (2002)
[13] A. Viljanen, Geophysical research letters, **24:6**, 631-634 (1997)
[14] V. A. Pilipenko et al., Trudy Kol'skogo nauchnogo tsentra RAN, **4-2,** 38 (2016)
[15] V. N. Ishkov, Vestnik ONZ RAN, **8**, NZ5001 (2016) (in Russian)
[16] V. N. Ishkov, Sb. «Solnechnaya i solnechno-zemnaya fizika–2013», 111-114 (2013)

Joint analysis of the ionospheric parameters and cosmic ray data during periods of magnetic storms 2015

Oksana Manrikova[1], *Yury* Polozov[1,*], and *Timur* Zalyaev[1]

[1]Institute of Cosmophysical Researches and Radio Wave Propagation, Mirnaya Str., 7, 684034, Kamchatka Region, Elizovskiy District, Paratunka 684034, Russia

Abstract. Method of joint analysis of ionospheric parameters and cosmic ray data is proposed. The method allowsto allocate anomalous changes in data before strong magnetic storms. The developed method is based on the application of wavelet transform, neural networks and classical autoregressive models. The application of the method has shown its effectiveness and the possibility of using in the problems of estimating space weather and predicting strong magnetic storms.
The research was supported by RSF Grant, project No 14-11-00194.

1 Introduction

The work is aimed at studying processes in the near-Earth space during periods of increased solar activity and magnetic storms. At present, many issues of energy transfer in the solar wind-magnetosphere-ionosphere system remain open, which determines the relevance of these studies. In the article we performed a joint analysis of ionospheric parameters and galactic cosmic ray data (GCR). It is known that the most significant changes in the parameters of GCR are caused by the ejections of the coronary mass and the following changes in the parameters of the interplanetary field and the solar wind [1,2]. GCR observations are used in a number of fundamental and applied studies related to monitoring and forecasting space weather [2,3]. The GCR variations have a complicated structure, their dynamics reflect the 11-year cycle and the 27-day solar rotation period [4] and there is a diurnal variation due to the asymmetry of the shape of the magnetosphere, which varies with time when the solar wind parameters change [5]. Traditional methods of data analysis are not sufficiently effective for their investigation [1,5-7]. The analysis of GCR in the work is based on the joint application of wavelet transform and neural network methods, which are currently gaining intensive development in this field [1, 6-12]. The method of

* Corresponding author: up_agent@mail.ru

joint application of mathematical devices for studying complex time series was first proposed by the authors in works [7, 10, 12-14] The method allows to study in detail the data structure, to determine typical variations and to allocate anomalies. Its application allowed us to isolate the anomalous effects in the dynamics of GCR (increases that occur 8-20 hours before the onset of magnetic storms) [7, 10, 11] arising on the eve of geomagnetic storms. These anomalies were first discovered by statistical analysis and described in [15]. The results of the analysis presented in this paper confirm the possibility of the appearance of the effect of an increase in GCR and show the effectiveness of the proposed technique for their detection..

A similar approach based on the combination of wavelet transform and neural networks was developed by the authors to study the dynamics of ionospheric parameters [13, 14, 16] (the data of the critical frequency of the F2 layer of the ionosphere (foF2) and PES data were analyzed). During the periods of disturbances, anomalous changes are observed in the recorded ionospheric parameters, which indicate the occurrence of anomalous processes in the ionosphere. It is known that the strongest ionospheric disturbances occur during solar events and geomagnetic storms, the study of which has important scientific and applied significance [17-21]. In this paper, based on the joint analysis of the ionosphere parameters and GCR data on the eve of magnetic storms, anomalous increases in variations in the intensity of cosmic rays and the increase in the electron density of the ionosphere that arise during these periods are likely to be associated with the approaching events.

2 Methods of data analysis

2.1 Modeling of the CR time series and isolation of abnormal changes.

1 On the basis of the multi-scale wavelet transform [22, 23], a time series of GCR data was obtained in the form:

$$f(t) = f_{a,(-6)}(t) + \sum_{j=-1}^{-6} f_{d,j}(t) \tag{1}$$

where $f_{a,(-6)}(t) = \sum_n c_{-6,n}\phi_{-6,n}(t)$, $f_{d,j}(t) = \sum_n d_{j,n}\Psi_{j,n}$, $\Psi_j = \{\Psi_{j,n}\}_{n\in Z}$ – wavelet-basis, $\varphi_j = \{\varphi_{j,n}\}_{n\in Z}$, basis, generated by scaling function, j–scale.

2. Approximation of the characteristic component of the time series of GCR based on the neural network:

$$\overline{\overline{c_{-m,n+1}}} = \phi_k^3\left(\sum_i \omega_{ki}\phi_i^2\left(\sum_l \omega_{il}\phi_l^1\left(\sum_{z=0}^{\gamma} \omega_{ln}c_{-m,n-z}\right)\right)\right) \tag{2}$$

where ω_{ki}, ω_{il}, ω_{ln}– weights of network ,$\phi_l^1 = \phi_i^2 = \frac{2}{1+e^{-2z}} - 1$;$\phi_k^3 = a * z + b$;$\gamma$ – dimension of the input vector of the network.

Estimates of the adequacy of this model are given in [24]. The architecture of the constructed neural network is shown in Fig.1.

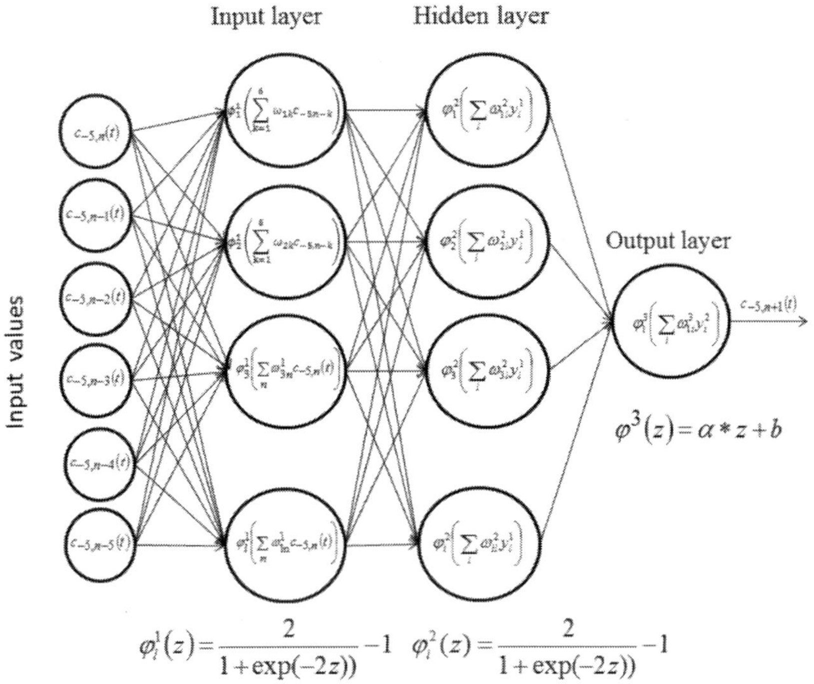

Fig.1. The architecture of a neural network that approximates the time series of a typical GCR component.

Performing the operation (2) allows to reproduce typical variations of the GCR (approximates the typical level of FCR variations). In the period of anomalous changes in the time series of the GCR, the absolute values of the errors of the trained NN increase, so the operation of their isolation can be based, for example, on checking the following condition:

$$|e(t)| > T_s, \tag{3}$$

where T_s – threshold value for anomaly. As the threshold T_s the threshold $T_s = 2.5\,\sigma_{calm}$, was used in the work, where σ_{calm} – is the root-mean-square error of the NN for the periods used in training..

2.2 Allocation of short-term anomalies in the CR variations.

Anomalies in recorded GCR variations may contain large-scale (trend) changes occurring during periods of prolonged Forbush effects, and may also contain short-term anomalous features characteristic of local GCR enhancements and depressions. The above neural network model (2) approximates the characteristic variations of GCR and on its basis large-scale GCR anomalies can be distinguished. To identify short-term anomalies, the authors developed computational solutions based on the continuous wave-transformation, first proposed in [25]:

1. Continuous wavelet transform of the data [22]:

$$(W_\Psi f_{b,s}) := |s|^{-\frac{1}{2}} \int_{-\infty}^{+\infty} f(t)\Psi\left(\frac{t-b}{s}\right) dt, \tag{4}$$

 DOI: 10.1051/e3sconf/20172001007

where $f \in L^2(R), s, b \in R, s \neq 0,$, Ψ – basis wavelet, parameter s characterizes the scale, b – time moment.

2. Application of threshold function:

$$\mathrm{P}^{\mathrm{CR}}_{T^l_s}(W_\Psi f_{b,s}) = \begin{cases} W_\Psi f_{b,s}, \text{ if } (W_\Psi f_{b,s} - W_\Psi f^{med,l}_{b,s}) \geq T^l_s \\ 0, \text{ if } |W_\Psi f_{b,s} - W_\Psi f^{med,l}_{b,s}| < T^l_s \\ -W_\Psi f_{b,s}, \text{ if } (W_\Psi f_{b,s} - W_\Psi f^{med,l}_{b,s}) < -T^l_s \end{cases} \quad (5)$$

where $W_\Psi f^{med,l}_{b,s}$–median value, calculated in a sliding time window of length l .$T^l_s = U * \sigma^l_s$– threshold function, where $\sigma^l_s = \sqrt{\left(\frac{1}{l} - 1\sum_{k=1}^{l}(W_\Psi f_{b,s} - \overline{W_\Psi f_{b,s}}\right)^2}$ – is the standard deviation calculated in a sliding time window of length l,$\overline{W_\Psi f_{b,s}}$- median value, U is a threshold coefficient.

The choice of the threshold coefficient U is based on the assessment of a posteriori risk, the results of experiments with real data and the results of statistical modeling were used in the estimates. The model data correspond to the neutron monitors data (the data of the Apatity stations, Cape Schmidt and Magadan were used), they contain a recurrence component of the variations and different-scale local features, and white noise. The results of estimating the efficiency of the algorithm for different values of the U coefficient on the model data are given in Table 1. In accordance with the estimates, U=2.5 was used in the work.

The choice of the analyzing wavelet Ψ was based on the following criteria: the number of zero moments, the smoothness of the wavelet and the size of the carrier. Taking into account the chosen criteria, the basic wavelet Coiflet of order 1 was used (has the least possible length of the carrier and ensures the detection of the feature).

Table 1. Estimation of the efficiency of the algorithm for different values of the threshold coefficient .

Analyzing wavelet	Percentage of isolated positive anomalies	Percentage of isolated negative anomalies	Percentage of false positives	U
coif1	89%	97%	27%	2,0
coif1	88%	94%	24%	2,1
coif1	86%%	91%	21%	2,2
coif1	85%	89%	19%	2,3
coif1	84%	86%	17%	2,4
coif1	83%	85%	15%	2,5

The length of the sliding time window l = 1440 counts, which corresponds to one day (it is determined taking into account the diurnal variations of the GCR).

The use of operation (5) allows one to fix the periods of anomalous increases and abnormal GCR decreases.

3. *To estimate the intensity of the anomaly at time* $t = b$, *we used the value :*

$$Y_b = \sum_s \mathrm{P}^{\mathrm{CR}}_{T^l_s}(W_\Psi f_{b,s}), \quad (6)$$

which in the case of a local increase in GCR is positive, and in the case of a local decrease in GCR, is negative.

2.3 Detailed analysis of ionospheric parameters.

1. The initial time series of ionospheric parameters is represented in the form of wavelet coefficients obtained on the basis of a continuous wavelet decomposition [23] (see representation 4).
2. On the basis of the computational solutions proposed in [18], we determine the values of wavelet coefficients that exceed a given threshold:

$$P_{Q_s}(W_\Psi f_{b,s}) = \begin{cases} W_\Psi f_{b,s}, \text{если} \left| W_\Psi f_{b,s} - W_\Psi f_{b,s}^{med} \right| \geq Q_s \\ 0, \text{если} \left| W_\Psi f_{b,s} - W_\Psi f_{b,s}^{med} \right| < Q_s \end{cases}, \quad (7)$$

Where threshold $Q_S = \theta * St_S$ – determines the presence of an anomaly on the scale s near the point ξ contained in a carrier $\Psi_{b,s}$, θ – threshold coefficient, $St_s = \sqrt{\frac{1}{\Phi - 1} \sum_{k=1}^{\Phi} \left(W_\Psi f_{b,s} - \overline{W_\Psi f_{b,s}} \right)^2}$, $\overline{W_\Psi f_{b,s}}$ и $W_\Psi f_{b,s}^{med}$ – mean and median values, defined in a sliding time window of length Φ.

In work to select the threshold Q_S, the criterion of the lowest error rate was used (the a posteriori risk was assessed and minimized). When assessing a posteriori risk, the geomagnetic activity index K and the level of solar activity were used to determine the state of the ionosphere. The parameter values θ for the area Kamchatka were determined: $2.5 \leq \theta \leq 3.5$ - for analyzing data during periods of high solar activity (parameter $f_{10.7} > 100$); $1.5 \leq \theta \leq 2.5$ - for analyzing data during periods of low solar activity (parameter $f_{10.7} \leq 100$).

3. The moments of exceeding the values of the wavelet coefficients show anomalous changes in the ionospheric parameters. An estimate of the intensity of the anomaly at time $t = b$ was fulfilled on the basis of the following formula:

$$I_b = \sum_s \frac{\left| P_{Q_s}(W_\Psi f_{b,s}) \right|}{\left\| W_\Psi f_{b,s} \right\|_2}, \quad (8)$$

where norm $\left\| W_\Psi f_{b,s} \right\|_2 = \sqrt{\sum_{N_s} \left(P_{Q_s}(W_\Psi f_{b,s}) \right)^2}$, N_s – length of series on scale s.

3 Results of data analysis

In processing, we used minute data from neutron monitors of Kingston stations (Australia) and ionospheric data of Paratunka station (Russia). To assess the state of the geomagnetic field, the Dst component was used.

Fig. 2 shows the results of data analysis for the period January 20-23, 2015 during the magnetic storm. According to space weather data [http://ipg.geospace.ru/] on January 21, due to the accelerated flow from the coronal hole, the speed of the solar wind increased to 550 km / sec, the southern Bz component of the IMF dropped to -15 nT, on January 22 the speed The solar wind remained within 550-400 km / s, Bz = -10 nT. Analysis of galactic cosmic ray data shows the appearance of a positive anomaly about 20 hours before the onset of the magnetic storm (shown in Fig. 2 in red). A few hours before the storm began, the level of GCR dropped (shown in Fig. 2 e in blue). The error of the neural network during the period of lowering level of GCR increased five-fold, and remained high for two days, which characterizes the occurrence of a deep and prolonged Forbush-lowering. During the period of lowering the level of GCR in the ionospheric parameters, a positive anomaly is observed (an abnormal increase in the electron concentration, shown in red in Fig. 2). During the main phase of the storm, the electron concentration remained elevated, the maximum anomaly reached at 05.00 UT on 22 January. Based on the results of processing events in the period 2010-2016. [26] such behavior of the ionosphere is not typical in the analyzed areas and can be observed during periods of moderate magnetic storms.

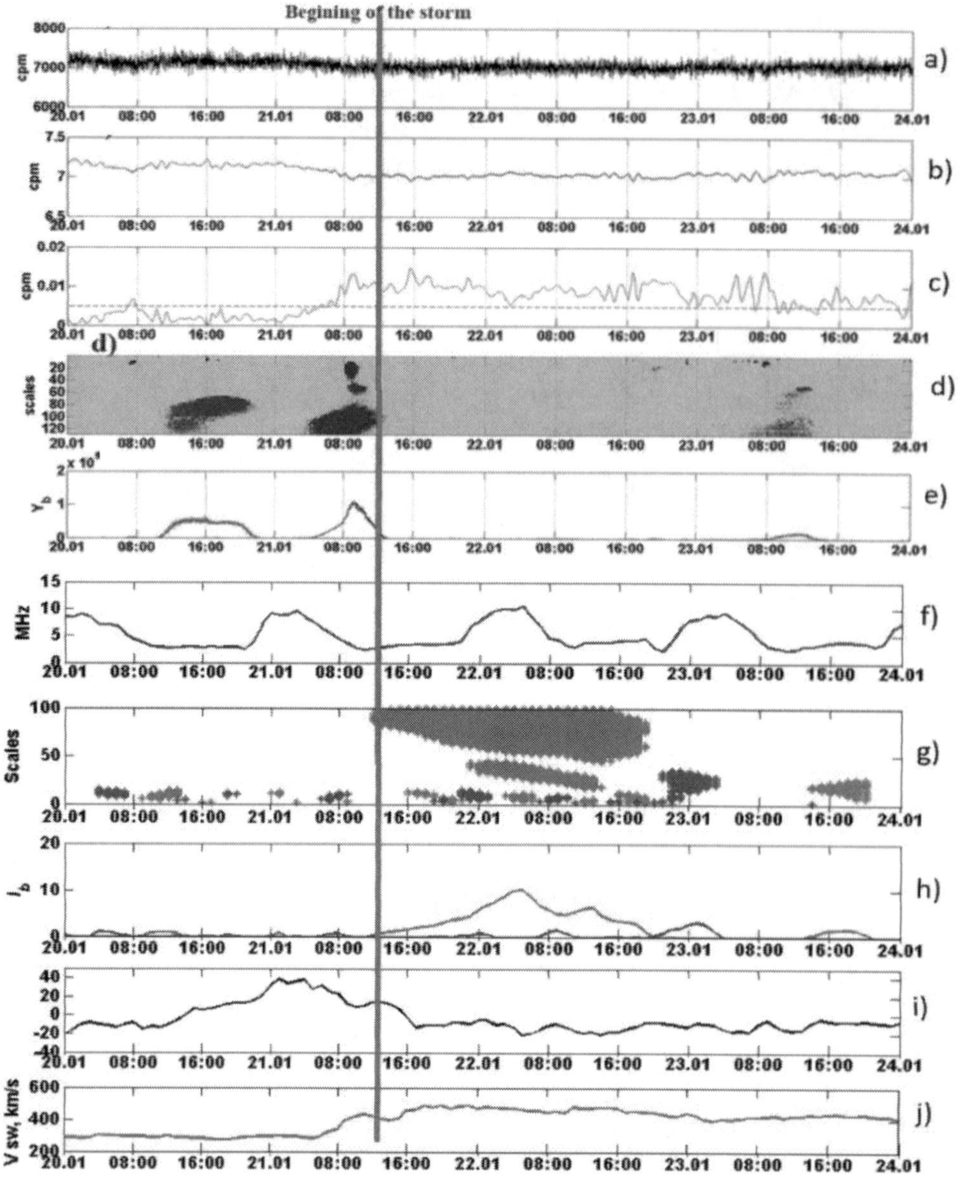

Fig. 2. Results of data processing for the period January 20-23, 2015 a) Data of the neutron monitor station Kingston; B) the quiet component of the neutron monitor data and its model, c) the neural

network error; D) isolated positive (red) and negative (blue) anomalies; E) intensity of positive (red) and negative (blue) anomalies; F) - ionospheric data; G) - isolated ionospheric anomalies; H) - the intensity of ionospheric anomalies; i) - Dst index; j) - solar wind speed..

Fig. 3 shows the results of data analysis for the period February 15-18, 2015. During this period, February 17, 2015, there was a magnetic storm. According to space weather [http://ipg.geospace.ru/] the effect of the solar wind velocity from the coronary hole in the evening time UT on February 16 led to a decrease in the Dst index to -25 nT. Approximately at 6:00 UT on February 17, the speed of the solar wind increased to 380 km/s and continued to grow until the evening of UT on February 17. During this period, the Dst-index continued to decline to -50 nT. The results of the analysis of ionospheric parameters show the occurrence of a prolonged positive anomaly that arose on the eve of the storm (in Fig. 1 it is shown in red). Maximum values of the intensity of the ionospheric anomaly reached at 04.00 UT on February 17, 2015. At the end of the day on February 18, the electron concentration of the ionosphere decreased and a negative anomaly arose (in Figure 1 it is shown in blue). Analysis of galactic cosmic ray data shows that during the period of an anomalous increase in the electron concentration in the ionosphere, an appreciable increase in their intensity was observed (arose 20 hours before the beginning of the magnetic storm, in Fig. 1 it is shown in red). At the time of the onset of the magnetic storm, a negative anomaly is observed in the GCR data, which is probably related to the occurrence of a short Forbush decrease of a small amplitude. As a result of the processing of events in the period 2010-2016, [26] a similar behavior of the ionosphere is typical during periods of increased solar activity and strong magnetic storms.

 DOI: 10.1051/e3sconf/20172001007

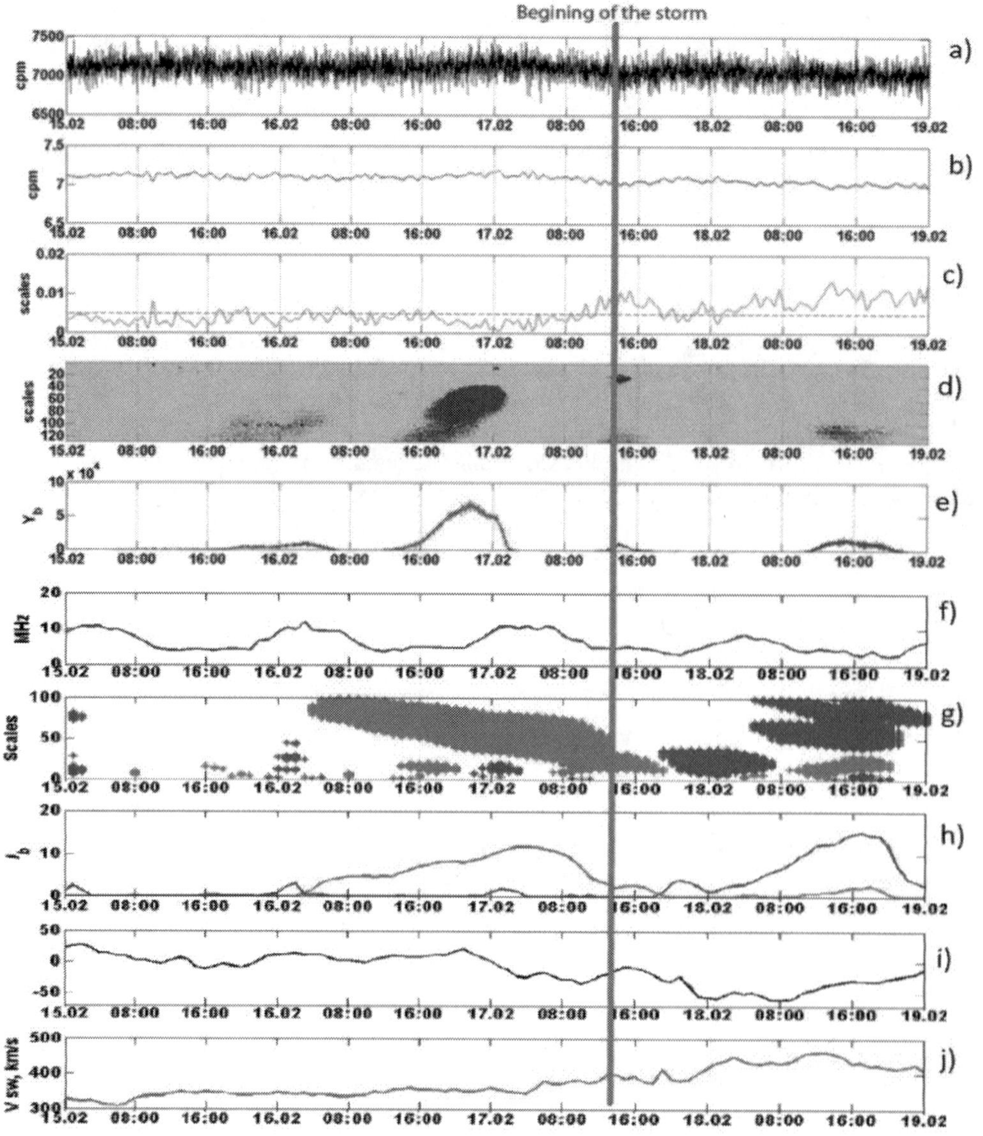

Fig. 3. Results of data processing for the period February 15-18, 2015a) data of the neutron monitor station Kingston; B) the quiet component of the neutron monitor data and its model, c) the neural network error; D) isolated positive (red) and negative (blue) anomalies; E) intensity of positive (red) and negative (blue) anomalies; F) - ionospheric data; G) - isolated ionospheric anomalies; H) - the intensity of ionospheric anomalies; i) - Dst index; j) - solar wind speed.

Conclusions

In the work, the fact of a possible anomalous increase in the intensity of GCR a few hours before the onset of strong magnetic storms was experimentally confirmed. In this case, these effects can be used as precursors of magnetic storms and be an additional factor in the tasks of space weather forecast. At the time of arrival of the shock wave, cosmic ray intensity decreases, typical to the Forbush decreases. An analysis of the ionospheric parameters showed that during periods of increasing galactic cosmic ray intensity in the analyzed regions, anomalous increases in the electron concentration in the ionosphere are typical. The results of [26 - 29] indicate repeated observations of this anomalous effect in the ionosphere on the eve of magnetic storms, but questions related to its nature and mechanisms remain open for the time being. The authors of the paper hold the opinion presented in [27] that such ionospheric anomalies are associated with a certain channel of energy penetration from interplanetary space and the magnetosphere. The study of the reaction of the ionosphere to anomalous changes in the intensity of galactic cosmic rays requires additional studies (using more representative statistics), and in the case of confirmation of the isolated effect, such pre-storm anomalies can serve as a signal for the forthcoming magnetic storm, which is of great practical importance.

The research is supported by a grant from the Russian Science Foundation (project No. 14-11-00194). The authors thank the institutions supporting the data recording stations that were used in the study, and also express their gratitude to the staff of the Institute of Terrestrial Magnetism, Ionosphere and Radio Wave Propagation RAS for the environment and database software that provide prompt information about the parameters of the state of outer space.

References

1. A. Vecchio, M. Laurenza, M. Storini, V. Carbone, Adv Astronomy, (2012)
2. L. M. Zelenyy, I. S. Veselovskiy, T. K. Breus, Plazmennaya geliogeofizika [Plasma geliogeofizika]. (Fizmatlit, Moscow, 2008)
3. I. N. Toptygin, Kosmicheskie luchi v mezhplanetnykh magnitnykh polyakh [Cosmicraysin the interplanetary magnetic field]. (Nauka, Moscow, 1983)
4. M. I. Tyasto, O. A. Danilova, V. M. Dvornikov, V. E. Sdobnov, Izvestiya RAN, ser. fizicheskaya [Izvestiya RAN, physical ser.], 73, (2008)
5. J. Kóta, A. Somogyi, Acta Physica Acad Sci Hung, 27, (1969)
6. P. Paschalis, C. Sarlanis, H. Mavromichalaki, Sol Phys, 182, (2013)
7. O. V. Mandrikova, T. L. Zalyaev, Tsifrovaya obrabotka signalov [Digital signal processing], 1, (2015)
8. V. I. Kozlov, V. V. Markov, Geomagnetism and Aeronomy, 47 (2007)
9. V. I. Kozlov, V. V. Markov, Geomagnetism and Aeronomy, 47 (2007)
10. V. I. Kozlov, V. V. Markov, Geomagnetism and Aeronomy, 48 (2008)
11. O. V. Mandrikova, T. L. Zalyaev, Mashinnoe obuchenie i analiz dannykh [Machine Learning Data Anal.], 1, (2014)
12. O. V. Mandrikova, I. S. Solovev, T. L. Zalyaev, Earth Plan. Space, 66, (2014)
13. O.V. Mandrikova, Yu.A. Polozov, V.V. Bogdanov, E.A. Zhizhikina A Journal of Software Engineering and Applications, 5, (2012)
14. O. Mandrikova, Yu. Polozov, I. Solovev, N. Fetisova, E3S Web of Conferences, 11, (2016)
15. A. V. Belov, J. W. Bieber, E. A. Eroshenko, P. Evenson, R. Pyle, V. G. Yanke, Adv. Space Res, 31, (2003)
16. O.V. Mandrikova, N.V. Glushkova, I.V. Zhivet'ev Geomagnetism and Aeronomy, 54 (2014)
17. D.G. Baishev, A.V. Moiseyev, R.N. Boroyev, S.E. Kobyakova, A.E. Stepanov, O.V. Mandrikova, I.S. Solovev, S.Yu. Khomutov, Yu.A. Polozov, A. Yoshikawa, K. Yumoto, Sun and Geosphere, 10, (2015)

18. O.V. Mandrikova, N.V. Fetisova, Y.A. Polozov, I.S. Solovev, M.S. Kupriyanov, Earth, Planets and Space, 67, (2015)
19. O.V. Mandrikova, N.V. Fetisova (Glushkova), R.T. Al-Kasasbeh, D.M. Klionskiy, V.V. Geppener, M.Y. Ilyash, Annals of Geophysics, 58, 2015
20. O.V. Mandrikova, Yu.A. Polozov, I.S. Solovev, N.V. Fetisova(Glushkova), T.L. Zalyaev, M.S. Kupriyanov, A.V. Dmitriev Pattern recognition and image analysis, 26, (2016)
21. M. Nakamura, T. Maruyama, Y. Shidama Journal of the National Institute of Information and Communications Technology, 56, (2009)
22. C.K. Chui, An introduction in wavelets (Academic Press, New York, 1992)
23. I. Daubechies, CBMS–NSF lecture notes nr. (SIAM, Philadelphia, 1992)
24. O.V. Mandrikova, T. L. Zalyaev, SCM`2017, 1, (2017)
25. T. L. Zalyaev, News of "LETI" [Izvestija «LETI»], (2015)
26. O.V. Mandrikova, N.V. Fetisova, Yu.A. Polozov, Proceedings of the 11th Intl School and Conference "Problems of Geocosmos", (2017)
27. A.D. Danilov, Advances in Space Research, 52, (2013)
28. D. Buresova, J. Laštovička, Adv. Space Res., 39, (2007)
29. L. Liu, W. Wan, M.-L. Zhang, B. Zhao, Ann. Geophys., 26, (2008)

Geospace monitoring for space weather research and operation

Tsutomu Nagatsuma[1,*]

[1] Space Environment Laboratory, Applied Electromagnetic Research Institute, National Institute of Information and Communications Technology, Japan, 184-8795, Tokyo, Koganei, Nukui-Kita, 4-2-1

Abstract. Geospace, a space surrounding the Earth, is one of the key area for space weather. Because geospace environment dynamically varies depending on the solar wind conditions. Many kinds of space assets are operating in geospace for practical purposes. Anomalies of space assets are sometimes happened because of space weather disturbances in geospace. Therefore, monitoring and forecasting of geospace environment is very important tasks for NICT's space weather research and development. To monitor and to improve forecasting model, fluxgate magnetometers and HF radars are operated by our laboratory, and its data are used for our research work, too. We also operate real-time data acquisition system for satellite data, such as DSCOVR, STEREO, and routinely received high energy particle data from Himawari-8. Based on these data, we are monitoring current condition of geomagnetic disturbances, and that of radiation belt. Using these data, we have developed empirical models for relativistic electron flux at GEO and inner magnetosphere. To provide user-friendly information , we are trying to develop individual spacecraft anomaly risk estimation tool based on combining models of space weather and those of spacecraft charging, Current status of geospace monitoring, forecasting, and research activities are introduced.

1 Introduction

It has been sixty years since the world first artificial satellite, Sputnik, was launched on Oct. 04, 1957. After that, space exploration is rapidly increasing, and more than several hundreds of artificial satellites are in orbit circulating around the Earth for practical purposes such as telecommunications, broadcasting, navigation, meteorology, etc. Thus, space assets are one of inevitable social infrastructure now. However, these space assets are vulnerable to space debris and space weather disturbances in geospace[1,2]. Based on the sixty years of space exploration, it was realized that the Sun is major driver of space disturbances around the Earth. The space environment around the Earth (geospace) dynamically varies depending on the solar wind conditions. To mitigate spacecraft anomalies, monitoring and forecasting of geospace is important. National Institute of Information and Communications Technology (NICT) is in charge of operational space

*Corresponding author: tnagatsu@nict.go.jp

weather service in Japan as regional warning centre of International Space Environment Service (ISES) [3].

Because the solar-terrestrial relations are quite complicated, and the part of the physical processes in the relations are still unknown, global network of ground-based and space-based observations needs to be preserved and improved. To contribute global space weather observation network, NICT have been operating regional ground-based observation network, and ground-based data reception facilities for space weather real time data stream from space-based observation. In this paper, NICT's activity about geospace monitoring for space weather research and operations are introduced.

2 Ground-based observation

Based on the international collaboration, NICT has constructed comprehensive ground-based networks for space weather monitoring along the Japanese meridian from the pole to the equator. These networks are used for operation and research of space weather forecasting [3]. For space weather monitoring, most of the data obtained from the networks are available in near-real time. These data are also useful for advanced research of space weather phenomena and for development of space weather simulations and models and its validation. Further, the data can be applied to future numerical space weather forecasting scheme based on data- assimilated simulation and modelling.

Current conditions of geomagnetic disturbances and ULF wave activities which contribute to acceleration of relativistic electron in geospace, are monitored by ground-based magnetometer networks and HF radar at King Salmon. Some of the magnetometers are operated by the international project Russian Auroral and Polar Ionospheric Disturbance Magnetometers (RapidMAG) [4]. Other magnetometers are operated under the research collaboration with IKIR in Russia. The HF radar at King Salmon is operated as a part of the Super Dual Auroral Radar Network (SuperDARN) [5].

EMIC waves play an important role for loss of relativistic electron in geospace. To monitor EMIC waves activity around sub-auroral region, we have installed induction magnetometers recently at sub-auroral region as part of project study of dynamical variation of Particles and Waves in the INner magnetosphere using Ground-based observations (PWING).

3 Space-based observation

Space-based observation is one of the key element for space weather monitoring and forecasting. However, there are limited number of operational satellite in orbit. Scientific satellites for basic research of solar-terrestrial physics contribute monitoring of space weather. Although NICT itself has no own spacecraft for space-based observations, two ground-based data reception facilities are operated for real-time space weather monitoring. One is for Advanced Composition Explorer (ACE) and Deep Space Climate Observatory (DSCOVR). The other is for Solar Terrestrial Relations Observatory (STEREO).

The L1 Lagrange point of the Sun–Earth system is one of the key locations for space weather monitoring. Because the L1 point is located about 1,500,000 km ahead of Earth toward the sun, the condition of the solar wind can be monitored with about one-hour leading time. NASA's Advanced Composition Explorer (ACE) was launched at the L1 point in 1997, and had the "real-time beacon mode" for broadcasting the solar wind and high-energy particle data in real time [6]. NICT has contributed to receiving the real-time beacon mode from the start of the ACE mission. In 2014, we renewed the spacecraft data reception system for ACE data acquisition because of the replacement of old facilities. This new data reception system can be applied to the coming real-time solar wind data from the Deep Space Climate Observatory (DSCOVR).

The Solar Terrestrial Relations Observatory (STEREO) is NASA's scientific mission for solar and interplanetary physics. A subset of the STEREO data is continuously downlinked in the real-time broadcast mode, called the Space Weather Beacon. The solar wind data (*e.g.*, velocity, density, temperature, and vector magnetic fields) and low-resolution images of the solar EUV, coronagraph, and interplanetary space are contained in the Space Weather Beacon [7]. The data obtained from the STEREO Space Weather Beacon mode is used for estimating the recurrence structure of the solar wind, propagating direction of the interplanetary CMEs, and so forth.

Monitoring of high-energy particle environments in geospace is necessary for satellite operation, because high-energy particles are one of the major causes of spacecraft anomaly [8]. When some anomaly happens in a satellite, the operator's quick understanding of the current status of the space environment is needed for anomaly triage [9]. We have collected many types of space-based observations from GOES operational space weather data including high-energy protons and electron data in near real time. In addition, the Space Environment Data Acquisition Monitor (SEDA), which measures high-energy electrons and protons, is on board Himawari-8, a Japanese meteorological satellite in GEO. SEDA have been collected in near real time since Nov. 2014 [10]. Near-real-time SEDA data is provided by Japan Meteorological Agency (JMA) to NICT for the monitoring and forecasting of the space environment around a geostationary orbit. NICT plays roles in the processing and analyses of SEDA data. We are operating a quick-look web page of SEDA data and data server for real-time SEDA data distribution. A registered user will be able to download SEDA data in near real time.

The Energization and Radiation in Geospace (ERG) project is JAXA's science program for studying the acceleration and loss of radiation belt particles based on a small satellite, ground-based space weather observation networks, and computer simulations [11]. ERG was successfully launched on Dec. 20, 2016. ERG is providing real-time space weather data on high-energy particles and magnetic field measurements for the real-time monitoring of geospace. NICT is collecting these data for monitoring and forecasting of the space environment in geospace.

4 Model and user-oriented information

Spacecraft charging is one of the major causes of spacecraft anomalies. There are two types of spacecraft charging. One is surface charging, the other is deep charging. The surface charging is caused by electrons whose energy range is from several tens of keV to several hundreds of keV. Those energy range of electron flux is enhanced due to substorm activity. The deep charging is caused by electrons whose energy range is more than several hundreds of keV. Those energy range of electron flux is enhanced due to variations of radiation belt. So we have been developed prediction model of geomagnetic disturbances including substorm activities, and radiation belt electron flux [12].

However, it is not easy to estimate risks of spacecraft anomalies only from space environment information. Because the threshold of charging condition depends not only on space environment, but also on materials and structure of the satellite itself. This means that the risk level of spacecraft anomaly is not the same for individual satellite. This means that nowcasting and forecasting information of space weather may not be sufficient for satellite operator. For this reason, we start developing a system to estimate the charging risk of individual satellites with different locations based on the combining forecast models of particle environment in geospace and multi-utility spacecraft charging analysis tool (MUSCAT)[13] and to promote the utilization of space weather forecast for satellite

DOI: 10.1051/e3sconf/20172001008

operators[14]. This activity is called as SECURES (Space Environment Customized Risk Estimation for Satellite) under Project for Space Terrestrial Prediction (PSTEP) (figure 1). We try to stimulate dialogues between satellite operator and space weather researcher and forecaster based on SECURES.

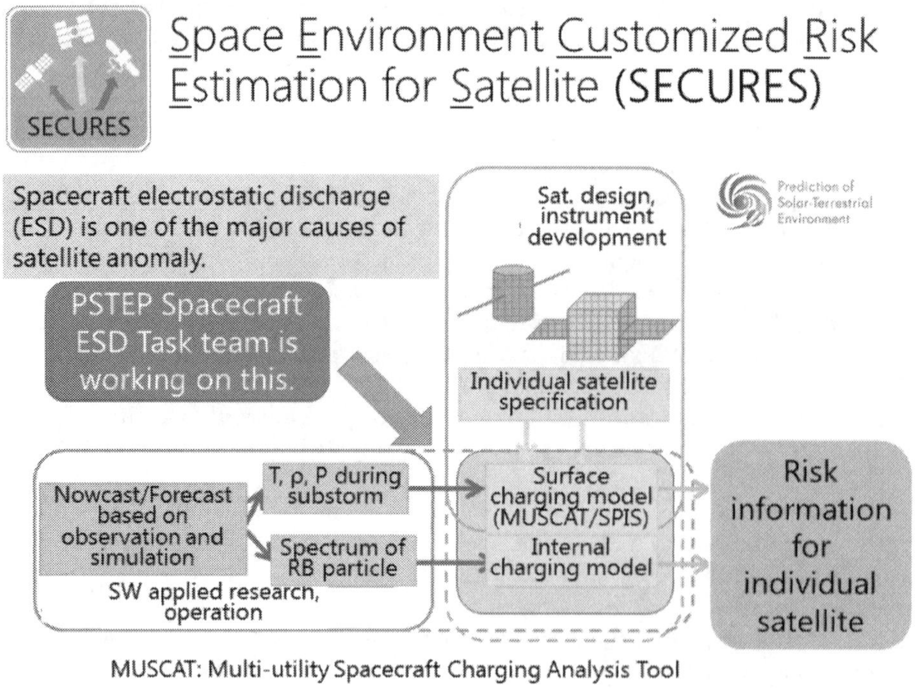

Fig. 1. Concept of SECURES (Space Environment Customized Risk Estimation for Satellite)

5 Summary

Geospace, a space surrounding the Earth, is one of the key area for space weather. Because geospace environment dynamically varies depending on the solar wind conditions. Many kinds of space assets are operating in geospace for practical purposes. Anomalies of space assets are sometimes happened because of space weather disturbances in geospace. Therefore, monitoring and forecasting of geospace environment is very important tasks for NICT's space weather research and development. To monitor and to improve forecasting model, fluxgate magnetometers and HF radars are operated by our laboratory, and its data are used for our research work, too. We also operate real-time data acquisition system for satellite data, such as DSCOVR, STEREO, and routinely received high energy particle data from Himawari-8. Based on these data, we are monitoring current condition of geomagnetic disturbances, and that of radiation belt. Using these data, we have developed empirical models for relativistic electron flux at GEO and inner magnetosphere. To provide user-friendly information, we are trying to develop SECURES, individual spacecraft anomaly risk estimation tool based on combining models of space weather and those of spacecraft charging. SECURES enables us to stimulate dialogues between satellite operator and space weather researcher and forecaster.

Acknowledgment

Part of this work was supported by JSPS KAKENHI Grant Number JP15H05813 and JP16H06286.

References

1. Lanzerotti L.J., Breglia C., Maurer D.W., Johnson, G.K.III, Maclennan C.G., Adv. Space. Res., **22**, 79-82, doi:10.1016/S0273-1177(97)01104-6 (1998)
2. Baker D.N., IEEE Trans. Plasma Sci., **28**, 2007-2016 (2000)
3. Nagatsuma T., Space Weather, **11**, doi :10.1002/swe.20050 (2013)
4. Takahashi K., Meng C.I., Kamei T., Kikuchi T., Kunitake M., Space Weather, **2**, S11003, doi:10.1029/2004SW000116 (2004)
5. Chisham G., Lester M., Millan S.E., Freeman M.P., Bristow W.A., Grocott A., McWilliams K.A., Ruohoniemi J.M., Yeoman T.K., Dyson P.L., Greenwald R.A., Kikuchi T., Pinnock M., Rash J.P.S., Sato N., Sofko G.J., Villain J.-P., Walker A.D.M. Surv. Geophys. **28**, 33-109 (2007)
6. Zwickl R.D. et al., Space Sci. Rev., 86, 633-648 (1998)
7. Biesecker D.A., Webb D.F., St. Cyr O.C., Space Sci. Rev., **136**, 45-65 (2008)
8. Bodeau, J.M., paper AIAA 2010-1608 presented at 48th AIAA Aerospace Sciences Meeting Including the New Horizons Forum and Aerospace Exposition, Orland, Fla., doi :10.2514/6.2010-2010-1608 (2010)
9. O'Brien T.P., Mazur J.E., Fennell J.F., Space Weather, **11**, 49, doi :10.1002/swe.20028 (2013)
10. Nagatsuma T., Sakaguchi K., Kubo Y., Belgraver P., Chastellain F., Muff R., Otomo T., Earth, Planets and Space, **69**:75 doi:10.1186/s40623-017-0659-6 (2017)
11. Shiokawa K., et al., Adv. Space Res., **38**, 8, 1861-1869 (2006)
12. Sakaguchi K., Miyoshi Y., Saito S., Nagatsuma T., Seki K., and Murata K.T., Space Weather, **11**, 79-80, doi :10.1002/swe.20020 (2013)
13. Hatta S., Kim J., J. Plasma Fusion Res., **88**, 92-95 (2012) (in Japanese)
14. Nagatsuma T., Aeronaut. Space Sci. Jpn., **65**, 96-99 (2017) (in Japanese)

 DOI: 10.1051/e3sconf/20172001009

Analysis of ionospheric parameters by the software system "Aurora"

Yury Polozov*, and *Nadezhda* Fetisova

Institute of Cosmophysical Research and Radio Wave Propagation FEB RAS, 684034, Kamchatsky krai, Paratunka, Russian Federation

Abstract. The paper presents methods of modeling and analysis of ionospheric parameters, which realized in the program system of complex analysis of geophysical parameters "Aurora". The methods allow to analyze of characteristic changes in the ionospheric parameters and allocate the anomalous features during periods of ionospheric disturbances. The algorithm parameters are adapted for analyzing the ionospheric data of the Paratunka station (Kamchatka) and based on results of the estimates (station data of Yakutsk, Gakona, etc. were analyzed). Methods can be applied for the mid-latitude region. The system is implemented in the public domain (http://aurorasa.ikir.ru:8580). The research was supported by RSF Grant, project No 14-11-00194.

1 Introduction

The realm of the research concerns the problems of the theory of direct experimental data processing. It is associated with the monitoring and the prediction of the state of near-Earth space. At the present time, the databases of various geophysical parameters (National Geophysical Data Center; MAGBAT; SuperMAG) were formed and provided with the means of primary processing and updating. However, the tasks of effective methods creating for the data analysis, the interpretation of the obtained results and their correspondence to model constructions remain largely open. The influence of solar activity on the magnetosphere and the Earth's ionosphere is complex, many aspects of which have not been sufficiently studied so far [1-6]. The most strong and complex ionospheric disturbances (irregularities) are formed during solar flares and geomagnetic storms. They manifest themselves as significant changes of electron concentration in comparison to some characteristic (calm) level and reflect in the ionospheric parameters [1, 2, 7-9]. Methods and computational algorithms developed by the authors for processing of the ionospheric data and the detection of ionospheric irregularities are described in the present paper. Ionospheric data of IKIR FEB RAS chain located in the northeast of Russia were used for the testing of software tools based on the methods and the algorithms. The software is presented in the paper.

* Corresponding author: up_agent@mail.ru

A large number of studies have been devoted the problems associated with the monitoring of ionospheric conditions and the detection of ionospheric irregularities [1-18]. The International Refence Ionosphere (IRI) model is the international standart of the Earth's ionosphere model [10-13]. It is based on a wide range of ground and space data. Its parameter estimation accuracy depends significantly on the availability of qualitative registered data for a particular region. The accuracy of the IRI model also depends on the level of solar activity, decreasing with a solar activity increase [8, 10-13]. As the IRI model allows to calculate the monthly averages of the ionospheric parameter data for calm conditions, its aplication do not effective for tasks of ionospheric disturbances detection [8, 10]. The recent developments of empirical models, based on neural networks, allow for a significant improvement of the forecast quality in comparison with the IRI model, as they are easy to implement automatically and flexible enough [7-9, 14, 15, 17]. However, these models belong to the "black box" model class. Therefore, for feature spatial description, long training samples with qualitative data are required, which are prone to over fitting and can lead to unexpected results with very noisy data. Also, the operation of systems, based on neural networks, in real time (or close to it) requires the availability of the current information on a complex of geophysical parameters, which is not always feasible [8, 9]. On the basis of the TEC measurements, Afraimovich E.L., Perevalova N.P., and Kosogorov E.A. [1, 7, 16] developed an approach and a software system for the global detection and the monitoring of ionospheric disturbances of natural and technogenic origin (such as, solar flares and eclipses, geomagnetic disturbances, earthquakes, missile launches, industrial explosions, etc.). TEC data are recorded using the GPS navigation system signals. As noted by the authors themselves, the method error increases significantly during ionospheric disturbances [1]. Also, the results of the ionospheric irregularities detection depend on the parameters setting of GPS/GLONASS ground-receivers net (spatial and temporal resolution, sensitivity). These parameters have influence with a type of detected irregularities (large-scale, medium-scale, small-scale, etc.), that can be registered by a specific network [7]. Therefore, the accuracy and the efficiency of the described methods depend largery on the availability of reliable and qualitative historical ionospheric data. Also, they are determined by the availability of the current geophysical parameters determining the state of the near-Earth space.

The methods described in the paper are based on a complex approach that combines the traditional methods with modern approaches of pattern recognition and artificial intellect, as well as information technologies and systems. The methods are realized in "Aurora" software system for complex analysis of geophysical parameters. The system is presented in open access on the website at http://aurorasa.ikir.ru:8580 (mirror is http://aurorasa2.ikir.ru:8580). Analysis of the ionospheric parameters is performed on the basis of multicomponent model (MCM) developed by the authors [17, 18]. MCM allows us to study the typical duirnal and season vatiations of ionospheric process and to detect anomalies. Ionospheric anomalies can occur during increased solar and geomagnetic activities, as well as during seismic events on Kamchatka. The MCM construction is described in the papers [17, 18] in detail. It is based on a joint application of the wavelet transform methods with traditional autoregressive models (ARIMA models). The MCM is theoretically justified [19]. Therefore, its advantage is the ability to obtaine the predicted data with a given confidence probability. Also, ARIMA methods, underlying of the MCM, allow obtaining quite accurate estimates with limited samples. The comparison of the MCM with IRI model, performed in the paper [17], showed that the MCM allowed a more accurate estimate of the predicted data, especially during the solar maximum. This confirms of the efficiency of the suggested approach.

Another approach, implemented in "Aurora" system's ionospheric component, is based on a combination of the wavelet transform and neural networks. Noise component is suppressed on the basis of ionospheric data preprocessing by wavelets. This operation allows improving the efficiency of neural networks for the ionospheric irregularities detection [7]. To more detailed study of the ionospheric parameter dynamics in "Aurora" system, the computational solutions based on the continuous wavelet transform (CWT) are used [17]. The CWT application allows the detection of different scale anomalies in the ionosphere and the estimation of their occurence time, duration, and intensity.

2 Methods for analysis of ionospheric parameters by the software system "Aurora"

2.1 Modeling of ionospheric parameter time variations based on MCM

Modeling of ionospheric parameters in the program system "Aurora" is performed on the basis of following operations:

2.1.1 Using the multiresolution wavelet decomposition (MRA) [17 , 20, 21], the f_0F2 time series is represented as components:

$$f(t)=\sum_k c_{-3,k}\phi_{-3,k}(t)+\sum_k d_{-3,k}\Psi_{-3,k}(t)+e(t)=f_1(t)+f_2(t)+e(t) \quad (1)$$

where $c_{-3,k}=\langle f,\phi_{-3,k}\rangle$, $\phi_{-3,k}(t)=2^{-3/2}\phi(2^{-3}t-k)$, ϕ is a scaling function, $d_{-3,k}=\langle f,\Psi_{-3,k}\rangle$, $\Psi_{-3,k}(t)=2^{-3/2}\Psi(2^{-3}t-k)$, Ψ is a wavelet basis, $j=-3$ corresponds to third level of the wavelet decomposition (decomposition level was determined by algorithm [17]), and $e(t)$ is noise.

The wavelet decomposition (Eq. (1)) is carried out by the orthonormal wavelet basis Daubechies of third order (the basis was determined by minimization of the approximation error) [17].

2.1.2. Using the ARIMA methods [19], the modeling of $f_1(t)$ and $f_2(t)$ components (see Eq. (1)) is performed as:

$$f_1(t)=\sum_{k=\overline{1,N_3^\mu}} s_{3,k}^1\phi_{-3,k}(t),\ f_2(t)=\sum_{k=\overline{1,N_3^\mu}} s_{3,k}^2\Psi_{-3,k}(t), \quad (2)$$

where $s_{3,k}^\mu=\sum_{l=1}^{p_3^\mu}\gamma_{3,l}^\mu\omega_{3,k-l}^\mu+e_{3,k}^\mu$ is an estimated μ th component ($\mu=1,2$), p_3^μ and $\gamma_{3,l}^\mu$ are the order and parameters of the μ th component autoregression, $\omega_{3,k}^\mu=\nabla\beta_{3,k}^\mu$, $\beta_{3,k}^1=c_{-3,k}$, $\beta_{3,k}^2=d_{-3,k}$, $e_{3,k}^\mu$ are the residual errors of the μ th component model, determined as the difference between the actual and modeled values, and N_3^μ is the length of the μ th component.

For the construction and the estimation of the model parameters for the $f_1(t)$ and $f_2(t)$ components, we used historical hourly data of the foF2 of Paratunka site (Kamchatka, Russia, Institute of Cosmophysical Research and Radio Wave Propagation FEB RAS (IKIR FEB RAS), 52° 58'N, 158° 15'E) from 1968 to 2013. During the MCM identification, we considered the dependence of the ionospheric parameters on seasonal variations of solar radiation, solar activity level (SA) and geomagnetic activity. To obtain models describing typical variations of the ionospheric parameters, the time intervals for a relatively calm geomagnetic field (sum of the daily K-indices $\sum K < 24$), without strong seismic events occurring on Kamchatka (without earthquakes of $\sum K \geq 12$, within a 300 km radius from the station), were used as estimates. The estimation of the model parameters was carried out separately for high and low levels of solar activity and for different seasons. The solar activity was estimated according to the average monthly radio radiation at a wavelength of f10.7 (for f10.7 < 100, the activity was considered low, while for f10.7 > 100, it was considered high). Modeling of the foF2 data in the program system "Aurora" is performed for winter (high and low SA) and summer (high and low SA) seasons. The following models are used:

- for winter (high and low SA):

$s^1_{3,k} = -0.62 \cdot \omega^1_{3,k-1} - 0.63 \cdot \omega^1_{3,k-2} + 0.36 \cdot \omega^1_{3,k-3} + e^1_{3,k}(t)$, $s^2_{3,k} = -0.97 \cdot \omega^2_{3,k-1} - 0.93 \cdot \omega^2_{3,k-2} + e^2_{3,k}(t)$

- for summer (high SA) :

$s^1_{3,k} = -0.50 \cdot \omega^1_{3,k-1} - 0.58 \cdot \omega^1_{3,k-2} + e^1_{3,k}(t)$; $s^2_{3,k} = -0.88 \cdot \omega^2_{3,k-1} - 0.80 \cdot \omega^2_{3,k-2} + e^2_{3,k}(t)$

- for summer (low SA):

$s^1_{3,k} = -0.83 \cdot \omega^1_{3,k-1} - 0.73 \cdot \omega^1_{3,k-2} + e^1_{3,k}(t)$, $s^2_{3,k} = -0.95 \cdot \omega^2_{3,k-1} - 0.86 \cdot \omega^2_{3,k-2} + e^2_{3,k}(t)$

where $s^1_{3,k}$ is the estimated component $f_1(t)$, and $s^2_{3,k}$ is the estimated component $f_2(t)$.

2.1.3. The models (see Eq. (2)) describe typical variations of the ionospheric parameters, during the anomalous changes, the model errors increase. Thus, in the program system the detection of anomalies is carried out by the estimation of the model errors:

$$\varepsilon_\mu = \sum_{q=1}^{Q_\mu} \left| e^\mu_{-3,k+q} \right| > T_\mu, \tag{3}$$

$$e^\mu_{-3,k+q} = s^{\mu,fact}_{-3,k+q} - s^{\mu,model}_{-3,k+q} \tag{4}$$

where $e^\mu_{-3,k+q}$ are the residual errors of the μth component model for point $k+q$, $\mu = 1,2$, q is time advance of the data, Q_μ is the length of the data time advance based on the μth component model, and T_μ is the threshold value of the μth component defining the presence of an anomaly. The threshold T_μ was determined for the Paratunka station (Kamchatka, Russia) as:

1. for winter : $T_1 = 1.37 / 1.22$ (high/low SA), $T_2 = 0.97 / 0.73$ (high/low SA).
2. for summer : $T_1 = 1.60 / 1.30$ (high/low SA), $T_2 = 0.88 / 0.80$ (high/low SA).

2.1.4 In the program system the wavelet reconstruction of the component models $f_1(t) = \sum_{k=1,N^\mu_3} s^1_{3,k}\phi_{-3,k}(t)$, $f_2(t) = \sum_{k=1,N^\mu_3} s^2_{3,k}\Psi_{-3,k}(t)$ is carried out to the initial resolution $j = 0$:

$$f^0_1(t) = \sum_k s^1_{0,k}\phi_{0,k}(t), \quad f^0_2(t) = \sum_k s^2_{0,k}\Psi_{0,k}(t),$$

where $s^\mu_{0,k} = \left\langle f_\mu, b^\mu_{0,k} \right\rangle$, $\mu = 1,2$, $b^1_{3,k} = \phi_{-3,k}$ is a scaling function, and $b^2_{3,k} = \Psi_{-3,k}$ is a wavelet basis.

The MCM errors, obtained after the wavelet reconstruction, are defined as:

$$E_{U_0} = \frac{1}{U_0} \sum_{\mu=1,2} \sum_{k=1}^{U_0} \left(e^{\mu}_{0,k+q}(t)\right)^2$$

where U_0 is the length of the observation time window for the resolution $j = 0$ (in the program, an user sets the value of U_0 on the basis of parameter «Error calculation window»).

Fig. 1 illustrates the results of the foF2 data analysis for Paratunka station, performed by the MCM. In the analyzed period (30.05.2013 – 04.06.2013) during the magnetic storm of June 1, 2013, anomalous changes occured in the foF2 data as evidenced by the significant increase of MCM errors (the standart deviation (SD) was 2.5 for the $f_1(t)$ component and 4.5 for the $f_2(t)$ component).

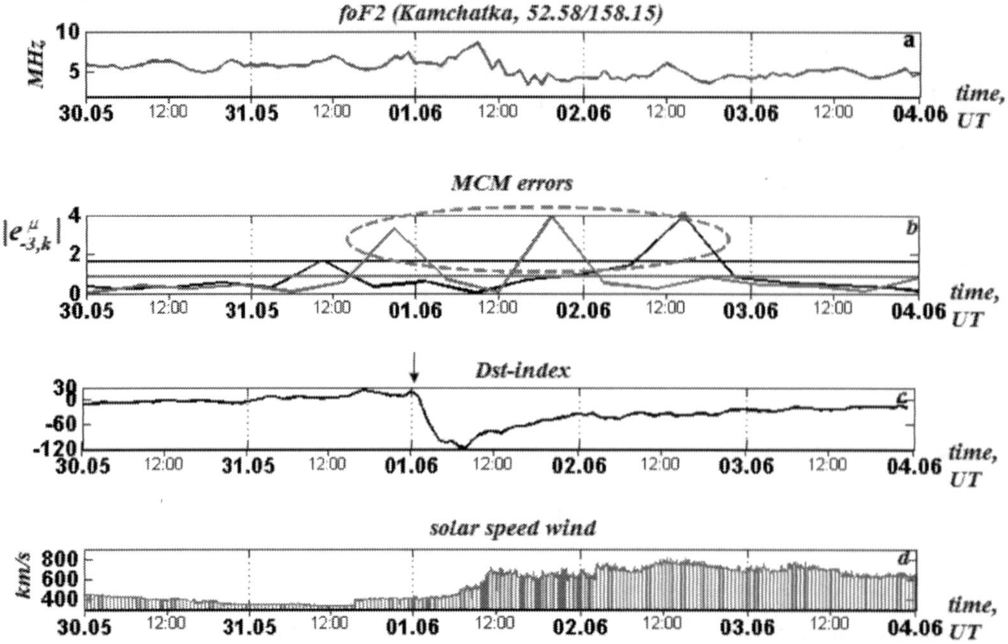

Fig.1. The application of MCM during the magnetic storm on June 1, 2013 (Kamchatka, Russia, time UT): **a** the foF2 data, **b** MCM errors ($f_1(t)$ component (black line) and $f_2(t)$ component (green line)) and their standard deviations, **c** Dst-index of geomagnetic activity, and **d** solar wind speed. Red dashed line on graph b indicates the increase of MCM errors during the magnetic storm. On the graph c the arrow indicates the beginning of the magnetic storm.

2.2 Approximation of ionospheric data based on the wavelet transform and neural network

Approximation of the ionospheric parameter time variations in the program system "Aurora" is performed by the following operations:

2.2.1 Using MRA the data time series is represented as $f_1(t)$ and $f_2(t)$ components (see Eq. (1)).

2.2.2 Using the inverse wavelet transform [21], the reconstruction of the initial resolution $j = 0$ is carried out for component $f_1(t) = \sum_k c_{-3,k}\phi_{-3,k}(t)$:

$$f_1^0(t) = \sum_k c_{0,k}\phi_{0,k}(t)$$

where $c_{0,k} = \langle f_1, \phi_{0,k} \rangle$, $\phi_{0,k}(t) = \phi(t-k)$. Modeling of the $f_1^0(t)$ component is performed on the basis of pre-trained neural networks (NN).

Hourly values of the foF2 from 1968 to 2010 were used for neural networks training. The data with significant gaps were not used in the training, and slight gaps in the data were filled with median values, which calculated for the corresponding hour. To approximation of the foF2 typical variations, the data for the periods without strong magnetic disturbances and seismic activity in Kamchatka were used during neural networks training. The construction of NN was performed separately for different seasons and different levels of solar activity (see 1.2). The quality criterion of the NN training was the condition:

$$E_A = \sqrt{e_A^2(l)} < \varepsilon_A,$$

where $\varepsilon_A > 0$ is predetermined small value, $e_A(l) = \hat{c}_{0,l} - c_{0,l}$ is NN error at time l, $c_{0,l}$ and $\hat{c}_{0,l}$ are desired and actual output value of the neural network.

The constructed neural networks carry out the advance of the data by the following transformation:

$$c_{-3,k+1}(t) = \varphi_3\left(\sum_i \omega_{ij}\varphi_2\left(\sum_q \omega_{qi}\varphi_1\left(\sum_k \omega_{kq}c_{-3,k}(t)\right)\right)\right),$$

where ω_{kq} are weight numbers of q neuron of the NN outpun layer, ω_{qi} are weight numbers of i neuron of the NN hidden layer, ω_{ij} are weight numbers of j neuron of the output layer, $\varphi_1(z) = \varphi_2(z) = \dfrac{1}{1+\exp(-z)}$, and $\varphi_3(z) = a*z+b$. Configuration of the constructed NN is represented on Fig. 2.

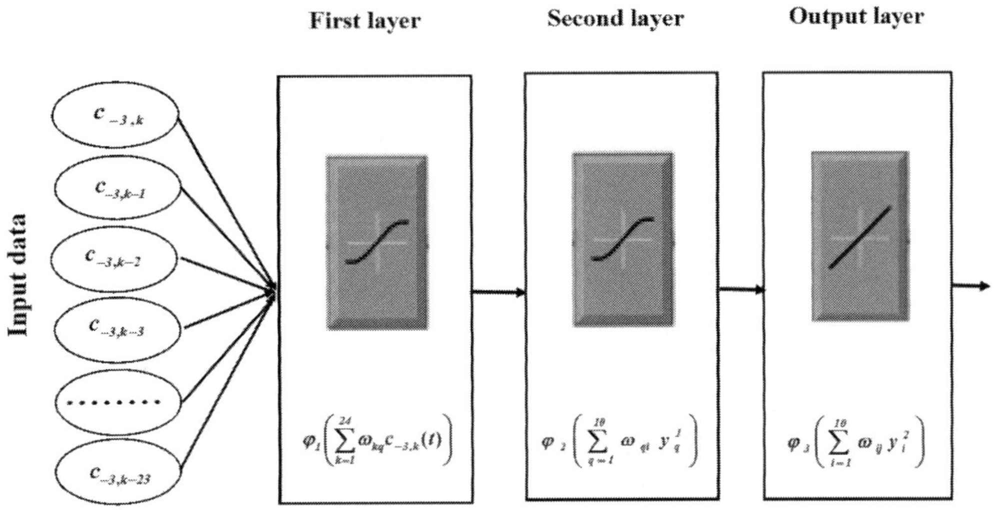

Fig. 2. Configuration of the neural networks

DOI: 10.1051/e3sconf/20172001009

2.2.3 The trained NN allows to simulate typical time variations of data. Thus, the procedure of anomalous changes detection may be based on the analyse of the NN errors $e_A(l)$.

Fig.3 illustrates the results of the foF2 data time variation approximation by the NN during the magnetic storm on March 17, 2013. The NN errors do not exceed 0.18 during calm geomagnetic conditions (Fig. 3, graphs c and d). During the magnetic storm, the NN errors significantly increase, indicating the disturbances of the ionospheric parameter typical variations (Fig. 3, graphs c and d).

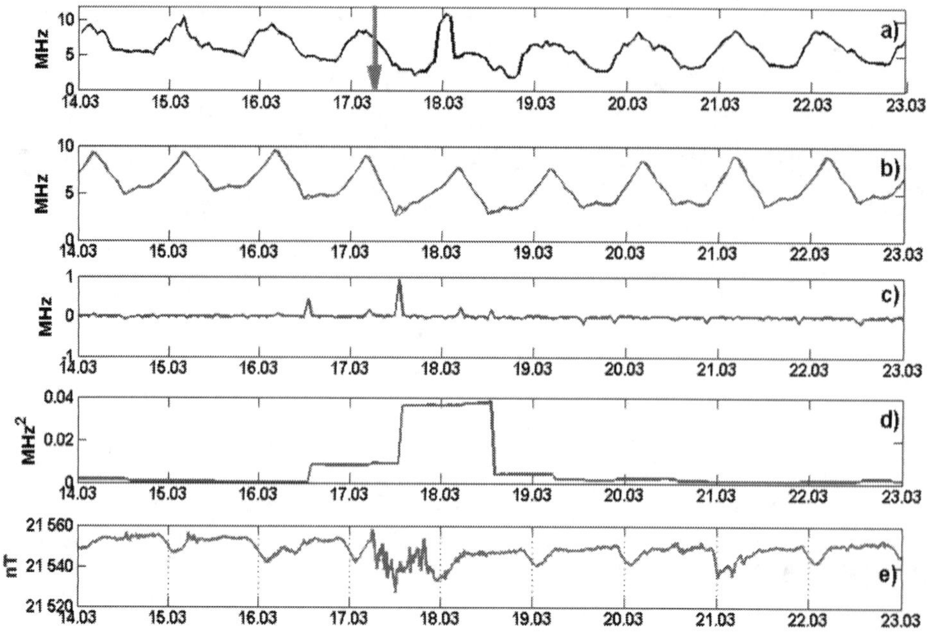

Fig. 3. The results of the foF2 data aproximation by the NN for March 14-22, 2013 : **a** the foF2 data (Paratunka station), **b)** the NN output (blue line), the NN aproximation (red line), **c)** the NN errors, **d)** – the variance of the NN errors, and **e)** H-component of the Earth's magnetic field for Paratunka station (it describes the stage of the geomagnetic field). The error variances were estimated in a moving time window with length of 24 samples (it corresponds to the day). The arrow indicates the beginning of the magnetic storm.

2.3 Anomaly detection in the ionospheric data based on the continuous wavelet transform

To detailed analysis of the ionospheric parameters in the system "Aurora", we use following computing solution:

1. Performance of the continuous wavelet transform of data:

$$W_\Psi f_{b,a} := |a|^{-1/2} \int_{-\infty}^{\infty} f(t)\Psi\left(\frac{t-b}{a}\right)dt,\ f \in L^2(R), a,b \in R, a \neq 0,$$

where a is a scale and Ψ is a basis wavelet.

2. Application of a threshold function for the obtained wavelet components $W_\Psi f_{b,a}$:

$$P_{T_a}(W_\Psi f_{b,a}) = \begin{cases} W_\Psi^+ f_{b,a}, & if\ \left(W_\Psi f_{b,a} - W_\Psi f_{b,a}^{med}\right) \geq T_a \\ 0, & if\ \left|W_\Psi f_{b,a} - W_\Psi f_{b,a}^{med}\right| < T_a \\ W_\Psi^- f_{b,a}, & if\ \left(W_\Psi f_{b,a} - W_\Psi f_{b,a}^{med}\right) \leq -T_a \end{cases} \quad (5)$$

where $T_a = U * St_a$ is the threshold detects the presence of an anomaly for an a scale (the largest possible scale of decomposition in the program $a_{\max} = 150$), $St_a = \sqrt{\frac{1}{\Phi - 1}\sum_{k=1}^{\Phi}\left(W_\Psi f_{b,a} - \overline{W_\Psi f_{b,a}}\right)^2}$, where $\overline{W_\Psi f_{b,a}}$ and $W_\Psi f_{b,a}^{med}$ are the average and median, which, taking into account the diurnal variation of the ionospheric data, are calculated in a moving time window of length Φ for each hour. U describes a value of the threshold. In the program, an user sets the values of the Φ and U parameters. The recommended value is $\Phi = 168$, which corresponds to a time interval of 7 days. Values of the U parameter for Kamchatka were determined by the estimation of the a posteriori risk, and were

$2.5 \leq U \leq 3.5$ for the data analysis during a high solar activity ($f_{10.7} > 100$);

$1.5 \leq U \leq 2.5$ for the data analysis during a low solar activity ($f_{10.7} \leq 100$).

The determined values of U may be recommended for the data analysis of middle latitudes stations.

3. The estimation of the intensity for detected anomalies:

$$I_b^+ = \sum_{a=1}^{a_{\max}} \frac{\left|P_{T_a}^+\left(W_\Psi f_{b,a}\right)\right|}{\left\|P_{T_a}^+\left(W_\Psi f_{b,a}\right)\right\|_2}, \quad (6)$$

$$I_b^- = \sum_{a=1}^{a_{\max}} \frac{\left|P_{T_a}^-\left(W_\Psi f_{b,a}\right)\right|}{\left\|P_{T_a}^-\left(W_\Psi f_{b,a}\right)\right\|_2}, \quad (7)$$

where $\|\cdot\|_2$ is Euclidian norm, $P_{T_a}^+\left(W_\Psi f_{b,a}\right) = \left|P_{T_a}\left(W_\Psi f_{b,a}\right)\right|$, if $\left(W_\Psi f_{b,a} - W_\Psi f_{b,a}^{med}\right) \geq T_a$, and $P_{T_a}^-\left(W_\Psi f_{b,a}\right) = \left|P_{T_a}\left(W_\Psi f_{b,a}\right)\right|$, if $\left(W_\Psi f_{b,a} - W_\Psi f_{b,a}^{med}\right) \leq -T_a$.

Fig. 4 illustrates an example of the aplication for the describing computing solutions. During analyzed period, the strong magnetic storm occured on August 3-4, 2010. On the eve of the magnetic storm, multiscale positive anomaly occured with about from 11.00 UT on August 2 to 10.00 UT on August 4 in the ionospheric data (positive anomaly is shown on Fig. 4 b, in red). The anomaly intensite reached a maximum at 23.00 UT on August 3, 2010 (see Fig. 4 c). In the period from 11.00 to 12.00 UT on August 4, a positive anomaly was replaced by a negative anomaly (a negative anomaly is shown on Fig. 4 b, in blue). Negative anomaly was reached the maximum intensity at 17.00 UT on August 4 and at 10.00 UT on August 5, 2010 (see Fig. 4 c).

 DOI: 10.1051/e3sconf/20172001009

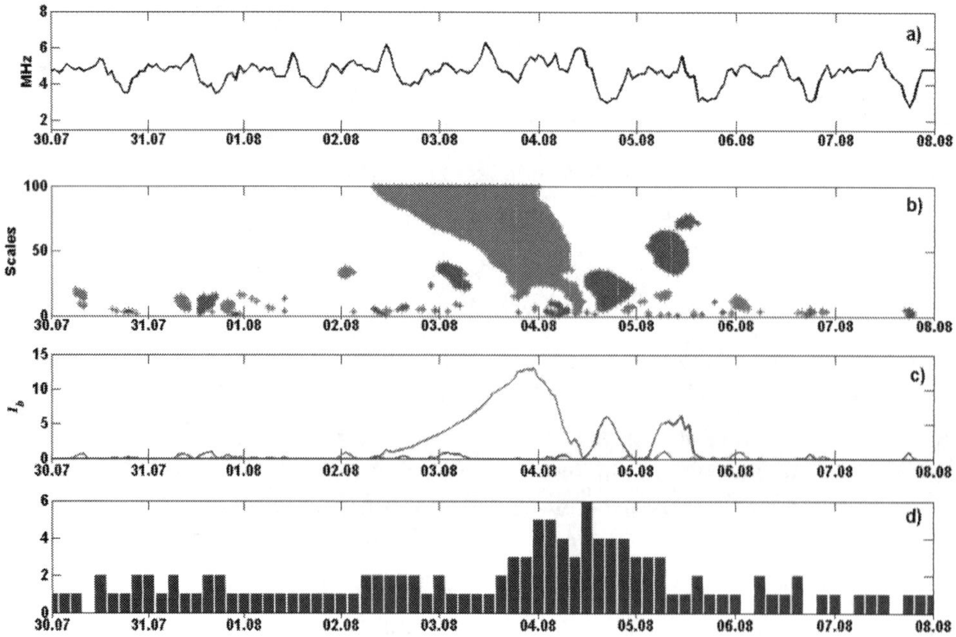

Fig. 4. The results of the foF2 data processing during the magnetic storm on August 3, 2010 : **a** the foF2 data (Paratunka), **b** detected anomalies, **c** the intensity of anomalies, and **d** K-index of the Earth's magnetic field (Paratunka).

3 Conclusions

The methods of the ionospheric parameter analisis, presented in the article, are realized in the ionospheric component of the program system "Aurora". The system is available to users on the website at http://aurorasa.ikir.ru:8580. The following functions are realized in the software system:

- Modeling of time variations by the MCM, based on the multiresolution wavelet decompositions and ARIMA methods;
- Approximation of the ionospheric data based on the wavelet transform and neural networks;
- Detection of ionospheric anomalies and the estimation of their intensity.

The algorithm parameters were adapted for the processing of the Paratunka station data (53.0 N, 158.7 E). Approbation of the algorithms was also performed and positive results were obtained for Magadan station (60.0 N, 151.0 E).

The paper was supported by RSF Grant No. 14-11-00194. The authors are grateful to the organizations carrying out the registration of ionospheric and magnetic data which were applied in the paper.

References

1. E.L. Afraimovich, N.P. Perevalova, GPS-monitoring verhnej atmosfery Zemli [GPS-monitoring of the Earth's upper atmosphere] (GU NU RVH VSNC SO RAMN, Irkutsk, 2006)

2. N. I. Budko, A.N. Zaitsev, A.T. Karpachev, A.N. Kozlov, B.P. Filippov, Kosmicheskaya sreda vokrug nas. [The space environment around us] (TROVANT, Troitsk, 2005)
3. O.V. Mandrikova, T.L. Zalyaev, I.S. Solovev, S.Y. Khomutov, D.G. Baishev AOGS13th (2016)
4. O. Mandrikova, I. Solovev, E. Zhizhikina, S. Khomutov, D. Baishev, 26th IUGG, A23p-059 (2015)
5. O.V. Mandrikova, E.A. Zhizhikina, E3S Web of Conferences, 11, (2016)
6. O.V. Mandrikova, I.S. Solovjev, S.Yu. Khomutov, D.G. Baishev, V.V. Geppener, D.M. Klionskiy, Pattern recognition and image analysis, 26, (2016)
7. O. Mandrikova, Yu. Polozov, I. Solovev, N. Fetisova, E3S Web of Conferences, 11, (2016)
8. N.P. Perevalova, Solar-Terrestrial Physics, 19, (2011)
9. M. Nakamura, T. Maruyama, Y. Shidama, Journal of the National Institute of Information and Communications Technology, 56, (2009)
10. K. Watthanasangmechai, P. Supnithi, S. Lerkvaranyu, T. Tsugawa, T. Nagatsuma, T. Maruyama, Earth Planets Space, 64, (2012)
11. D. Bilitza, B.W. Reinisc, Advances in space research, 42, (2007)
12. O.S. Oyekola, P.R. Fagundes, Earth, Planets and Space, 64, (2012)
13. M.V. Klimenko, Earth, Planets and Space, 64, (2012)
14. R.G. Ezquer, J.L. López, L.A. Scidá, M.A. Cabrera, B. Zolesi, C. Bianchi, M. Pezzopane, E. Zuccheretti, M. Mosert, Journal of Atmospheric and Solar-Terrestrial Physics, 107, (2014)
15. X. Zhao, B. Ning, L. Liu, G. Song, Advances in Space Research, 53, (2014)
16. R. Wang, C. Zhou, Z. Deng, B. Ni, Z. Zhao, Journal of Atmospheric and Solar-Terrestrial Physics, 92, (2013)
17. E.L. Afraimovich, E.A. Kosogorov, L.A. Leonovich, Earth, Planets, and Space, 52, (2000)
18. O.V. Mandrikova, N.V. Fetisova, Y.A. Polozov, I.S. Solovev, M.S. Kupriyanov, Earth, Planets and Space, 67, (2015)
19. O.V. Mandrikova, N.V. Fetisova (Glushkova), R.T. Al-Kasasbeh, D.M. Klionskiy, V.V. Geppener, M.Y. Ilyash, Annals of Geophysics, 58, 2015
20. G. Box, G. Jenkins, Time series analysis: Forecasting and control (Holden-Day, San Francisco, 1970)
21. S. Mallat, A wavelet tour of signal processing (Academic Press, London, 1999)
22. C.K. Chui, An introduction in wavelets (Academic Press, New York, 1992)
23. A.D. Ageev, Neiromatematika: uchebnoe posobie dlya vyzov [Neuromathematics: Textbook for high schools] (IPRZhR, Moscow, 2002)

Two types of geomagnetic storms and relationship between Dst and AE indexes

Lyudmila P. Shadrina[1, *]

[1] *Academy of sciences of Sakha (Yakutia) Republic, Yakutsk, Russia*

Abstract. The study of the relationship between Dst and AE indices of the geomagnetic field and its manifestation in geomagnetic storms in the XXIII solar cycle was carried out. It is shown that geomagnetic storms are divided into two groups according to the ratio of the amplitude of Ds index decrease to the sum of the AE index during the main phase of the storm. For the first group it is characteristic that for small depressions of the Dst index, significant amounts of the AE index are observed. Most often these are storms with a gradual beginning and a long main phase associated with recurrent solar wind streams. Storms of the second group differ in large amplitudes of Dst index decrease, shorter duration of main phase and small amounts of AE-index. Usually these are sporadic geomagnetic storms with a sudden commencement caused by interplanetary disturbances of the CME type. The storms of these two types differ also in their geoeffects, including the effect on human health.

1 Introduction

In this paper, the analysis of geomagnetic storms from 1996 to 2007, during the XXIII solar cycle was fulfilled. This solar cycle was actively discussed in the literature [1-11] mainly because unusually intense geomagnetic storms with Dst index decreases more than 400 nT during the maximum of the cycle (2000-2004) were observed. All of them are devoted to a comprehensive study of these storms – the analysis of solar phenomena, the identification of interplanetary perturbations, ground manifestations in the geomagnetic field, in optical phenomena, propagation of waves, etc. The main attention was paid to studies of extra storms in 2001, 2003 and 2004, and only in [10] the moderate storms were considered. In contrast to these works here are the results of a statistical analysis of geomagnetic disturbances from the point of view of dividing storms into two types – sporadic and recurrent – and not only and not so much intense and extra storms but moderate and weak one are studies.

It is known that sporadic storms are associated with the passage of solar wind disturbed structures, caused mainly by solar flares, with the propagation of coronal mass ejections (CME). The reason for the second type of storms are recurrent high-speed fluxes from coronal holes [8, 9, 11]. So, usually the division of storms into two types is determined mainly by the recurrence of events: geomagnetic storms refer to recurrent ones, if they are repeated with a periodicity of 27 days – the time of rotation of the Sun around its axis.

* e-mail: lushadr@mail.ru

2 Data

In papers [12-14] it was proposed to determine the type of storm using the ratio of ground based indices of geomagnetic activity – AE and Dst. The basis for determining the magnitude of these indices is the horizontal H component of the geomagnetic field. To calculate the AE index, magnetograms of the H components of 12 observatories uniformly distributed over the longitude in the northern hemisphere at auroral or subauroral latitudes, between 60 ° and 70 ° are used. The AE index is defined as the difference between the AL and AU indices (the largest negative and positive variations of the H component) and characterizes the intensity of the auroral zone currents (auroral electrojet). In a quiet time, the value of this index is tens of nT, and during storms and substorms it increases to several hundred and more than a thousand nT. It was shown in [14, 15] that for two types of storms there is a difference in the behavior of auroral activity: during storms of sporadic type, a chain of substorms with a sharp start and a short duration are observed. The depth of modulation of the AE-index is average, up to several hundred nT, there are few intensifications. During recurrent storms, a high level of background perturbations is observed, frequent and intense surges of the AL-index lasting several minutes and a small depth of modulation of the AE-index.

The Dst-index is an axisymmetric component of the disturbed magnetic field relative to the geomagnetic dipole and is determined on the basis of measurements of the magnetic field at four equatorial stations. Its decrease by 30 nT and more, up to 400 nT is a manifestation of the geomagnetic storm, its main phase. According to the magnitude of this decrease storms are dividing into weak ones (30-50 nT), moderate (50-100 nT), strong (100-200 nT) and extra storms (more than 200 nT) [5]. The main reason for this variation is the development of a ring current located at a distance of 2-6 Earth radii due to the injection of charged particles with an energy of 1-300 KeV/nucleon from the tail of the magnetosphere.

The amplitude of Dst-index decrease for the storm main phase is proportional to the number of particles injected into the ring current, and thus is a measure of ring current increase, a measure of the energy input to the interior of the magnetosphere during a storm. The energy characteristic of auroral activity is the sum of the AE index over the same time interval [16, 17]. On this basis, the parameter $\boldsymbol{k}$ is proposed in [12], which is calculated by the formula: $\boldsymbol{k} = \Delta \text{ Dst} / \Sigma \text{ AE}$, where Σ AE is the sum of the AE index during the main phase of the storm T, and Δ Dst is the amplitude of the decrease of the Dst index, the difference between the maximum and minimum value of Dst for the same time interval. It is shown in [12-14] that the parameter $\boldsymbol{k}$ differs significantly for two types of storms. Storms of sporadic (flare) type have $\boldsymbol{k} < 0.010$, for recurrent storms $\boldsymbol{k} > 0.015$. Thus, the value of k suggests to determine the type of storm – sporadic or recurrent.

3 Results and discussion

The result of comparison of statistical characteristics of the two types of storms during 12 years of the XXIII solar cycle are shown in Table 1. In the first column there are Wolf numbers – the number of sunspots characterizing solar activity. One can see the maximum cycle occurs in 2000 and remains high for three years, until 2002. The type of storm was determined by the parameter $\boldsymbol{k}$. For the sporadic storms its annual average varies from 0,0151 to 0,0260 and has a mean value 0.0199, for the recurrent ones – from 0,0064 to 0,0081 and a mean value is 0.0076. The duration of the main phase of sporadic and recurrent storms (Ts and Tr) is significantly different, 10,0 and 16,2 hours, respectively. The average storm intensity for sporadic type is noticeably higher than for recurrent one (respectively, 83.3 nT and 54.7 nT). Auroral activity is almost twice as high for recurrent storms (Σ AE = 7676 nT) than for sporadic ones (Σ AE = 4454 nT). So, the division of storms on two types according to the parameter k made it possible to pick out different quantitative characteristics of the storms.

Table 1. Characteristics of the geomagnetic storms in XXIII solar cycle

Year	Wolf	Sporadic storms				Recurrent storms			
		Ts	Δ Dst	Σ AE	*k*	Tr	Δ Dst	Σ AE	*k*
1996	8,6	8,6	48,1	3100	0,0160	14,0	40,6	5519	0,0079
1997	21,5	9,0	80,9	4321	0,0214	17,2	58,3	7658	0,0080
1998	64,3	9,2	86,3	4863	0,0187	16,0	58,3	7580	0,0076
1999	93,3	10,4	94,0	4709	0,0201	17,4	50,0	8748	0,0064
2000	119,6	10,0	100,3	4487	0,0260	15,5	57,3	8033	0,0075
2001	111,0	9,5	110,9	5123	0,0235	17,8	59,1	8317	0,0074
2002	104,1	10,3	80,1	4302	0,0215	17,8	56,2	7975	0,0077
2003	63,6	9,5	104,6	5411	0,0180	16,7	58,6	9085	0,0071
2004	40,4	12,1	79,0	5241	0,0151	14,0	51,5	6780	0,0078
2005	29,8	7,0	82,1	3881	0,0223	15,3	60,7	7575	0,0081
2006	15,2	11,8	74,2	4435	0,0172	14,5	50,5	6712	0,0080
2007	7,5	12,4	58,9	3575	0,0187	17,8	55,6	8135	0,0073
Mean		**10,0**	**83,3**	**4454**	**0,0199**	**16,2**	**54,7**	**7676**	**0,0076**

Fig. 1 shows a diagram of the quantitative characteristics of storms. Curve shows solar activity (Wolf numbers), solid line – the total number of storms (N), columns with a rare hatching – quantity of sporadic storms (Ns), with a frequent hatching – quantity of recurrent storms (Nr). In the cycle under consideration it should be noted an appearance of a well-known pattern – two maxima of geomagnetic activity – at growth phase (1998 and 2000) and at the decay phase of solar activity (2003 and 2005).

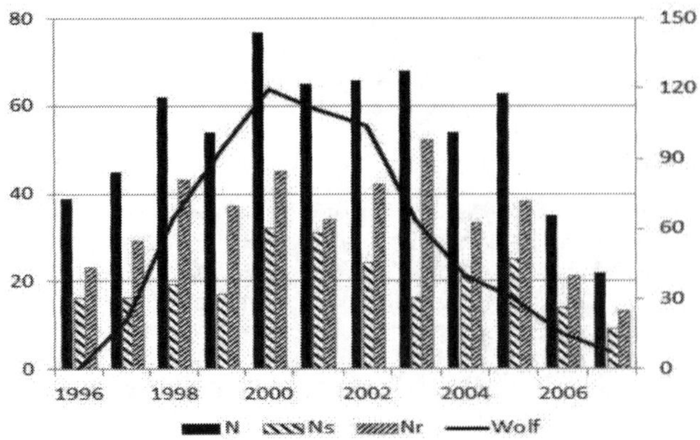

Figure 1. The total number of geomagnetic storms (N), number of sporadic (Ns) and recurrent (Nr) storms in XXIII cycle of solar activity (Wolf).

The ratio of sporadic and recurrent storms, against the expected, persists throughout the cycle: recurrent storms number are always greater than sporadic storms number (Fig. 1). Although it is generally believed that the number of sporadic storms is greater in the phase of solar activity growth, in the phase of recession – less than recurrent [10]. If do not take into account weak storms with ΔDst <50 nT then in the maximum of the cycle (2000 and 2001) and in the minimum (1996, 2004-2007) number of sporadic storms are greater than recurrent ones. Table 2 shows the total number of storms (N), the number of all sporadic (Ns) and recurrent (Nr) storms, as well as the number of moderate, strong and extra sporadic (Ns-sm) and recurrent (Nr-sm) storms, i.e . with the exception of weak storms.

Table 2. The number of two types of storms and regression coefficients

Year	N	Sporadic				Recurrent			
		Ns	Ns-sm	***a***	***b***	Nr	Nr-sm	***a***	***b***
1996	39	16	5	61	168	23	4	62	3008
1997	45	16	12	37	969	29	14	72	3442
1998	62	19	14	48	725	43	18	91	2290
1999	54	17	12	17	3150	37	12	74	5060
2000	77	32	24	24	2074	45	19	103	2454
2001	65	31	26	27	2069	34	20	72	4041
2002	66	24	17	43	753	42	20	138	183
2003	68	16	11	34	1830	52	23	57	5752
2004	54	21	15	25	2562	33	8	107	1267
2005	63	25	19	21	2195	38	18	91	2037
2006	35	14	11	55	364	21	7	117	799
2007	22	9	6	60	67	13	5	82	3553
Mean	**650**	**240**	**172**	**37,7**	**1410,5**	**410**	**168**	**88,8**	**2823,8**

The total number of storms in this cycle was 650, including 240 sporadic and 410 recurrent storms, i.e. Nr > Ns. If we exclude weak storms, we will get approximately equal number of two types of storms, respectively, Ns-sm = 172 and Nr-sm = 168. The number of moderate and strong sporadic storms at the maximum of the cycle (2000-2001) is greater than recurrent storms number, Ns-sm > Nr-sm, which agrees with the generally accepted opinion.

In paper [12] it was proposed the parameter ***k*** on the basis of the analysis of the geomagnetic data of 1967-1972. It was shown that two point clouds can be distinguished on the dot diagrams of Σ AE versus Δ Dst, each of which has its own linear regression dependence. Dot diagrams of 1996-2007 (see Fig. 2) also shows two groups of points for two type of storms for every year. And for each group of points, a linear relationship ΣAE from ΔDst can be find. This was more clearly manifested in 1999-2006. The coefficients ***a*** and ***b*** of the linear regression dependence $\Sigma AE = \boldsymbol{a} * \Delta Dst + \boldsymbol{b}$ were calculated for each group of points (see Table 2). Using the solar cycle average values, we can write the regression relationships:

Sporadic storms: $\Sigma AE = 38 * \Delta Dst + 1410$

Recurrent storms: $\Sigma AE = 89 * \Delta Dst + 2824$

DOI: 10.1051/e3sconf/20172001010

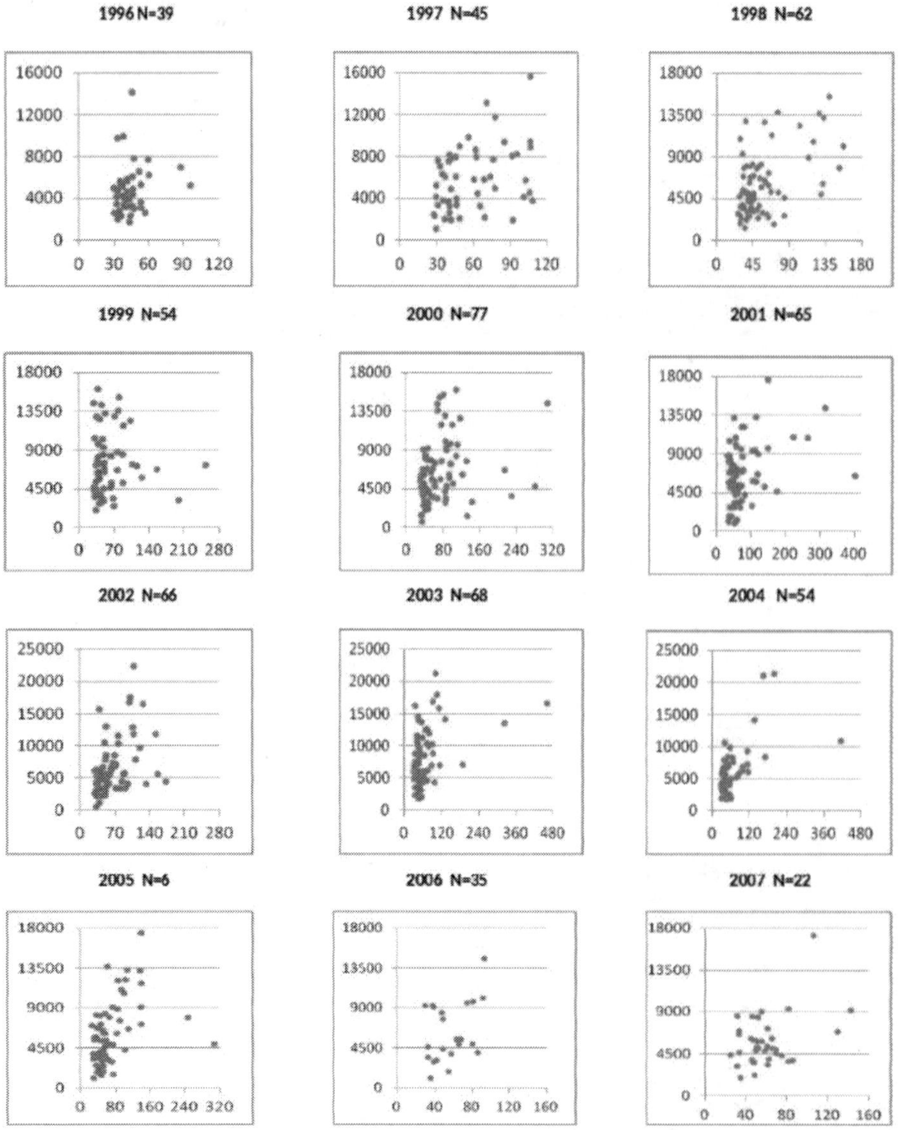

Figure 2. Point diagrams of the dependence of ΣAE (vertical) from ΔDst (horizontal).

And one more feature of sporadic and recurrent storms – the differ in their terrestrial manifestations. In particular, it is shown in paper [18] that an increase in one of the parameter of a human cardiograms is associated with storms. During sporadic storms, the greatest number of patients with such changes occur on the first day of the geomagnetic storm, and for recurrent storms a significant increase in this parameter occurs on the 2nd day before the storm's beginning.

4 Conclusion

Thus, the fulfilled analysis showed:

1. sporadic and recurrent storms differ by the ratio of the energy characteristics of geomagnetic activity of auroral ΣAE and low-latitude ΔDst, each of which has its own linear dependence;
2. the parameter k = ΔDst / ΣAE can be used as a numerical indicator of storm type, it differs significantly for sporadic and recurrent storms: in the XXIII solar cycle average value of k for sporadic storms was 0.0199, and for recurrent storms 0.0076;
3. sporadic storms have a shorter main phase, from 8 to 12 hours, compared with recurrent storms, which have duration of the main phase from 14 to 18 hours;
4. sporadic and recurrent storms differ by their effects on human health;
5. in spite the fact that in the maximum of XXIII solar cycle there were extra storms with an intensity of more than 400 nT, the general statistical regularities persist: moderate and intense storms are more often sporadic, and weak ones are recurrent.

Acknowledgements. This work was made in the frame of project of the Program of Complex Scientific Researches in Republic Sakha (Yakutia) 2016- 2020, Russia and partly supported by RFBR grants № 15-45-05090-p_east_a, and № 15-42-05085-p_ east_a.

References

[1] Yu.I. Yermolaev, M.Yu. Yermolaev, Cosmic Investigation, **41:2**, 105-109 (2003) (in Russian)
[2] N. Srivastava, P. Venkatakrishnan, J. Geophys. Res., **109**, doi:10.1029/2003JA010175 (2004)
[3] J. Zhang et al., J. Geophys. Res., **112**, doi:10.1029/2007JA012321 (2007)
[4] E. Echer et al., J. Geophys. Res., **113**, doi:10.1029/2007JA012744 (2008)
[5] D. Xu et al., Planetary Space Sci., **57**, 1500-1513 (2009)
[6] R.P. Kane, Planetary and Space Sci., **58**, 392-400 (2010)
[7] C. Oprea et al., Ann. Geophys., **31**, 1285-1295 (2013)
[8] N.S. Nikolaeva, Yu.I. Yermolaev, I.G. Lodkina, Geomagnetism and Aeronomy, **51**, 51-67 (2011)
[9] I.G. Richardson, H.V. Cane, Sp. Weather, **9**, doi:10.1029/2011SW000670 (2011)
[10] E. Echer, B.T. Tsurutani, W.D. Gonzalez, J. Geophys. Res., **118**, 385-392 (2013)
[11] A.A. Petrukovich, Yu.I. Yermolaev, N.A. Aismont, Proc. 11-th conf. "Physics of plasma in solar system", Moskow (2016) (in Russian)
[12] L.P. Shadrina et al., Magnetospheric Phenomena, **13**, 33-37 (1989) (in Russian)
[13] L.P. Shadrina, Solar-Terrestrial Physics, **1**, 107-108 (2002) (in Russian)
[14] L.P. Shadrina, V.G. Vasileva, Issledovaniya po geomagnetizmu, aeronomii i fizike Solnza, (ISZF SB RAS, Irkutsk), **112,** 163-167 (2001)
[15] M.G. Gelberg, S.Z. Kershengolz, S.V. Sharaeva, Geomagnetism and Aeronomy, **40**, 26-32 (2000)
[16] S.I. Soloviev et al. Geomagnetism and Aeronomy, **48**, 306-319 (2008)
[17] T.I. Pulkkinen et al., Adv. Space Res., **30**, 2231-2240 (2002)
[18] L.P. Shadrina, S.N. Samsonov, V.I. Manykina, Yakut Medical Journal, **43**, 108-111 (2013)

Observations of electromagnetic radiation of stand EISCAT on Kamchatka

Vladimir Sivokon[1], *Irina* Agranat[1,*]

[1] Institute of Cosmophysical Research and Radio Wave Propagation FEB RAS,, 684034, Kamchatka region, Elizovskiy district, Paratunka, Mirnaya str., 7, Russia

Abstract. In experiments on active influence on an ionosphere powerful short-wave complexes – heater stands are used. As a result of such influence the set of the physical processes [1] which studying represents scientific interest is formed. One of them is formation artificial field-aligned irregularities. Studying properties natural field-aligned irregularities on Kamchatka [2], authors have come to a conclusion about possibility of realisation of the received approaches in their studying, on short-wave lines of the big extent. In work results of supervision on Kamchatka of displays of active influence on an ionosphere in Tromse (Norway) are resulted.

1 Introduction

Interest in the fissile impact on an ionosphere does not weaken for a number of reasons, including, applied character. Need of signaling of management on a global scale on pilotless devices is known, for example. Use of the ionospheric wave duct possessing minimum losses on long-distance trasses can be one of solutions of this problem. But in the presence at the similar channel of a number of advantages it is necessary to resolve an issue of its efficient exaltation. One of ways of "powering" of the channel is scattering on the simulated small-scale inhomogeneities of an ionosphere which are formed under the influence of a potent heating wave. In this case the effectiveness of exaltation is defined by type of scattering and its spatial configuration. As a rule, it is considered that aspect scattering when falling an electromagnetic wave on the magnetofocused inhomogeneities of an electron concentration scattering happens in a cone which top is in the place of their arrangement takes place. But if to consider set of the magnetofocused radiators as the system of reradiators it is possible to draw a conclusion that the effectiveness of exaltation of the ionospheric channel depends on their dimensional orientation and situation in it.

For studying of topology of these inhomogeneities there are various methods, including, methods of a trial wave [3] and aspect scattering [4]. In case of the former the modified area of an ionosphere is periodically irradiated with auxiliary radiation and on change of its characteristics assessment of variations of its properties is made. In a method of aspect scattering the auxiliary transmitter remote from the venue on rather big removal is used, and the dispelled signal is accepted in point located so that trajectories of a direct signal and

* Corresponding author: agranat@ikir.ru

disseminated on the modified area made a corner, the close to 90 degrees. It allows to carry out space selection of direct and dispelled signals and as radiation happens continuously, to estimate speed occurring in the modified area of movements of plasma on the doppler shift of frequency. As the auxiliary transmitter the broadcasting station is, as a rule, used. This approach is inapplicable to studying of the magnetofocused inhomogeneities as polarizable coordination - polarization of a wave at such stations horizontal is not carried out.

As the magnetofocused inhomogeneities are extended along power lines of a magnetic field of Earth, their dimensional orientation is defined by properties of a magnetic field in the venue of experiments. For Tromse, according to the Kyoto university, the magnetic dip is equal 78.230, and the magnetic declination makes 8.740. Therefore, inhomogeneities will be focused at an angle 78.23 degrees in the vertical plane. Follows from the same data that in the horizontal plane the true direction on a geographical pole will differ by 8.74 degrees. We will return to these parameters a bit later, and now we will consider separate nuances of the fissile impact on an ionosphere.

For increase in effectiveness of impact on an ionosphere there is a number of technologies. Let's remind those from them which are essential to our observations: the selection polarizable exaltation of characteristic waves and effect of a magnetic zenith.

When falling a wave of the arbitrariest polarization on an ionosphere there is a formation of two types of waves: ordinary and unusual [5]. Waves differ in polarizable characteristics and if on the transferring party to create the electromagnetic field coinciding in this parameter with one of waves, then this wave will be mainly excited. Ordinary and unusual waves have different refractivities and extend in an ionosphere with different absorption indexes. Respectively extent of indignation of plasma and amount of the formed ionospheric inhomogeneities will be waves of various polarization, various at radiation. Therefore, information on type of the polarization applied in a heating experiment is important for our observations.

The effect of a magnetic zenith consists in essential increase in effectiveness of influence at the radiation of electromagnetic waves not upright up, and along power lines of a magnetic field of Earth. Therefore, proceeding from parameters of a magnetic field of Earth, the maximum of a directional diagram of the radiating antenna will be focused at an angle 78.23 degrees in the vertical plane, and at an angle 188,74 degrees in the horizontal plane.

2 Method of investigations

Let's estimate probability of direct reception of a heating wave on Tromse-Petropavlovsk-Kamchatsky Route. For this purpose we use the VOACAP program, including its on-line version and technical parameters of transmitting antennas of the website EISCAT [8]. The VOACAP program estimates distance between Tromse and Petropavlovsk-Kamchatsky at 6021 kilometer, the azimuth at the same time makes 24,6 degrees. The difference between a principal direction of radiation of a transmitting antenna and an azimuth will make 188,74-24,6 =164,14 degrees. Directional properties of transmitting antennas depend on quantity of elements in a lattice and the used frequency, the figure 1.

According to data, the presented head of experiments Blagoveshchenskaya N.F., in experiments used an antenna lattice number 2. Follows from the schedule that directional diagram width in the vertical plane on the level of 3 dB will be from 15 to 10 degrees depending on frequency. In experiments the frequency of 5421 kHz for which width of a directional diagram is about 10 degrees, fig. 1 was used. Therefore, emitted energy is generally concentrated within 73-83 degrees in the vertical plane. Meanwhile, calculations with use of VOACAP show, fig. 2 that for reception of a heating wave on Kamchatka

 DOI: 10.1051/e3sconf/20172001011

radiation has to occur within 3-10 degrees in the vertical plane.

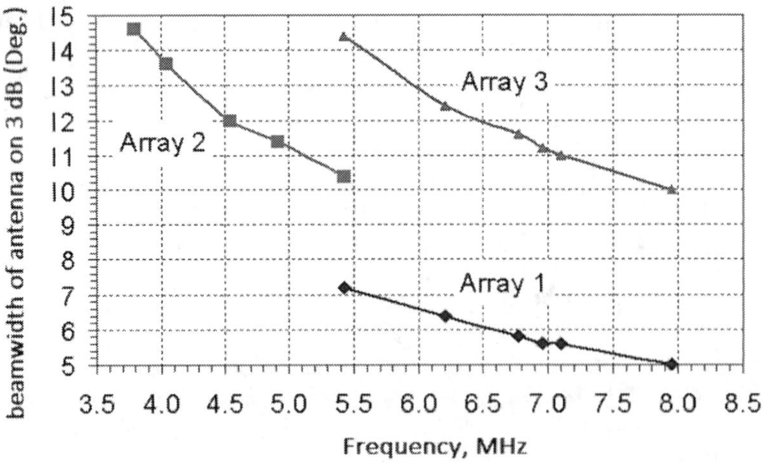

Fig. 1. Directional properties of transmitting antennas of the heating EISCAT stand.

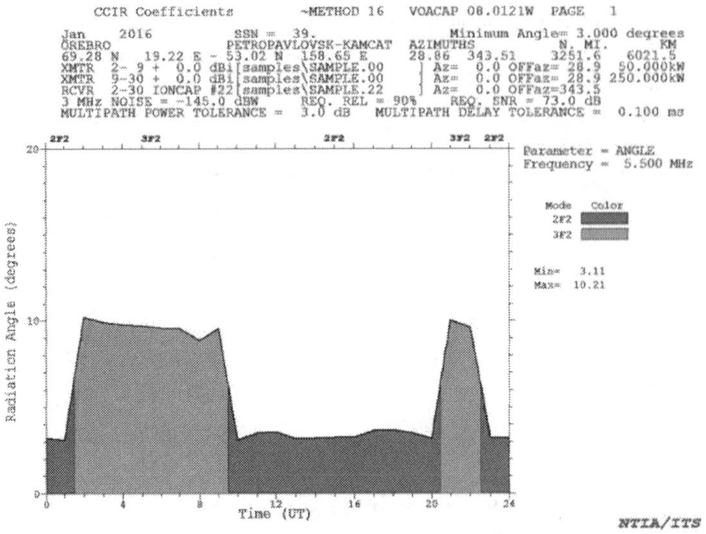

Fig. 2. Estimated values of a corner of radiation in the vertical plane for Tromse-Petropavlovsk-Kamchatsky Route.

From the carried-out analysis it is possible to draw a conclusion that reception of the heating wave extending in a wave guide ionosphere-Earth is improbable.

For definition of conditions of distribution in an interlayered wave guide we use the most probable model in which distribution happens in channel between layers of E and F. Information on a possibility of existence of such condition can be obtained from ionograms of the place of influence, fig. 3

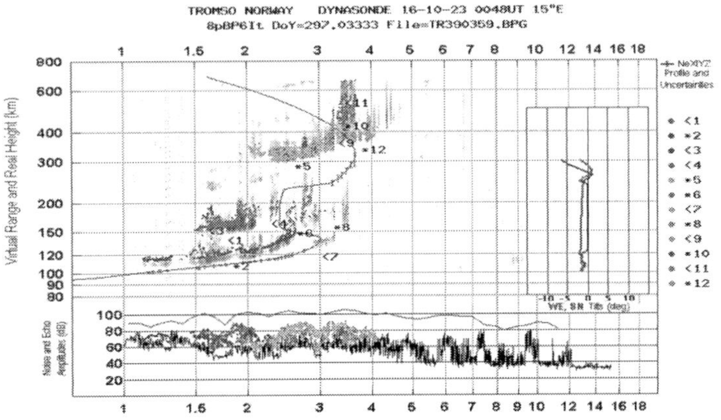

Fig. 3. Ionogram to Tromsø on October 23, 2016 00.48 UT.

In fig. 3 reflections at the level of a layer of E and F are visible that can demonstrate existence of conditions for formation of an interlayered ionospheric wave guide in the venue of impact on an ionosphere.

The majority of the radio route of Tromse-Petropavlovsk-Kamchatsky passes in polar and subpolar latitudes (fig. 4) of which the serious violations of a radio communication caused by magnetic storms are characteristic.

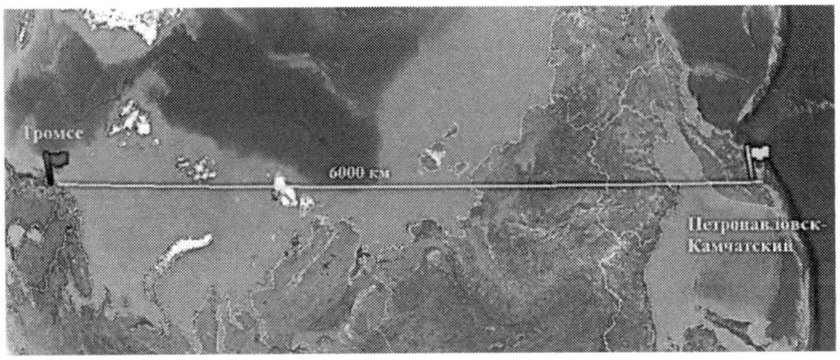

Fig. 4. Radio route of Tromse-Petropavlovsk-Kamchatsky.

Besides in the conditions of the indignant magnetic field of Earth in polar regions the moving ionospheric agitations which can significantly affect results of observations are formed. Using data of the university of Kyoto, Japan, we will estimate a condition of a magnetic field of Earth during experiments. As the parameter in which assessment is made we use Dst variations as unlike the planetary magnetic index Kr, Dst allows to estimate not only extent of indignation of a magnetic field, but also an indignation phase. Earlier it was shown that at the selection polarizable exaltation of characteristic waves in an ionosphere the phase of indignation of a magnetic field plays an important role [6]. As a result of processing of magnetic data it was received distribution on fig. 5.

 DOI: 10.1051/e3sconf/20172001011

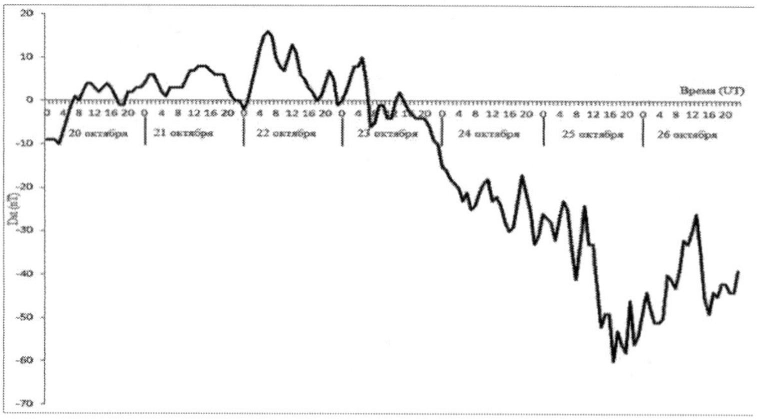

Fig. 5. Variations of a magnetic field of Earth during the period from October 20 to October 26, 2016.

Experiments were made after 12 clocks of world time, and for this time slice, as appears from fig. 5, for reception and the analysis of heating radiation are optimum on October 20-23, 2016.

We will give spectrograms of radiations as an example of the made observations at a frequency of 5423 kHz on October 21 at 14 hours 13 minutes, fig. 6 and 23 of October at 12 hours 43 minutes fig. 7

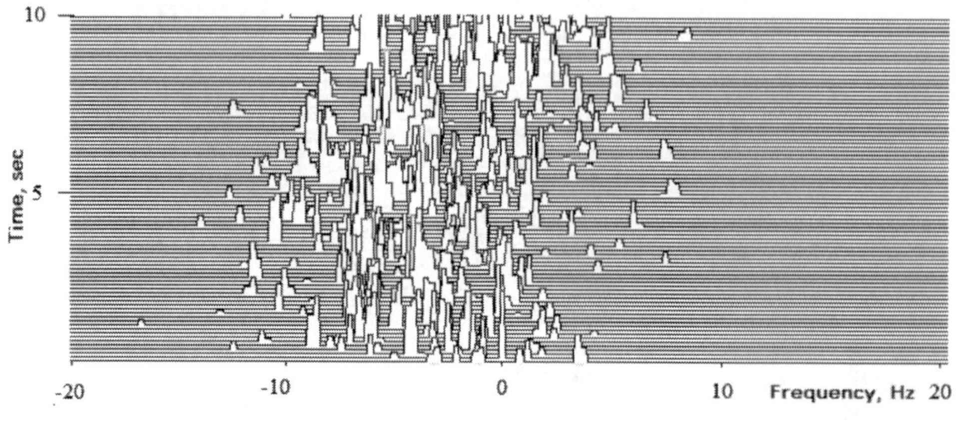

Fig. 6. Spectrogram of radiation on October 21, 2016.

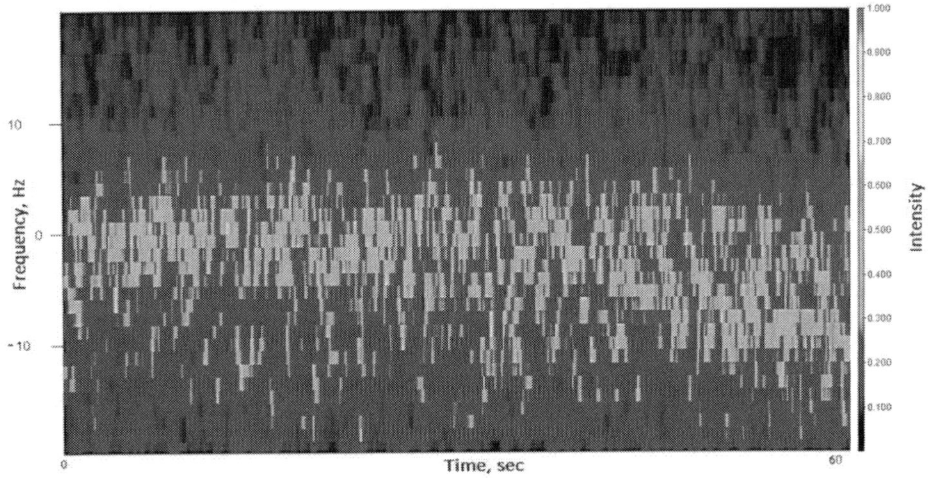

Fig. 7. Spectrogram of radiation on October 23, 2016.

For descriptive reasons ranges of radiations are presented in different formats, and it is apparent that in both cases the broadening of a range and its variation in time, mainly towards the negative detuning of frequency takes place.

Follows from data of the website EISCAT that stability of frequency of the master oscillator used in experiments makes the Hz. Stability of frequency of the radio-receiving device makes the Hz. Then, if to proceed from parameters of the least stable device – the radio-receiving – the absolute instability of frequency caused by technical means will be 0,05423 Hz that is tens times less than the detuning recorded in an experiment.

The shift on frequency can be caused by existence of the moving ionospheric agitations which main mechanism of emergence are geophysical indignations in auroral area. As for the analysis the data obtained in instants when the magnetic field of Earth was quiet are used it is possible to consider that the probability of manifestation of the moving ionospheric agitations in a frequency shift is improbable.

3 Results of investigation and discussion

When carrying out experiments with use for diagnostics of a method of aspect scattering [4,7] frequency shifts of the diagnostic transmitter of the same order, as well as recorded by us were received. In general, they correspond to the data obtained by us in these observations and in work [2]. But, as, it is considered that these indignations consist of the magnetofocused inhomogeneities [8], it is necessary to specify effectiveness of scattering on them. For this purpose we will use the drawing provided in work [8], fig. 8.

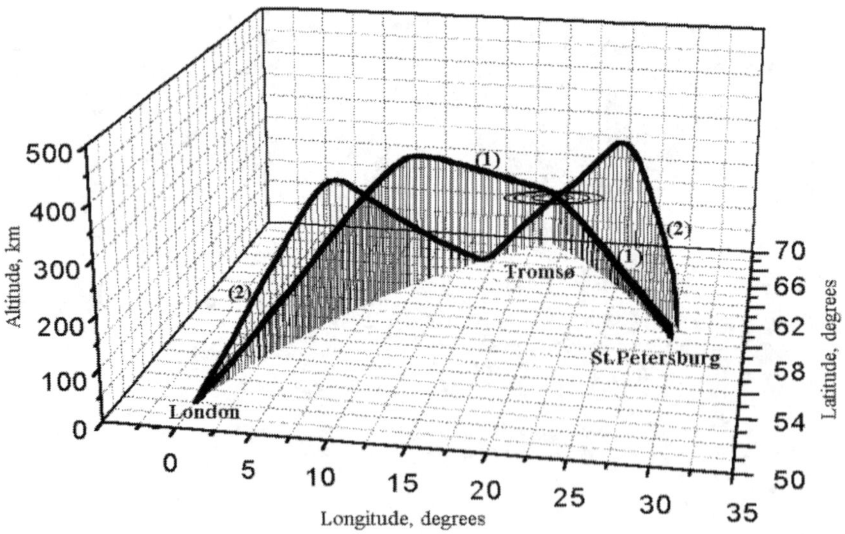

Fig. 8. Geometry of aspect scattering.

Let's remind that the transmitting antenna in London has horizontal polarization. The magnetofocused inhomogeneities over Tromse are located at an angle 78 degrees. In the figure 8 two fashion 1 and 2 are shown, each of which has refraction in an ionosphere, i.e. when falling on area of influence, the wave will be elliptically polarized. Therefore, unlike observations on Kamchatka [2], in experiments [9] it is not possible to provide completely polarizable coordination with the magnetofocused inhomogeneities. As the frequency of a heating wave and a frequency shift is known, it is possible to determine the speed of inhomogeneities which sizes for thirty second intervals of the experiments made on October 21 at 14 hours 13 minutes, are shown in fig. 9.

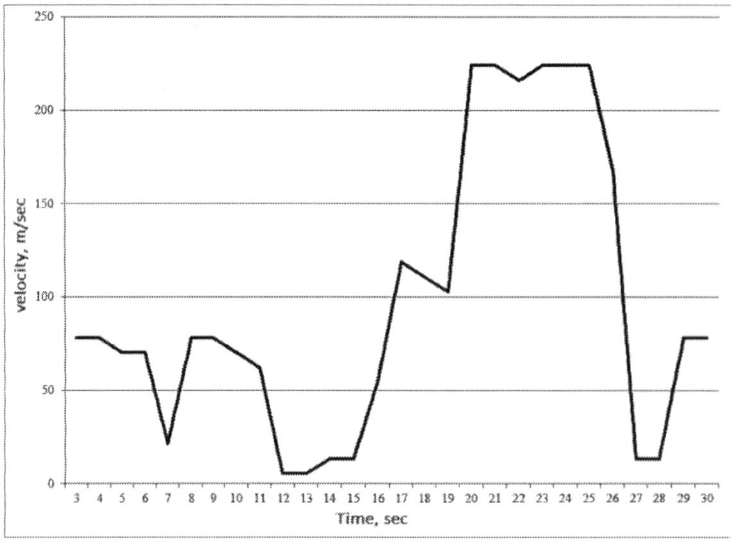

Fig. 9. Speed of the magnetofocused inhomogeneities

4 Conclusions

The possibility of distant selection diagnostics of properties of the magnetofocused inhomogeneities without use of auxiliary transmitters is shown. Observations showed existence in the radiation of the frequencies shift which size is comparable to the sizes received with use of a method of aspect scattering. Peculiar properties Route distribution of a signal allows to allocate as the most probable interlayered ionospheric waveguide. The specifics of exaltation of an interlayered ionospheric wave guide allow to assume that the frequencies shift is caused by driving of the magnetofocused inhomogeneities.

References

1. Gurevich A.V. Uspehi fizicheskih nauk, **177**, 11, 1145-1177 p. (2007)
2. Sivokon V.P., Kalugin I.A., Kobylkin V.S., Popov A.V. Vestnik KamchatGTU, **37,** 14-18 p.
3. Grach S.M., Komrakov G.P., Schvartz M.M., Yurishev M.A. Izvestiya vyschih uchebnyh zavedeniy. Radiofizika, **41,** 8, 966-977 p
4. Blagoveshchenskaya N. F., Borisova T. D., Kornienko V. A., Robinzon T.R., Yoman T.K., Frolov V.L., Ritveld M.T. Solnechno-zemnaya fizika, **2,** 12, 206–209 p. (2008)
5. K.Devis. *Radiovolny v ionosphere*, 504 p. (1973)
6. Sivokon V.P., Druzhin G.I. Geomagnetism i aeronomiya, **46,** 4, 521-524 p. (2006)
7. Blagoveshchenskaya N. F., Borisova T. D., Kornienko V. A., Moskvin I. V., Rietveld M. T., Frolov V. L., Uryadov V. P., Kagan L. M., Yampolski Yu. M., Galushko V. L., Koloskov A. V., Kasheev S. B., Zalizovski A. V., Vertogradov G. G., Vertogradov V. G., Kelley M. C. 24, 2333–2345 p. (2006)
8. Borisova T. D., Blagoveshchenskaya N. F., Moskvin I. V., Rietveld M. T., Kosch M. J., Thid´e B. Geophys., 20, 1479–1486 p. (2002)

Variations of atmospheric electric field and meteorological parameters in Kamchatka in 1997-2016

Sergey Smirnov[1,*]

[1] *Institute of Cosmophysical Research and Radio Wave Propagation FEB RAS, Paratunka, Russia*

Abstract. Analysis of seasonal and annual variations of aero-electric field at a mid-latitudinal observatory Paratunka in Kamchatka was carried out for 1997-2016. Stable seasonal intervals of the highest and the lowest values are observed. Changeability of the annual trend of aero-electric field in the near ground air layer at the observatory located in an active geodynamic region is shown. A large positive trend was changed by a smooth negative one. It is likely to be associated either with radon emanation intensity change in the observatory region or with volcanic activity change in Kamchatka.

1 Introduction

It is known that meteorological processes and relief have the greatest impact on the dynamic characteristics of atmospheric electricity in the near ground layer. Thus, it was interesting to consider the climate features on Kamchatka peninsular in detail.

"Paratunka" observatory ($\varphi = 52°58,3'N, \lambda = 158°14,9'E$) is located in the South of Kamchatka peninsular (Fig. 1), to the west from Avacha Bay, in a valley protected by a ridge of small mountains, at the height of 50 m above the sea level.

2 Climate and weather in the South of Kamchatka

The main climate-forming process in the Far East and in Kamchatka is the atmospheric circulation. It contributes the most into the weather state on the whole. Kamchatka location at the eastern edge of Eurasia, for which significant temperature-pressure contrasts, active cyclone activity, reconstruction and change of the general direction of atmospheric circulation meridional components are characteristic, determine the complicated and changeable weather here.

The main climatic centers of atmosphere impact in the northern part of the Pacific Ocean and in Asia are the Aleutian depression and the Siberian maximum in winter, the northern Pacific maximum and the Asian depression in summer, the interaction of which determined the character of the atmospheric circulation in Kamchatka and the seas surrounding it. One of the peculiarities of atmospheric circulation over the South of the peninsular is the active cyclonic activity, especially in autumn-winter period (Fig. 2) [1]. In summer, the cyclonic activity significantly decreases, however, this fall is determined not mainly by the change of cyclone number, but by the decrease of their depth. Recurrence

*Corresponding author: sergey@ikir.ru

 DOI: 10.1051/e3sconf/20172001012

Fig. 1. Location of "Paratunka" observatory on Kamchatka peninsula.

of pressure formations and the circulation character, associated with them, determine the air mass transfer to the south-eastern coast. It is obvious that marine moderate air prevails here within the whole year. The continental moderate air, which is sometimes noted in the north-east of Kamchatka, is just the transformed marine moderate air coming from the Sea of Okhotsk. The Arctic air comes to the region of the observatory in winter from the northern regions of the Bering sea (marine) or from Kolyma regions (continental) significantly transforming on the way.

Fig. 3 illustrates a histogram of atmospheric pressure distribution at the observatory according to a sensor located at the height of 60 meters above the sea level for 2003 and 2004. Stability of readings for May and September indicates that climate changes in a standard way within the summer-winter period. Less predictable are the winter months. The pressure annual variation is completely determined by the circulation processes over the South of Kamchatka and the waters of the Pacific Ocean, the Bering and the Okhotsk seas surrounding the peninsular. Minimum of the average monthly pressure falls on December -January, the period of maximal development of the cyclonic activity in the Aleutian region. This variation refers to the oceanic type, for which small amplitude of annual variation and the maximum within a warm period are typical.

The climatic summer, in contrast to the calendar one, comes when the average day temperature exceeds 10 degrees by centigrade. Stable summer but not necessarily sunny weather in the region of the observatory is observed only in August. Average monthly temperatures for 2003-2009 are shown in Fig. 4, for 2010–2016 in Fig. 5.

Air average annual temperature at the observatory for 2003–2016 is illustrated in Fig. 6. The average over the whole period was 2.7°C, that corresponds to the value in the paper [1]. It is lower than that in St. Petersburg and in Stockholm located more to the North, at the latitude of more than 60°.

 DOI: 10.1051/e3sconf/20172001012

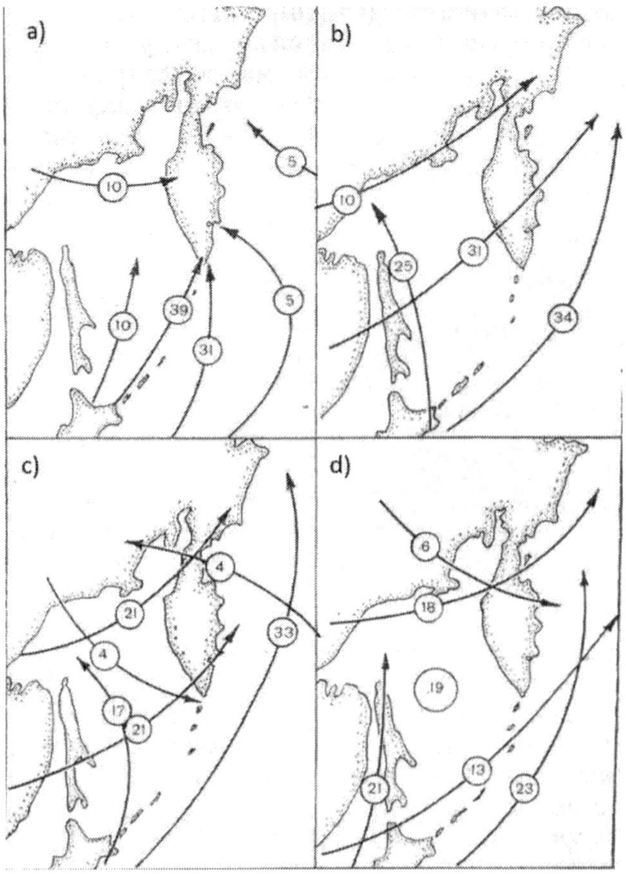

Fig. 2. Typical trajectories of cyclones affecting the weather in the South of Kamchatka and their recurrence [1].

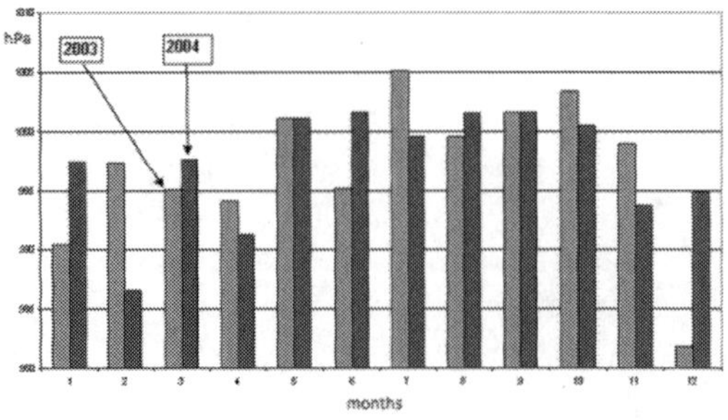

Fig. 3. Histogram of pressure distribution by months for 2003 and 2004.

Wind is horizontal motion of air relatively the earth surface. Wind sensor was mounted on a water tower at the height of 25 meters. Measurements are recorded every 10 minutes. Fig. 7 shows the graphs of wind strength in the region of Paratunka observatory for 2003 and 2004.

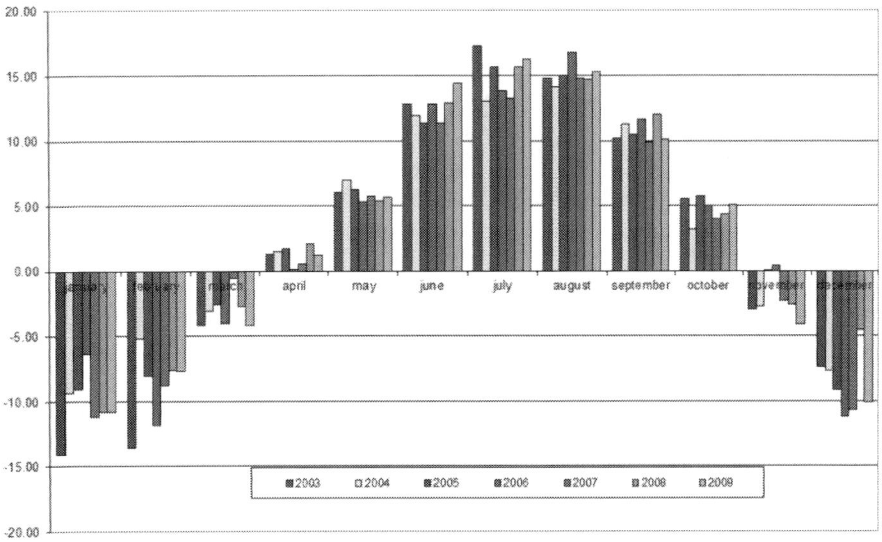

Fig. 4. Annual variation of average monthly temperature for 2003–2009.

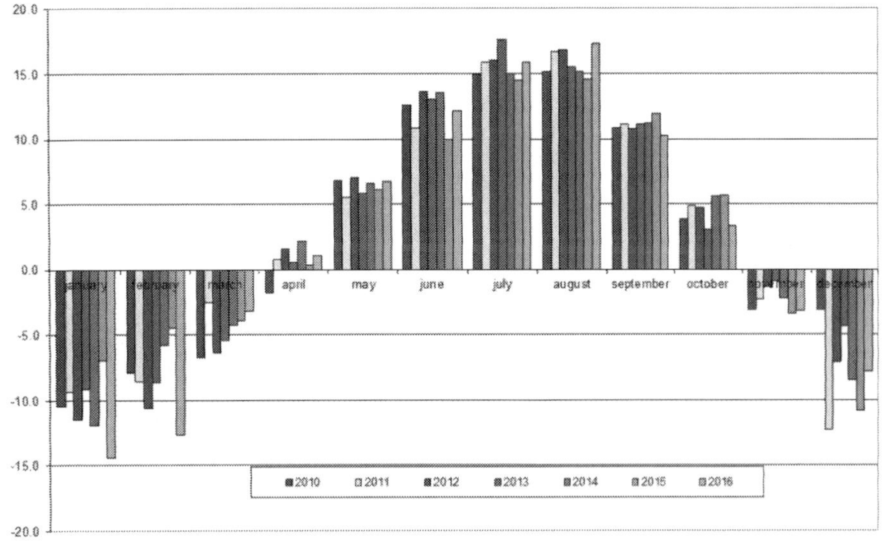

Fig. 5. Annual variation of average monthly temperature for 2010–2016.

Wind is associated with a whole complex of hydro-meteorological characteristics. This relation is clearly seen in the coastal regions. In winter time, fluxes from the sea determine warm weather with precipitation and winds from the mainland (north-western, northern and western in Paratunka) promote dry, cold weather.

Wind regime by the ground surface is formed under the influence of three factors: atmospheric circulation, pressure gradient and the relief. The first two factors are the reason of inter-annual,

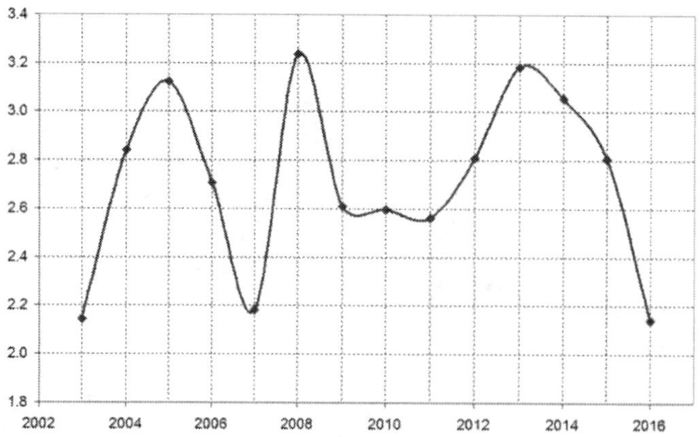

Fig. 6. Average annual temperature for 2003–2016.

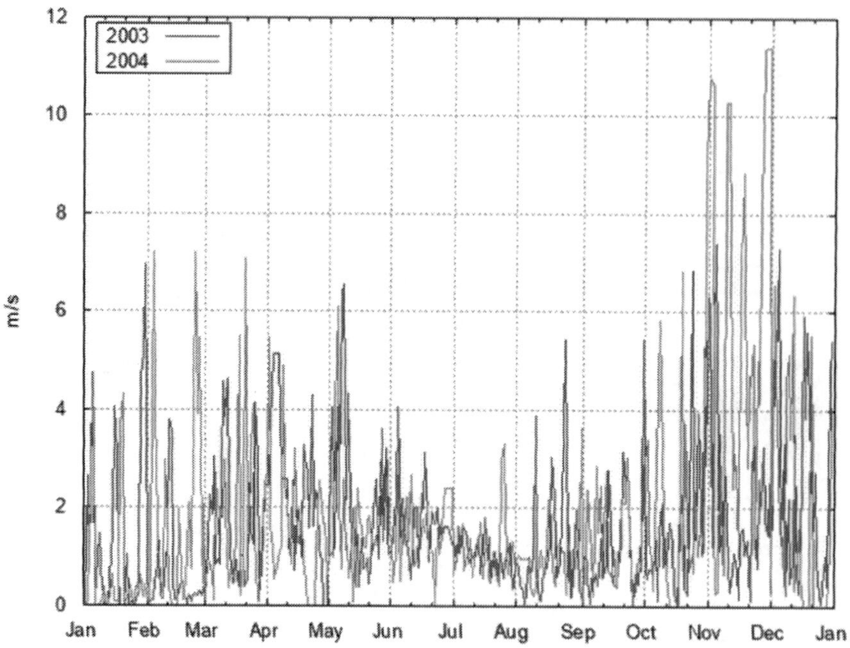

Fig. 7. Wind strength in the region of Paratunka observatory for 2003 and 2004.

seasonal and nonperiodic changes in wind regime. The third one, the relief, manifests itself in the change of wind direction and velocity in the boundary layer, i.e. it introduces some error into the wind regime. Fig. 8 illustrates a wind rose in the region of Paratunka observatory. The highest are the northern and the north-western winds.

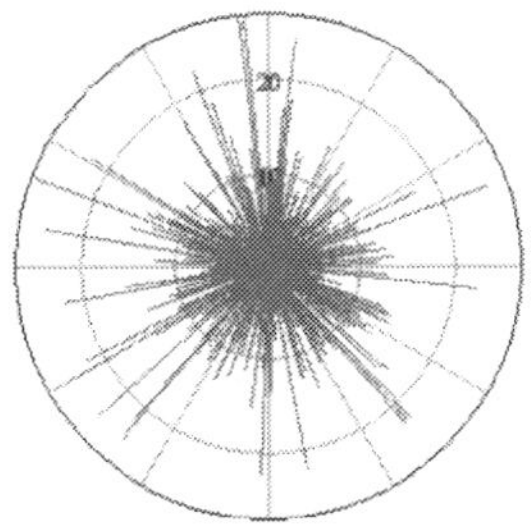

Fig. 8. Wind rose in the region of Paratunka observatory.

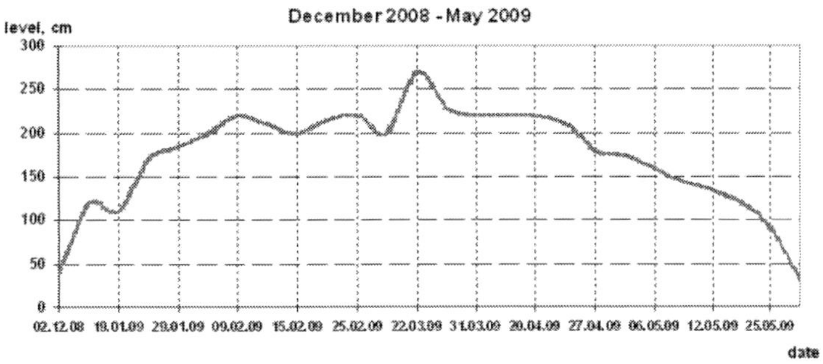

Fig. 9. Snow level in winter 2008-2009.

3 Precipitation effect on electric field strength

The strongest meteorological factor affecting the electric field intensity is precipitation. An example of field behavior when precipitation type changed as snow-rain-snow is shown in Fig. 10. The graph of unipolarity (the relation of electrical conductivity caused by positive ions to the electrical conductivity caused by negative ions $K = \lambda_+/\lambda_-$) during the rain on August 8, 2009 is shown in Fig. The graph of unipolarity during the snowfall on February 26, 2010 is illustrated in Fig. 12. As it is clear from the graphs, these two phenomena are associated with the excess of negative ions (rain) in the first case, and with the excess of positive ions (snow) in the second case. However, there are cases when during strong precipitation they are accompanied by strong sign-variable oscillations. The snow covers the ground in the region of the observatory 7 months a year, from November to the end of May (Fig. 9).

Thunderstorm is an atmospheric phenomenon when multiple electric strokes (lightning) accompanied by thunder occur between separate clouds or between a cloud and the ground. Lightning activity is determined by the processes of atmospheric circulations and mainly by the local physical-geographical conditions: relief, water proximity. It is characterized by the number of days with lightning and lightning duration.

In the result of insufficient heating, lightning in Kamchatka is quite a rare phenomenon within the warm time of the year. Since 1937 to 1982 (45 years) only 22 lightning strokes were registered. They all, except for two, occurred in summer time. In 1980, there were two lightning strokes and both in winter (in March and in December).

In most cases, summer lightning in the South of Kamchatka occurs during cold front passage at the background of relatively high temperatures. Winter lightning is usually associated with occlusion

 DOI: 10.1051/e3sconf/20172001012

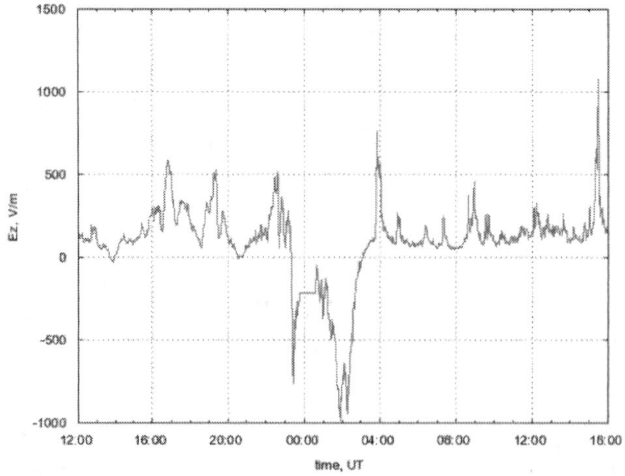

Fig. 10. Electric field behavior in the conditions of precipitation change snow-rain-snow.

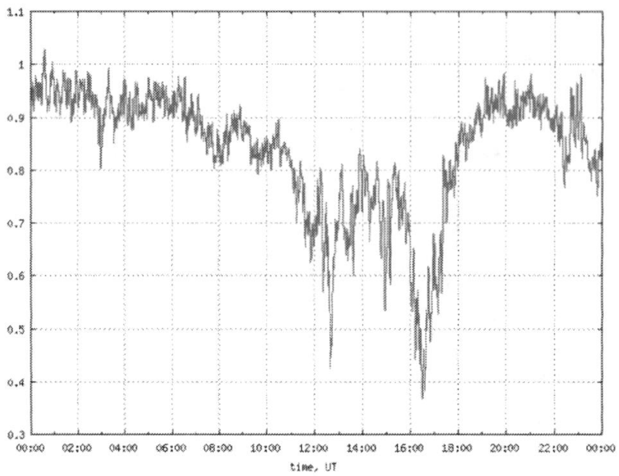

Fig. 11. Graph of unipolarity during the rain on August 8, 2009.

fronts and advection of warm, very humid air mass from the Pacific Ocean. That was just the case, when winter lightning strokes were observed in Petropavlovsk-Kamchatkskiy in March and December 1980.

Average annual number of days with lightning in Paratunka is less than 0.5, the maximal number is 3. The average duration of lightning does not exceed 0.5 of an hour per a day with lightning and 0.2 of an hour on the whole for a year. In the annual variation, lightning is the most probable in August (the average number of days with lightning is 0.2) and in the diurnal variations, it is from 15 to 18 hours [1].

Hail is a rarer phenomenon than lightning. The average annual number of days with hail is 0.2, the maximal number is 3. Hail is noted from August to November. By the way, hail was observed

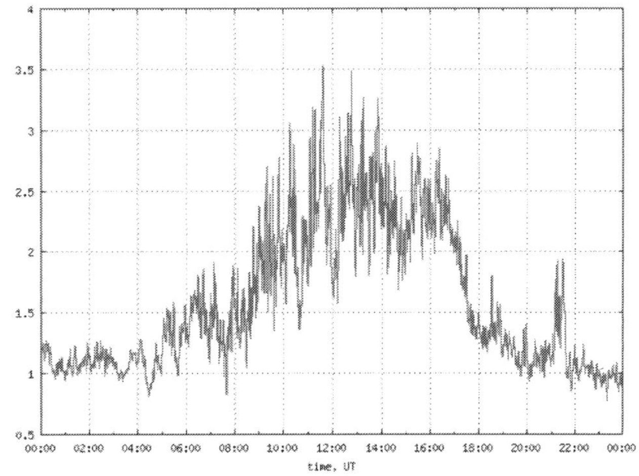

Fig. 12. Graph of unipolarity during the snow on February 26, 2010.

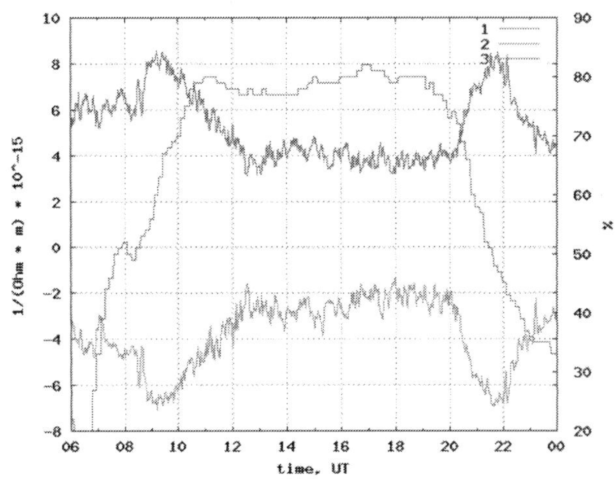

Fig. 13. Fog effect on May 31, 2010 1 - air electric conductivity is caused by negative ions, and 2 - that caused by positive ions with multiplier (-1), left scale, 3 - air humidity, right scale.

only once in August and in November. In September and October, the average number of days with hail is less than 0.1 and the maximal is 2.

In summer-autumn period, fogs are observed in the region of the observatory. Its effect manifests in the air electric conductivity fall and, consequently, in the increase of the electric field level. Conductivity decreases in fog in the result of attachment of light ions to fog particles so that ion mobility decreases. An example of thick fog on May 31, 2010 is illustrated in Fig. 13. The time of fog development can be judged from the graph of air humidity increase (Fig. 13, curve 3). Whereas, the air electric conductivity caused by negative ions (Fig. 13, curve 1) and positive ions (Fig. 13, curve 2) decreased in a bay-like manner.

Electric conductivity behavior can be explained by the following formula:

DOI: 10.1051/e3sconf/20172001012

$$\lambda = e\Sigma(n_{i+}u_{i+} + n_{i-}u_{i-}), \tag{1}$$

where n is the particle concentration and u is their mobility. Water molecules join to the charged particles of aerosol ions that sharply decreases their mobility. Mobility decrease causes electric conductivity fall.

4 Seasonality

Seasonal mode of radon emanation is determined by the value of gas diffusion through the ground. The diffusion is affected by the ground freezing depth in winter time. The depth and the duration of freezing can be judged from the snow cover thickness.

The measurements are carried out on a flat forest clearing, on a plain. The distance to the nearest mountains is not less than 10 km. The snow thickness at the observation point reaches 270 cm (Fig. 9). Fig. 14 shows the distribution of average monthly values of field intensity and the snow level in the vicinity of "Paratunka" observatory. Snow in the region of observations remains for 7 months (from November to the end of May). The higher the level of snow is and the deeper the ground freezing is, the less radioactive gas emanation is. Thus, the electric field intensity increases.

Electric field measurement point in summer time is mounted at the height of 3 meters. The reduction factor which reduces the field value to the ground level is determined by a special procedure. In order to do that, a second sensor is installed even with the ground and the readings from the two sensors are compared for fair weather days. The reduction factor for Paratunka observatory equals 0.343. In winter time, the snow level almost reaches the level of the sensor. Thus, the reduction factor over a season should change within the range from 0.3 to 1. However, for the whole period it is excepted to be a constant of 0.343. That is reflected in the seasonal variation.

Temperature seasonal variation affects the ion mobility that manifests in the air electric conductivity.

It is very difficult to separate the three seasonality factors, the ground freezing factor, the ion mobility change factor and the measurement height point factor.

5 Electric field secular variation

20-year observations of potential gradient variations on the days without precipitation showed two intervals where the trend was positive and than negative (Fig. 15).

Fig. 16 illustrates a graph of M/lg(R) for the closest earthquakes where M is the magnitude, R is the distance to the observatory. Based on Dobrovol'skiy formula, $R = 10^{0.43M}$ [2]. If the value M/lg(R) is more than 2.4, the observation point is within the earthquake preparation zone. In terms of this factor, the following earthquakes were the most significant: October 31, 1998. ($\varphi = 52.95°N, \lambda = 158.26°E$, M=5.5, 2.6 km from the observatory), M/lg(R) = 13.3 and July 29, 2000 ($\varphi = 52.81°N, \lambda = 157.96°E$, M=5.4, 26.5 km from the observatory), M/lg(R) = 3.8. Preparation of these two earthquakes was accompanied by a positive trend of the electric field with the angular coefficient of 0.14 and the root-mean-square deviation (RSD) of 307 V/m. We can suppose that radioactive gas outflow regime changed at that time. After July 2000, the trend became negative with the angular coefficient of -0.16, RSD was 293 V/m.

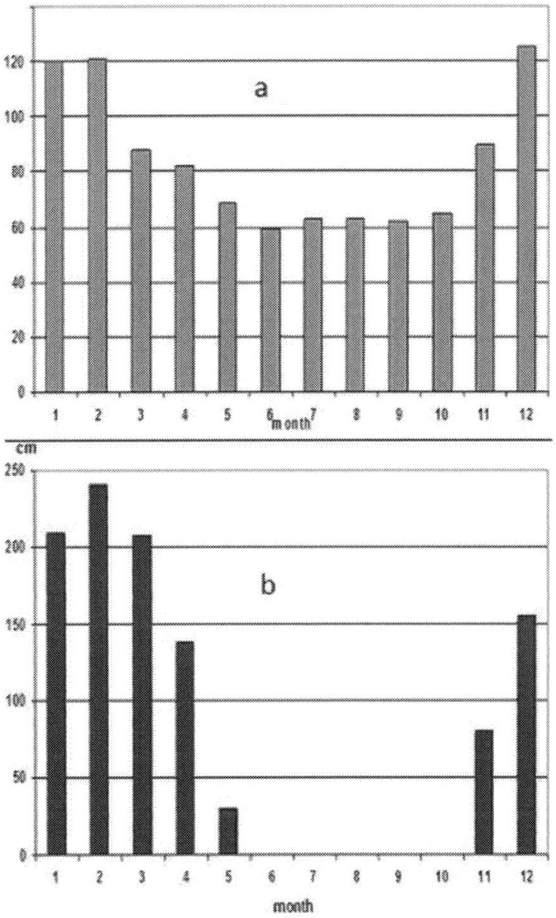

Fig. 14. Histograms of month distribution of electric field intensity (1997–2009) (a) and of snow level (2008–2009) (b) at "Paratunka" observatory.

6 Discussion of the results

Light ion mobility $u_{1,2}$ depends on pressure P(z) and air temperature T(z) and is determined by the expression [3]:

$$u_{1,2} = u^0_{1,2} \cdot \frac{P(0)}{P(z)} \cdot \frac{T(z)}{T(0)}, \tag{2}$$

where P(0) = 1013 mbar, and T(0) = 273°K.

In stationary conditions, ion concentration is determined by the balance equation:

$$\frac{dn_{1,2}}{dt} = 0 = q - \alpha n^2_{1,2} - \beta n_{1,2} N, \tag{3}$$

where q is the ion formation intensity, α is the recombination factor, β is the factor of attachment of light ions to aerosol particles, N is the aerosol particle concentration.

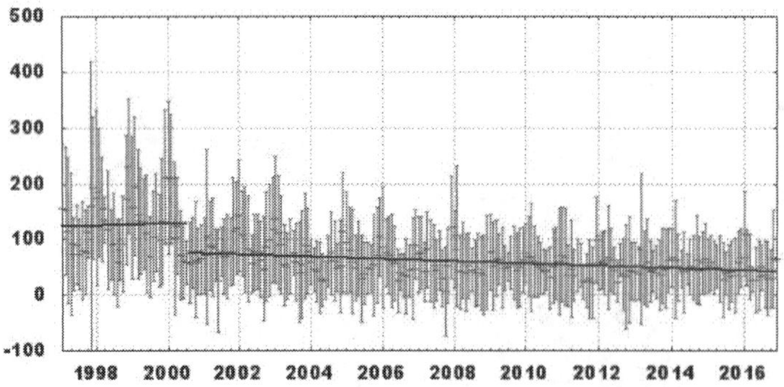

Fig. 15. Potential gradient, root-mean-square deviations and trend lines.

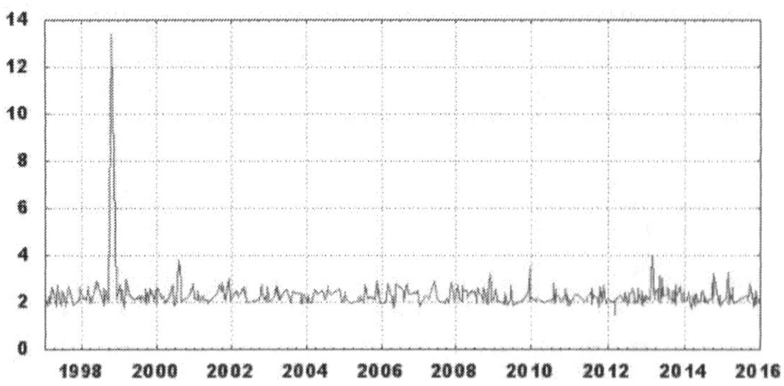

Fig. 16. Graph of earthquakes M/lg(R), where M–magnitude, R–distance to the observatory.

In stationary cases, excluding the effects of global electric circuit, convective and turbulent processes, we can consider the relation of electric field intensity with air electric conductivity as $E \propto \lambda^{-1}$. In its turn, λ, according to equation (1), is proportional to ion concentration (n) and mobility (u). The concentration is $n \propto q^{1/2}$, and the mobility for large averaging is proportional to temperature $u \propto T$. Thus, seasonal variation of electric field intensity is determined by two processes. In terms of ion mobility, the field intensity is associated with air temperature $E \propto T^{-1}$. And in terms of ion concentration, it is associated with seasonal variation of radioactive gas emanations $E \propto q^{-1/2}$.

Kamchatka is one of the most active geodynamic regions. Electric field annual variation can be determined by the annual variation of radioactive gas emanations from one side, and by air pollution of volcanic outbursts, from the other side.

We consider equation (3) more in detail. For example, for positive ion concentration:

$$\frac{dn_+}{dt} = 0 = q - \alpha n_+ n_- - \beta n_+ N_- - \gamma n_+ N_0, \qquad (4)$$

where N_- is the concentration of negatively charged heavy ions, N_0 is the neutral particle concentration, α, β, γ are the corresponding recombination factors. Analogous equations can be written for

n_-, N_-, N_+ and N_0. In the stationary state and in still air, $dn_+/dt = 0$. Thus,

$$n_+ = q/(\alpha n_- + \beta N_- + \gamma N_0) \tag{5}$$

We can take into account that $\beta N_- + \gamma N_0 >> \alpha n_-$ and $\beta \approx \gamma$. We denote $N_- + N_0 = N$. On the assumption that approximately $n_+ = n_-$ and $u_+ = u_-$, the dependence of electric conductivity on heavy mixture concentration is estimated as follows:

$$\lambda \sim 1/N \tag{6}$$

29 active volcanoes are located in Kamchatka. Field intensity annual variation, except for natural radioactivity, can be determined by volcanic outburst mixture in the air $E \propto N$.

7 Conclusions

Based on the analysis of meteorological and climatic parameter effect on the Earth electric field in the near ground atmosphere in Kamchatka, we can conclude that:

1. Precipitation has the greatest impact on electric field intensity variations. To carry out field and observatory measurement, one has to control it as a separate type of observations or by signal form and amplitude.
2. Electric field seasonal variation in the conditions of absence of precipitation is determined by three factors, they are: air temperature affecting the ion mobility, ionization by natural radioactivity affecting the ion concentration in the air, measurement height point change determined by snow cover thickness.
3. Two trends were discovered in the electric field secular variation measured in the conditions of absence of precipitation over the 20-year observation period at Paratunka observatory. From 1997 to July 2000, the trend was positive and then to January 2017, the trend was negative. Natural radioactivity and contaminants in the air from volcano eruptions may affect the secular variation.

References

[1] Kondratyuk V. I. *Climate of Petropavlovsk-Kamchatksiy* (Leningrad, Gidrometeoizdat, 1983) 150
[2] Dobrovolsky I.P., Zubkov S.I. Pure Appl. Geophys. **117** 1025–1044 (1979)
[3] Brikar Dzh. *Radioactivity and pollution effects on atmospheric electricity elements. Problems of atmospheric electricity* (Leningrad, Gidrometeoizdat, 1969) 68–105.

Monitoring of Space and Earth electromagnetic environment by MAGDAS project: Collaboration with IKIR – Introduction to ICSWSE/MAGDAS project

Akimasa Yoshikawa[1,*], *Akiko* Fujimoto[1], *Akihiro* Ikeda[1], *Teiji* Uozumi[1], and *Shuji* Abe[1]

International Center for Space Weather Science and Education, Kyushu University, Japan

Abstract. For study of coupling processes in the Solar-Terrestrial System, International Center for Space Weather Science and Education (ICSWSE), Kyushu University has developed a real time magnetic data acquisition system (the MAGDAS project) around the world. The number of observational sites is increasing every year with the collaboration of host countries. Now at this time, the MAGDAS Project has installed 78 real time magnetometers – so it is the largest magnetometer array in the world. The history of global observation at Kyushu University is over 30 years and number of developed observational sites is over 140. Especially, Collaboration between IKIR is extended back to 1990's. Now a time, we are operating Flux-gate magnetometer and FM-CW Radar. It is one of most important collaboration for space weather monitoring. By using MAGDAS data, ICSWSE produces many types of space weather index, such as EE-index (for monitoring long tern and shot term variation of equatorial electrojet), Pc5 index (for monitoring solar-wind velocity and high energy electron flux), Sq-index (for monitoring global change of ionospheric low and middle latitudinal current system), and Pc3 index (for monitoring of plasma density variation at low latitudes). In this report, we will introduce recent development of MAGDAS/ICSWSE Indexes project and topics for new open policy for MAGDAS data will be also discussed.

1 What is ICSWSE?

As a strategy of enhancement and sustainability of our project, Kyushu university reorganized the Space Environment Research Center (SERC) of Kyushu University was re-organized on 01 April 2012. On that date it became the "International Center for Space Weather Science and Education (ICSWSE) ". To allow for easy pronunciation, the short name of the Center is "iku-sei", which means "to nurture" in Japanese.

The purpose for this re-organization is to allow space weather research to continue on a more global basis, and to establish a permanent international institution (center) for space weather science and education. This new center was also specifically requested as a part of the International Space Weather Initiative (ISWI) in the "Abuja ISWI Resolu-

* Corresponding author: yoshi@geo.kyushu-u.ac.jp

tion" which was unanimously approved by the participants of the "UN/Nigeria Workshop on ISWI" (Abuja, Nigeria, 2011).

The purposes of ICSWSE are:

- expansion and promotion of Space Weather as a field of science;
- closer investigation of changes of climate/geospace/disasters;
- examination of the medical and biological aspects of man working in space;
- realizing international space weather research and education.

The new center is well-placed to achieve these purposes because of the strong human network created by the MAGDAS Project and its associated Magnetometer Data Archives (http://www.icswse.kyushu-u.ac.jp/data/index.php

There are various research activities taking place in ICSWSE. The ICSWSE is composed of 6 research divisions: Geospace atmosphere coupling lab, Geospace fundamental theory lab, Geospace hazard simulation lab and Space medicine lab. The new ICSWSE will engage in the following: Enlarge the scope of its space weather research; evolve internationally as the Center for Space Weather Science and Education; conduct more detailed investigation of the links between geospace changes, climate changes, and natural disasters; start new research in the biomedical aspects of geospace.

2 The 30 years history of geomagnetic observations by Kyushu University

For space environment monitoring, for this 30 years, ICSWSE has developed a real time magnetic data acquisition system (the MAGDAS project) around the world. The number of observational sites is increasing every year with the collaboration of MAGDAS host countries. Now at this time, the MAGDAS Project has installed 78 real time magnetometers – so it is the largest magnetometer array in the world.

Here we introduced 30 years history of geomagnetic observations by Kyushu University.

The Equatorial Magnetometer Network (EMPN) project was started in 1985 and continued to 1995 at the Kyushu University. The P-I of this project is Prof. T.-I. Kitamura. This network had total of 44 stations, especially along the magnetic equator. In this project, we developed and manufactured the magnetometer system ourselves. Researchers stayed observation site for a few months, and took care the system, then brought back data with us.

While the 210 Magnetic Meridian (MM) magnetic observations project started in 1990 and continued to 1995 at Nagoya University. The P-I of this project was first Prof. M. Tanaka and it taken over by Prof. K. Yumoto. Stations densely distributed along 210MM and this project becomes the foundation of collaboration between Russian academy and Japan. This network had total of 34 stations.

The Circum-pan Pacific Magnetic Network (CPMN) Project was started in 1996. The EMN and 210MM were integrated into the CPMN by Prof. Yumoto at Kyushu University. After that, some other stations were appended. Total of 93 stations belong this network.

Real-time monitoring of geomagnetic field, the MAGDAS/CPMN Project, started in 2005. Now at this time, the MAGDAS Project has installed 78 real time magnetometers and total of 140 stations were installed all over the world including non-real time stations. The P-I of this project is now taken over by Prof. A, Yoshikawa.

3 Capacity building activity in ICSWSE

The installation and maintenance of MAGDAS is strongly supported by the cooperation of local collaborators. Many students as well as foreign partners attended field works. This MAGDAS activity has become the foundation of our capacity building activities.

Consistent with the goals of the ISWI, the ultimate goal of our Core-to-Core research program is to:

- strengthen the space-weather-related research abilities of young scientists in the Asia region and Africa region;
- expose young Japanese scientists to more internationally-oriented science activities, such as doing field work in Asia and Africa.

The primary agenda, therefore, is to conduct Capacity Building for young scientists in Japan and Asia/Africa regions through a wide-variety of approaches:

- conducting overseas ISWI/MAGDAS Schools;
- implementing student exchanges;
- installing MAGDAS in "missing areas";
- bringing Asians and Africans to study at Japanese universities.

Through this myriad of approaches, a human network for space weather research is established.

With the support of JSPS, one of the capacity building activities we have at ICSWSE is through trainings. We invite our MAGDAS hosts for training at ICSWSE. We have conducted four trainings so far. Our trainees were from the Philippines, Malaysia, Indonesia Vietnam, Peru, Ivory Coast, Mozambique, Nigeria and Egypt. We teach our trainees how to install and maintain our magnetometers in their respective countries. We also teach them how to monitor how to use a data for their research. At the end of the MAGDAS training, we awarded each of the trainees a certificate of completion. For more information, access the Research Note, Space Research Today, Volume 188, December 2013, Pages 18-20, ISSN 1752-9298, http://dx.doi.org/10.1016/ j.srt.2013. 11.007. (http://www.sciencedirect.-com/science/article/pii/S1752929813000601)

We also organized international MAGDAS School with collaboration of UN/OOSA, ISWI, COSPAR, and SCOSTEP. The schools were held at Egypt (2009), Nigeria (2011), Indonesia (2012), Cote d'Ivoire (2013), Peru, Malaysia, Indonesia (2014), Japan (Fukuoka: 2015) and Nigeria, Russia, Peru, Malaysia (2016).

The topics is the UN/JAPAN Space Weather Workshop held at Fukuoka, Japan 20015 March. In this workshop, participants were 160 people from 32 countries and 9-science session and 4 panel discussions were organized. The recommendations and observations of this workshop were officially reported at UN COUPUS and UN general assembly.

We finally repot an education through MAGDAS project in Kyushu University. We dispatched over 80 graduate students to over sea fieldwork during 2010 to 2016. After 1996, 19 Japanese students received Ph.D and after 2012, 6 foreign student received Ph.D through MAGDAS project. They have strong relationship to ICSWSE activity through Offering collaboration, Convening of International MAGDAS workshop in each country, and Student exchange and Researcher exchange.

4 Research activity of MAGDAS project in ICSWSE and collaboration with IKIR

Geomagnetic field variations on the ground are generated by reflecting: "solar terrestrial coupling" electromagnetic disturbances across the magnetosphere and ionosphere, "horizontal coupling" electromagnetic energy is distributed from polar to equatorial ionosphere and vice verse and "atmospheric vertical coupling" propagation of lower atmospheric disturbances to upper atmosphere via gravity wave and tidal waves. By producing space index by using geomagnetic field variations, we can investigate influences to Earth system caused by solar eruptions, relation between solar activity and climate phenomena and/or climate change, global circulation of energy and material of upper atmosphere, and relations between seismic activity and solar activity. Super Multipoint Geomagnetic field Network Observationimportant tool for exploring of frontier of solar terrestrial coupling process. By using MAGDAS data, ICSWSE produces many types of space weather index, such as EE-index (for monitoring long tern and shot term variation of equatorial electrojet), Pc5 index (for monitoring solar-wind velocity and high energy electron flux), Sq-index (for monitoring global change of ionospheric low and middle latitudinal current system), and Pc3 index (for monitoring of plasma density variation at low latitudes). Since 2012, MAGDAS project produced 56 papers in the international refereed Journal.

Collaboration between IKIR is extended back to 1990's. Now a time, we are operating Flux-gate magnetometer and FM-CW Radar. It is one of most important collaboration for space weather monitoring. In June 2017, we visited IKIR and fixed the problem of FM-CW Radar system, and the data successfully begun to obtain. We now plan to measurement of Poynting flux by using magnetic field data obtained MAGDAS and electric field data from FM-CW radar. By developing this new type of Poynting flux measurement system, we expected to understand global circulation of electromagnetic energy and materials through ionosphere. So the collaboration between IKIR and ICSWSE become key element for succession of this new trial.

Analysis of the Earth's magnetosphere states using the algorithm of adaptive construction of hierarchical neural network classifiers

Sergey Dolenko*, *Vsevolod* Svetlov, *Igor* Isaev, and *Irina* Myagkova

D.V.Skobeltsyn Institute of Nuclear Physics, M.V.Lomonosov Moscow State University, 119191 Moscow, Russia

Abstract. This paper presents analysis of the results of clusterization of the array of increases in the flux of relativistic electrons in the outer radiation belt of the Earth by two clustering algorithms. One of them is the algorithm for adaptive construction of hierarchical neural network classifiers developed by the authors, applied in clustering mode; the other one is the well-known k-means clusterization algorithm. The obtained clusters are analysed from the point of view of their possible matching to characteristic types of events, the partitions obtained by both methods are compared with each other.

1 Introduction

The dynamics of the system "interplanetary space –solar wind –magnetosphere of the Earth" can be described by a multivariate time series, which includes the parameters of the interplanetary magnetic field (IMF) (IMF vector modulus and its three components); velocity, density and temperature of the plasma of the solar wind (SW); geomagnetic indices: Dst, Kp and AE. The outer Earth's radiation belt (OERB) is a part of the inner magnetosphere, and it is also a complex dynamic system, whose parameters (e.g., the intensity of the particle flux and the position of the boundaries) can greatly vary depending on the variations of IMF and SW components (in particular the speed of SV), as well as on geomagnetic perturbation.

The level of perturbation of the Earth's magnetosphere is estimated by using geomagnetic indices – Dst, Kp, AE-AL. The amplitude of planetary perturbations during geomagnetic storms, embracing the entire magnetosphere, is characterized by the amplitude of Dst-variation, which represents the maximum deviation of the variation of Earth's magnetic field from the quiet level, averaged over the values measured at the control chain of magnetic stations located at lower latitudes, i.e. near the equator [1]. Information about the experimentally measured values of Dst variation is provided by the branch of the World Data Center located in Kyoto (Japan) (http://wdc.kugi.kyoto-u.ac.jp/dstdir/index.html [2]). Geomagnetic disturbances can be caused as by reaching the Earth's orbit coronal mass ejections (CMEs), associated with solar flares, as by high-speed SW streams emerging from

* Corresponding author: dolenko@srd.sinp.msu.ru

coronal holes. A necessary condition for the development of a magnetic storm is the presence of Southern (negative) IMF component, which enables energy transfer from the solar wind to the magnetosphere [3].

Disturbances of Earth's magnetic field (geomagnetic perturbations), called magnetic storms, are one of the major manifestations of space weather. At present, as space weather one usually understands the processes and phenomena occurring in the Sun, in the solar wind, the Earth's magnetosphere and ionosphere, which can affect the operation of onboard and ground-based technological systems, as well as the health and well-being of people. Experimental studies show that geomagnetic disturbances have a direct influence on the state of the near-earth space (NES) [4], e.g., its radiation state, because the flux of relativistic electrons (RE) in the OERB typically increases by an order of magnitude and more (see [5-6] and the references therein). Extremely high RE flux can cause failures in electronic circuits of the apparata located onboard the spacecraft [7]. In addition, strong magnetic storms can cause disruption of telegraph lines and radio communications, pipelines, power lines and power grids [1].

From the foregoing it follows that the detailed study of processes occurring in the magnetosphere of the Earth, is a very important task – both from a fundamental point of view with the aim of understanding the processes of particle acceleration in the NES, and to improve methods of forecasting the values of its parameters, including those using selection of its different states. The latter can be accomplished using data clusterization algorithms.

2 Algorithm of adaptive construction of hierarchical neural network classifiers

In their laboratory, the authors of this study developed the algorithm of adaptive construction of hierarchical neural network classifiers (HNNC), which is a method of construction of a neural network tree, allowing one to reduce the solution of a complex problem of classification to a sequential or parallel solution of several problems of smaller complexity. The approach consists in an iterative analysis of the responses of a neural network and in adjusting its desired answers in the process of training. As the result of such adjustment, "similar" classes are combined into one, in which the same output neuron corresponds to patterns of both merged classes. Thus, at each level of the hierarchy similar classes form clusters (groups), and the task of the model is to determine the number of the corresponding cluster. In fact, at the same time the algorithm solves the problem of clustering of the plurality of the desired original classes and the problem of classification of patterns as belonging to the created clusters. Thus by reducing the number of clusters and by strong simplification of the desired separating surface the requirements on the complexity of the used model are weakened, which in turn leads to the fact that this approach can be applied even on small sample sizes.

Recursive application of the algorithm allows building adaptively a hierarchical system of classification; the number of clusters at each level of the hierarchy, i.e. the number of classes recognized by each neural network, remains small. The algorithm for constructing the hierarchical neural network classifier, first presented in [8-9], was studied and described in [10-13]. This algorithm can be used as in the classification mode, as in clustering mode [14]; the latter variant was applied in the present study.

3 Data sources and preparation

In this study, the SW and IMF data from the ACE (Advanced Composition Explorer) spacecraft located at the Lagrange L1 point, obtained by the SWEPAM and MAG instruments respectively, were used [15]. Processes in the chain "IMF – SW – magnetosphere of the Earth" are described using multivariate time series, which includes the parameters of IMF – module of the IMF vector and its three components; the parameters of the SW – velocity, density and temperature of SW plasma, and the geomagnetic indices Dst and Kp obtained from the page providing data for geomagnetism (Geomagnetic Data Service) [16], and the RE flux of the outer ERB (flux of electrons measured in the experiments at the spacecraft of the GOES series [17]). The data range used was from October 22, 1997 (the beginning of data receipt from ACE) until March 2016.

Analysis of the states of the outer radiation belt of the Earth was performed on the basis of data on the flux of RE in the geostationary orbit. Each of the clustered patterns represented a description of an event of increasing of the flux of RE of the OERB with energy > 2 MeV. In the computational experiment, the results of which are presented below, each event is described by the values of the following attributes:

count – duration of the event in hours (minimum considered event duration 12 hours);

e0 – the average value of the RE flux over the last 3 hours prior to the event;

el_max – the maximum value of the RE hourly average flux during the time of the event;

el_min – the minimum value of the RE hourly average flux during the time of the event;

el_sum_average – average RE flux during the time of the event (integral RE flux over the time of the event, normalized to the duration of the event);

spd_max – maximum hourly SW velocity during the event;

Bz_maxabs – maximum modulo hourly average value of the z-component of the IMF;

B_magn_max – the maximum hourly average value of the IMF modulus during the event;

Kp_mean – the mean value of the geomagnetic index Kp during the event;

Dst_max – maximum modulo hourly value of the geomagnetic index Dst during the event;

AE_mean – the mean value of the geomagnetic index AE during the event.

Clustering of events was performed by the developed algorithm with the following parameters: 5 neurons in the hidden layer, learning rate 0.01, time-out before the first class merging – 1 epoch, time-out between class mergings – 1 epoch, the threshold of neuron "voting" – 0.02. Clusterization into 5 clusters was obtained. As an alternative, we considered the division into the same number of clusters obtained by using the k-means algorithm.

4 Results and discussion

Due to the fact that the distribution for most of the considered input features has a shape far from Gaussian, the values of their mean and standard deviation are insufficient for a meaningful analysis of the clustering results. Fig. 1 (a, b) shows the histograms for several of the most informative of these input characteristics for the entire dataset and for each cluster individually. The left column shows the clustering obtained using the developed HNNC algorithm, the right column is the clustering obtained using the k-means algorithm.

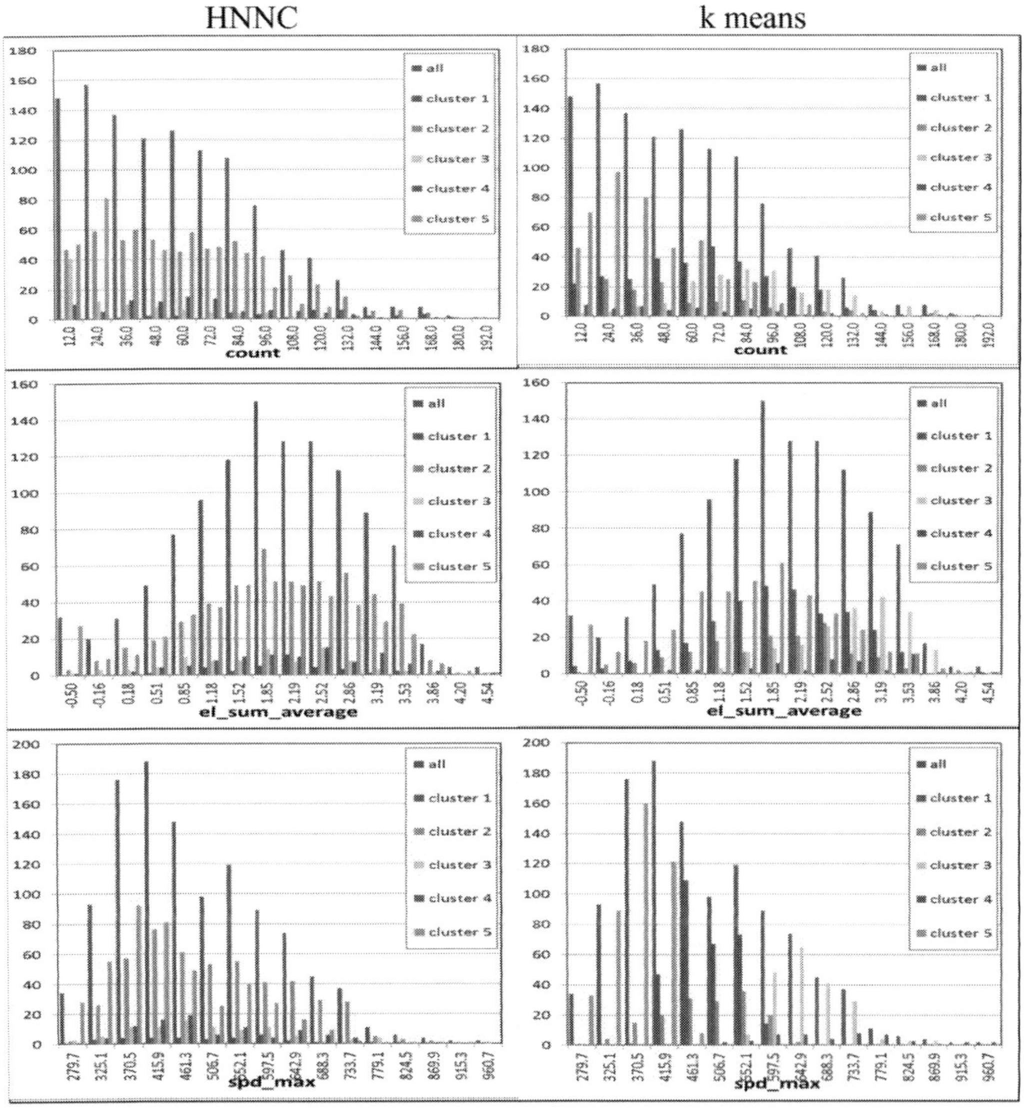

Fig. 1a. Histograms of distribution of several input features, for the entire dataset and for each cluster: 1) count – duration of the event in hours; 2) el_sum_average – average RE flux during the time of the event; 3) spd_max – maximum hourly SW velocity during the event. Left – clusterization, produced by the developed HNNC algorithm, right – clusterization, produced by the k-means algorithm.

 DOI: 10.1051/e3sconf/20172002001

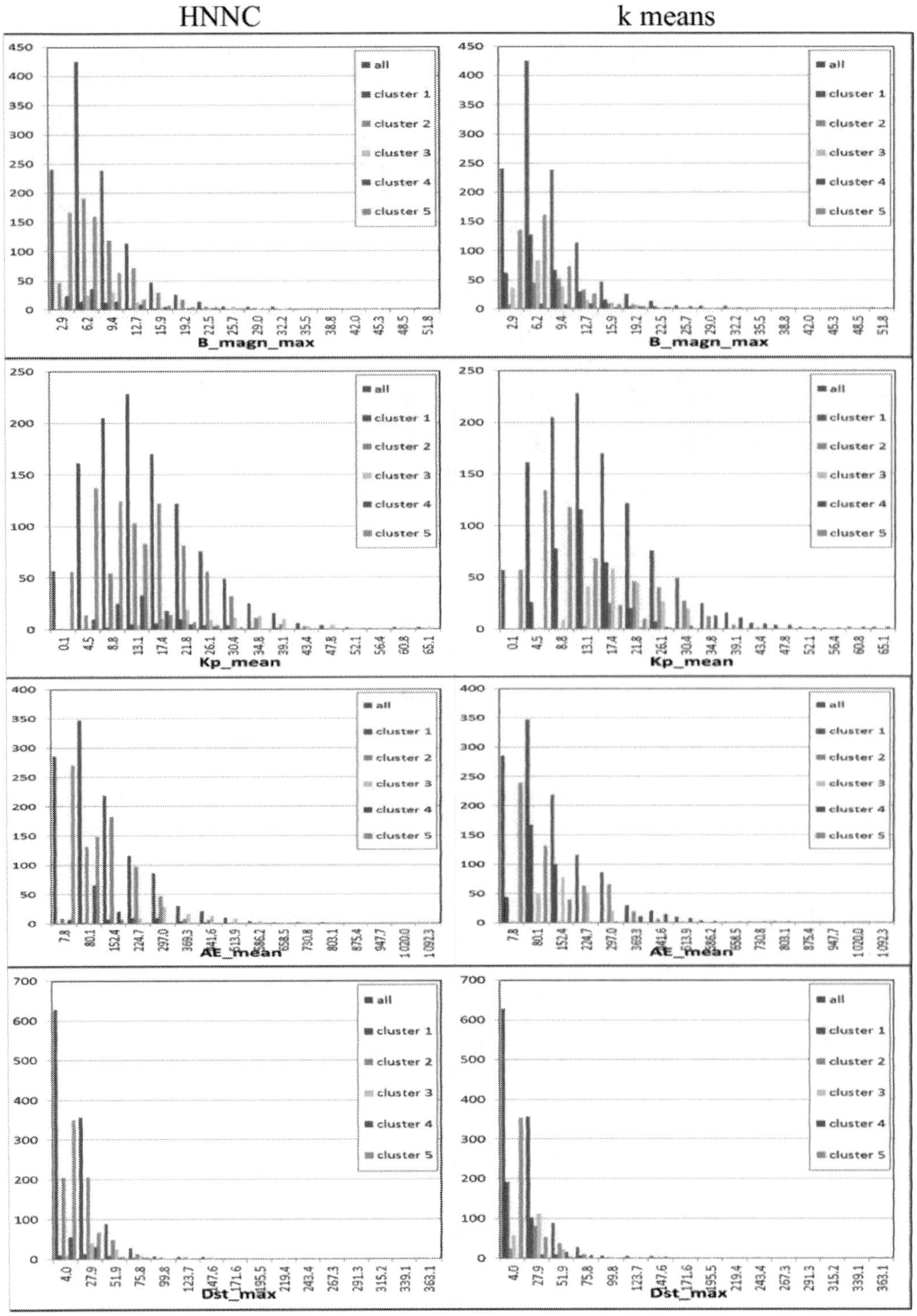

Fig. 1b. Histograms of distribution of several input features, for the entire dataset and for each cluster: 1) B_magn_max – the maximum hourly average value of the IMF modulus during the event; 2) Kp_mean – the mean value of the geomagnetic index Kp during the event; 3) AE_mean – the mean value of the geomagnetic index AE during the event; 4) Dst_max – maximum modulo hourly value of the geomagnetic index Dst during the event.

From the above histograms it can be seen that both algorithms identified in cluster five (the green bars) short increases of the electron flux of medium intensity, occurring at a relatively low SW velocity and at small but non-zero amplitudes of geomagnetic indices. It can be assumed that the cause of such disturbances of the magnetosphere and of variations of the electron flux in the OERB caused by them may be slow coronal mass ejections coming to the Earth's orbit.

Further, from the analysis of the histograms obtained by the k-means method, it is clear that the events most protracted and having the most intense electron flux were selected by this method to the cluster number 3 (the yellow bars). They correspond to the maximum SW velocity and to relatively large values of the Kp index and of the amplitude of the IMF vector B. By comparison of the histograms we can assume that this cluster included the increases of the electron flux caused by powerful coronal mass ejections associated with solar flares.

The next one by the intensity of the events (and also by the SW velocity) is cluster 2 (depicted by orange bars). It is seen that in this cluster there are significantly less events than in clusters 3 and 1. The histogram of the distribution of AE-index shows that this cluster includes events with maximum AE index, allowing one to suggest that for the events of cluster 2 the responsible processes are sub-storm processes which are characterized by large amplitude of the AE index.

Cluster 1 (dark red bars), larger by the number of events than clusters 2 and 3, includes events of medium intensity, emerging at SV with velocity less than for clusters 2 and 3. Most importantly, they occurred when the amplitude of the Dst index during the events of this cluster was low, in contrast to the amplitude of the Kp index, i.e. geomagnetic disturbances, characteristic for this cluster, can be characterized as high-latitude. Presumably this could be the disturbances caused by the arrival of high speed streams of solar wind from coronal holes.

Cluster 4 (grey-green) is too small, while the distribution of events in almost all parameters follows the distribution of cluster 2. The reason why the algorithm highlighted some of these events into a separate cluster may be the subject of further research.

As for the results shown by the developed algorithm, from the analysis of the histograms it follows that the clusters 1, 2, 3 according to the k-means algorithm were merged into a single cluster numbered 2; formally, the clustering result with HNNC was more "rough". On the other hand, the analysis of histograms shows that with the HNNC method, the events are well separated by the intensity of the Kp and AE indices, that is, the events caused by high-latitude perturbations were highlighted. A more detailed study of the obtained results and improving the performance of HNNC should also be continued.

5 Conclusions

Thus, in this study the array of the increases of the flux of relativistic electrons of the outer radiation belt of the Earth was split into 5 clusters characterizing the states of the Earth's magnetosphere different from the physical point of view, by the developed HNNC algorithm and by the k-means algorithm. The obtained clusters were analyzed from the point of view of their possible matching to distinctive types of events, and the two obtained clustering results were compared with each other.

This study was supported by the Russian Foundation for Basic Research (RFBR), grant No. 15-07-08975-a.

References

1. L.L. Lasutin, *World and polar magnetic storms* (MSU, 2012)
2. Geomagnetic Equatorial Dst Index Home Page.
   URL: http://wdc.kugi.kyoto-u.ac.jp/dstdir/index.html
3. B.T. Tsurutani, W.D. Gonzalez, F. Tang et.al., J. Geophys. Res. **A8**, 8519 (1988)
4. I.N. Myagkova, Yu.S. Shugay, I.S. Veselovsky, and O.S. Yakovchouk, Solar Syst. Res., **47**(2), 141 (2013).
5. R.H. Friedel, W.G.P. Reeves, T. Obara, J. Atmos. Solar. Terr. Phys. **64**, 265 (2002).
6. R. Kataoka, Y. Miyoshi, Ann. Geophys. **25**, 1335 (2008)
7. N. Iucci, A.E. Levitin, A.V. Belov et al., Space Weather, **3**, S01001 (2005)
8. S.A. Dolenko, Yu.V. Orlov, I.G. Persiantsev, J.S. Shugai, E.K. Eremin, The 4th Open Russian-German Workshop on Pattern Recognition and Image Analysis, Valdai, Russia, March 4–8, pp. 45–49 (1996).
9. S.A. Dolenko, Yu.V. Orlov, I.G. Persiantsev, Ju.S. Shugai, and E.K. Eremin, Patt. Recogn. Image Anal. **7**(1), 24 (1997).
10. S.A. Dolenko, Yu.V. Orlov, I.G. Persiantsev, Ju.S. Shugai, E.K. Eremin, Proceedings Fifth Int. Conf. on Artificial Neural Networks, Churchill College, University of Cambridge, UK, July 7-9, 1997. IEE Conference Publication, **440**, 285 (1997).
11. S.A. Dolenko, Yu.V. Orlov, I.G. Persiantsev, Yu.S. Shugai, Patt. Recogn. Image Anal. **8**(2), 122 (1998).
12. V.A. Svetlov, I.G. Persiantsev, J.S. Shugay, S.A. Dolenko, Opt. Mem. Neural Netw. (Inform. Opt.), **24**(4), 288 (2015). DOI: 10.3103/S1060992X15040062
13. V.A. Svetlov and S.A. Dolenko, Opt. Mem. Neural Netw. (Inform. Opt.), **26**(1), 40 (2017). DOI 10.3103/S1060992X17010076
14. V.A. Svetlov and S.A. Dolenko, *XIVth All-Russian scientific conference "Neurocomputers and their application"*. Theses, p.55 (M., MGPPU, 2016).
15. World Data Center in Kioto (Geomagnetic Data Service).
    URL: http://wdc.kugi.kyoto-u.ac.jp/wdc/Sec3.html (accessed 21.05.2017)
16. Caltech, Advanced Composition Explorer Science Center.
    URL: http://www.srl.caltech.edu/ACE/ASC/ (accessed 21.05.2017)
17. GOES (Geostationary Operational Environmental Satellite) Project site.
    URL: http://rsd.gsfc.nasa.gov/goes/ (accessed 21.05.2017)

Magnetic observations at Geophysical Observatory Paratunka IKIR FEB RAS: tasks, possibilities and future prospects

Sergey Y. Khomutov *

[1] *Institute of Cosmophysical Research and Radio Wave Propagation FEB RAS, 684034, Kamchatskiy Kray, Paratunka, Mirnaya st., 7, Russia*

Abstract. Continuous magnetic measurements at Geophysical Observatory "Paratunka" (PET) of IKIR FEB RAS are performed since 1967. In the new millennium analogue magnetometers were modernized to digital, the technologies of absolute observations were changed, the data processing was completely transferred to computers, and the status of INTERMAGNET observatory was obtained. Currently, the observatory uses the following magnetometers: (a) for absolute observations – DIflux LEMI-203 (theodolite 3T2KP) and Mag-01 (theodolite Wild-T1), Overhauser magnetometers POS-1 and GSM-19W; (b) for variation measurements – fluxgate magnetometers FGE-DTU, FRG-601 and MAGDAS (installed under international agreements of IKIR), vector magnetometers dIdD GSM-19FD and POS-4 with Overhauser sensors and coil systems, scalar magnetometer GSM-90 and induction magnetometer STELAB. During Spring-Autumn season dIdD also is installed at remote station "Karymshina" at distance of 15 km from Observatory. There is monitoring system for monitoring of conditions in which magnetic observations are performed, including the semi-professional weather stations Davis Vantage Pro2 and WS2000 and a network of digital temperature sensors DS19B20 located at various points in magnetic pavilions and outdoor. All measurements are synchronized with the UTC. The results of observations are collected by the IKIR data server from the recorders and loggers, including in real-time. Specialized software was developed (based on MATLAB and Octave packages), which allows automatic and semi-automatic processing of data, the comparison of the results from different magnetometers and presenting final data in formats, defined by international standards, including INTERMAGNET. Significant efforts of observatory staff are direct to archive (raw) magnetic data, a significant part of which has not been entirely processed, is not presented in international data centers and is still not available to the scientific community. Digital images of analog magnetograms have been obtained, the digitization of which allowed to fill the gaps in the hourly data available in the WDC for 1991, 1996-1997. The state of the set of the instruments and databases shows that the observatory "Paratunka" can solve most of the problems that arise in the geomagnetic field researches, including the provision of data for the development of regional and global field models, ground support of magnetometers located on satellites, support of the surface (sea) and aeromagnetic surveys and

*e-mail: Khomutov@ikir.ru
This work was supported by the Grants of RSF No.14-11-00194 and RFBR No.16-55-45007

 DOI: 10.1051/e3sconf/20172002002

drilling, providing the data for researches of processes in the lithosphere, magnetosphere and ionosphere over a wide frequency range, and also for a number of applied problems.

1 Introduction

Magnetic observatories (MOs) perform long-term standard observations that provide information about the magnetic field, and are an important element of the experimental base of scientific and applied research. The distribution of MOs over the Earth and over time is very heterogeneous: some MOs perform observations during hundreds of years, some MOs have been created in recent years; in some regions the density of the MO network is high, but also there is no observatories over large area. Only a few active MOs existed on the territory of the USSR until about the middle of the last century. However, the results of the International Geophysical Year showed that this network is unsatisfactory. In accordance with the Goverment decision the new network of Joint magneto-ionospheric stations (KMIS) was designed and in the late 60s of the last century it was deployed. KMIS in Kamchatka later became the Geophysical Observatory (GFO) Paratunka, the international code PET (IAGA). In the last century magnetic measurements on the GFO were carried out using Bobrov's quartz variometers with recording on a photographic tape in accordance with IAGA standards. The result of these regular measurements is the database of hourly data on the total field vector (www.wdc.bgs.ac.uk/catalog/master.html, WDC for geomagnetism) and also the database of digital images of analog magnetograms (www.wdcb.ru/stp/magnetogr_list.en.html, WDC for Solar-Terrestrial Physics).

The modernization of GFO Paratunka began at the end of the 1990s due to international cooperation. The digital magnetometers for vector and scalar measurements and new instruments for absolute observations were installed, the infrastructure of the observatory developed, new methods for digital processing of magnetic data were introduced. The observatory gradually achieved the modern standards of measurements and in 2013 got the status of the observatory of the INTERMAGNET network (www.intermagnet.org).

The goal of this report is to show the current state of magnetic measurements on the GFO Paratunka, including hardware, software and methodological support, and to note the some results and the possibility of their application in various scientific and applied tasks. The actuality of this report is due to there is no detailed description of observatory over all its history in publications.

2 Location, measurement conditions and infrastructure

The Institute of Cosmophysical Research and Radio Wave Propagation (IKIR) FEB RAS and Observatory Paratunka as department of IKIR are located about 25 km to south-west from Petropavlovsk-Kamchatsky, to west from Avacha Bay, on the northern outskirts of the village Paratunka. In a significant area around observatory (within a radius of 10 km) there are no industrial facilities and other sources of industrial disturbance. The main consumers of electricity, i.e. potential sources of noise, are large number of sanatoriums and recreation centers in the vicinity of the village Paratunka.

Figure 1 shows the main infrastructure objects of institute and observatory and some measurement systems. The building of ionospheric station 2, technical building 11, garage with workshops 10 and magnetic huts 3, 4, 5 were built in 1965-1967 as KMIS objects. At the same time, there were built two three-storey apartment houses 12 for KMIS staff. The main building 1 of the institute was built later, in the late 80's. On the meadow 7 the sensors of the STELAB induction magnetometer are installed (buried in the ground), the area around the sensors is cleared of trees to reduce wind effects. On the meadow 8 there is an electric field meter Pole-2 for measuring the vertical gradient of the potential of the atmospheric electric field. The meadow is cleared from the trees to avoid distortion of the field distribution around the sensor Pole-2. There is a special remote mark 9 (target) on the western wall of residential house. The azimuth of target (209°48'56") is determined every year by the astronomical observations of the Sun. This mark is necessary to obtain a magnetic declination D using the results of the absolute observations in pavilion 4. The visibility of the target is provided by a special glade 300 m long. Along the direction to main target at a distance of about 200 m from the absolute pavilion, an additional mark was installed. This additional target is located at guarded area of IKIR and used during absolute observations in bad weather conditions. There are several old abandoned buildings and two wells providing residential houses and the institute with water. Main well is located near 50 m to the northeast of the institute building, cased with a steel pipe to a depth of about 100 m, with a submersible pump. Figure 2 shows a panoramic photograph of the magnetic pavilions (variation, absolute and for POS-4).

Figure 1. The location of the main buildings and objects near Observatory (from www.google.ru/maps/). 1 – main building of institute; 2 – building of ionospheric station (working and office rooms, lidar, ionosonde Parus with vertical dipole transmitting antenna oriented in SE-NW direction); 3 - pavilion for variation measurements (magnetometers FGE, GSM-90, MAGDAS, FRG-601); 4 – pavilion for absolute observations (magnetometers LEMI-203, Mag-01H, POS-1); 5 – technical pavilion; 6 – pavilion for magnetometer dIdD GSM-19FD; 7 – sensors of induction magnetometer STELAB; 8 – area with electric field meter Pole-2; 9 – azimuthal target at wall of house, the its visibility from absolute pavilion is provided by glade; 10 – garage and workshops; 11 – service building; 12 – two three-store buildings. The POS-4 pavilion is placed between pavilions 3 and 4. At the left of photo there is federal road between Paratunka and Petropavlovsk-Kamchatsky. The framing shows the map of central part of Kamchatka with Petropavlovsk-Kamchatsky (▲) and Paratunka (★).

Figure 2. The main pavilions of Observatory Paratunka: variation (left), POS-4 (center) and absolute (right). Photo was made from roof of technical pavilion at April, 2016.

The pavilion for absolute magnetic observations (absolute pavilion)

The absolute pavilion is the most important at the observatory because the absolute observations are supplied the reliability and stability of magnetic field long-time series. Especially high requirements for magnetic cleanliness are met to this pavilion and to the nearby area. The old absolute pavilion was built in 1967. It had the size of 3x3 m and was constructed of wooden beam and there were four larch pillars for magnetometer installation. For forty years there was a destruction of both the walls and the buried part of the pillars. This made problems during observations and reduced the accuracy of the results.

In 2009 the new absolute pavilion with wall from a wooden beam was installed at the place of old one, the size is 3x3 m. Two important points were taken into account. (1) There is a high magnetization of the underlying rocks, which led to significant vertical gradient of field. (2) The snow level during winter season reaches 1-2 m above the absolute hut window used for observation of remote target 9 (see Fig. 1) was reached. Bulldozer is need to clear the glade from the snow, but this causes the significant disturbances of magnetic field. Therefore the new absolute pavilion was raised on piles, height of working floor is about 1.5 m. The piles are an asbestos-cement pipes filled with nonmagnetic crushed stone. Two pillars were installed in the pavilion. Bottom part of pillars (below the floor level) are also made of asbestos-cement pipes, top part of pillars from about 0.5 m below the floor and inside the pavilion are made from glass blocks (see Fig. 3e). The working height of the pillar is about 1.5 m, cross section is 40x40 cm, a thick glass plate is glued on the upper surface. The long-term stability of the main pillar is quite high: the astronomical measurements of the remote target azimuth over last three years have shown that the azimuth changes do not exceed 10", that is close to the astronomical observation error. The possible pillar movements across the direction to the target are no more than 1-2 cm. The pavilion has large area of the side due to a high pointed roof. This causes the oscillations of pavilion and pillar during strong wind, but dos not effect the results of absolute observations. The position of pillars in pavilion provide the visibility of remote target from both pillars.

For heating of the pavilion in the cold season, a heater with a Nichrome spiral and a mechanical thermoregulator are used, during warm season the heating is turned off. The temperature in the pavilion is controlled by the digital temperature sensor DS18B20 [1], the results of measurements are recorded to computer every 10 seconds. Also here are the DIflux magnetometers LEMI-203 (Fig. 3e)

Figure 3. Magnetometers of the Observatory Paratunka: (a) the sensors of the fluxgate magnetometers Magdas (in front) and FRG-601 (at the back) in the variation pavilion; (b) - the suspended sensor block of the fluxgate magnetometer FGE in the variation pavilion; (c) – sensor of magnetometer dIdD GSM-19FD on a low pillar in a special pavilion; (d) - pavilion of the magnetometer POS-4 with a sensor on the pillar; (e) - DIflux LEMI-203 on the main pillar in the absolute pavilion with the window to observe the remote target (at the back); (f) – sensor of magnetometer POS-1 on the main pillar in the absolute pavilion.

and Mag-01H and the Overhauser scalar magnetometer POS-1 (Fig. 3f) for absolute determinations (D, I) and F, respectively.

The change of the pavilion led to the jumps in the long-term series of magnetic data. For example, the declination has changed by 1', the horizontal component by 2 nT, the vertical component by 20 nT. The possible reason is the little shift of the absolute observation point under conditions of sufficiently high spatial gradients of field.

The pavilion for variation magnetic measurements (variation pavilion)

The variation pavilion was built in 1967 according to a typical KMIS project: a size is about 6x6 m with a lobby, the material of the main walls is a wooden balk trimmed from the outside with boards. The gable roof has as light slope. Inside the pavilion there is a chamber of planks, which provides thermal insulation of the internal space with pillars. The four pillars are made from the lime-cement mixture, three of which (main) are oriented in the direction of "north-south" and one with the orientation "east-west". The eastern main pillar is place for fluxgate sensors of the component magnetometer FGE-DMI (Fig. 3b), the western main pillar is place for the fluxgate sensors of MAGDAS and FRG-

 DOI: 10.1051/e3sconf/20172002002

601 magnetometers (Fig. 3a). Scalar Overhauser magnetometer GSM-90 is installed on the roof of the inner chamber. Recorder Magdalog is placed at the wall inside the camera, GSM-90 console and the electronics of the FGE and GSM-90 magnetometers are installed at outside wall of pavilion.

Heating of the variation pavilion is carried out by a heater with a Nichrome spiral and an additional low-power heater, which is switched on in winter as a background heater. Both heaters are installed between the outer wall and the inner chamber, the temperature stability is maintained by a mechanical temperature regulator, also located between outer and inner walls. There are four digital temperature sensors DS18B20 in the pavilion: one between outer and inner walls and three in the inner chamber (near the floor, at the level of pillars and near the ceiling). The data are recorded every 5 seconds. In order to provide the regulation of temperature with limited heaters capacity, the level of the temperature of the switching of the heater raises up to +(18-20)°C during the summer and decreases up to +(12-13)°C during the winter. Because the limited heating of the pavilion the high stability of temperature is also restricted. Therefore, the absolute magnetic observations are made very often, generally twice per day. The baseline values obtained make it possible more effectively control the temperature effects in variometer records. For example, Figure 4 (left panel) shows the minute values of the temperature of recorder Magdalog installed in inner chamber for 2013-2016. It is seen that temperature conditions in pavilion was noticeably stabilized as the results of the improvement of the thermal insulation of the outer wall of the pavilion in November 2014. The points of the switching of the temperature control device to summer (June-July) and winter (October-November) modes are also visible.

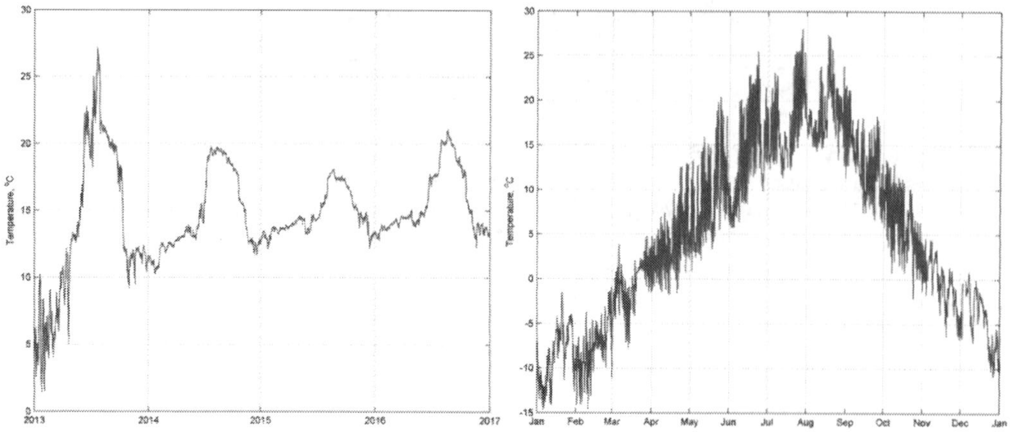

Figure 4. The temperature in inner chamber of the variation pavilion inside the recorder Magdalog for 2013-2016 (left panel) and in pavilion of dIdD for 2016 (right panel).

Pavilion for recording and auxiliary equipment (technical pavilion)

The technical pavilion is built simultaneously with other magnetic pavilions and is a wooden plastered building about 6x6 m in size with a gable roof. It contains recording computers, control electronics and data acquisition systems for some magnetometers, power supply systems, including uninterruptible power supplies, communication devices, GPS systems and antennas for timing synchronization, etc. There are also spares and temporary workplace for service staff. In winter, a small heater is used to support the temperature conditions. The digital sensor DS18B20 is installed to control the temperature. All the recording computers in the technical pavilion are connected to the local network and

DOI: 10.1051/e3sconf/20172002002

connected to the main building of the institute by means of a fiber-optic line. This ensures the on-line monitoring of the state of measuring systems and online access to data from personal computers of staff of Observatory.

The pavilion for the magnetometer dIdD GSM-19FD

The pavilion for the magnetometer dIdD GSM-19FD was built in 2011 at some distance from other magnetic pavilions, because dIdD creates significant disturbations of the magnetic field around itself. The experiments with installation of dIdD in the variation pavilion at Observatory Paratunka and also at other magnetic observatories of IKIR shown that effect can reach up to 10 nT at distances of the 1-2 m and tenths of nT at distance of up to 5 m ([2], [3]).

The pavilion is a low building about 1x2 m in size from a timbers with expanded polystyrene plates 10 cm thick on the inner side of the walls and the roof. The roof is gable and steep. The entrance door is missing — the access inside provides the attic hole, which is closed by a lid fixed with brass bolts. The floor is also absent and the foam polystyrene boards are used for thermal insulation. The pillar is made of glass blocks, has a cross section of 40x40 cm and is buried at 0.5 m. The height of pillar above the ground is about 0.2 m. The dIdD sensor is mounted on the pillar (Fig. 3c), the console is also inside the pavilion. The magnetometer is controlled by standard shielded twisted-pair cable via RS232 protocol. Power 220 V for the console is supplied from the technical pavilion by cable. The digital temperature sensor DS18B20 is installed inside the pavilion. Active heating is not used, therefore the indoor temperature is close to the outdoor temperature, but more smoothed due to thermal insulation (see Fig. 4, right panel). Because of the small height the pavilion is almost entirely covered with snow during winter and spring, this improves the stability of thermal conditions inside.

The pavilion for the magnetometer POS-4

The pavilion for the magnetometer POS-4 was built in 2015 at a distance from other magnetic pavilions (place is selected between pavilions 3 and 4, see Fig. 1 and 2). The reason is the same as for the pavilion of the dIdD, because the vector magnetometer POS-4 is close to the dIdD by technology: a scalar Overhauser sensor is placed inside the ring system. Therefore, it causes field disturbations in the vicinity during operation (the magnitude of the effect was not estimated). The pavilion about 2.0x2.5 m in size with a wall height of about 1.7 m is built of timbers without a foundation. The gable roof made of chipboard has a height in the center of about 0.5 m, that makes it possible to work with a magnetometer quite comfortably. On the inside of the walls and roof fixed panels of polystyrene 10 cm thick. The door is a plate insulated with expanded polystyrene and in the closed position is fixed by four brass bolts. The floor is absent, but insulating panels are laid on the ground (Fig. 3d).

The pillar is composite, the underground part is cast from a sand-cement mixture to a depth of about 1 m, the overgound part with height of 1 m and a section of 40x36 cm is made of glass blocks with a thick glass plate on top. Continuous measurements with POS-4 during autumn 2015 - spring 2016 showed that due to freezing and defrosting of the ground the pillar tilts, creating significant trends of non-magnetic origin in the results [4]. Therefore, in the autumn of 2016 the lower part of the pillar was enlarged up to 90x80 cm in cross-section, its lateral planes were thermally insulated with expanded polystyrene and isover, and between the pillar and the ground an air gap of 10-20 cm was left insulated from the top. POS-4 measurements during the winter season of 2016-2017 shows, that this update proved to be quite effective against freezing and defrosting.

Communication and power supply of the magnetometer is provided by cable from the instrument kit, protocol is RS232. In contrast to the pavilion of dIdD, the pavilion of POS-4 does not have 220 V power. Heating is not used, the temperature is controlled by three sensors DS18B20: on the cover of the magnetometer electronics to control of possible overheating in summer and on the north wall at a height of about 1 m. The third sensor after updating of pillar was lowered into the air gap between pillar and ground to a depth of about 0.6 m. Because of the considerable dampness in the gap and the risk of flooding with groundwater during the spring 2017 it was reinstalled to the north wall of the pavilion at floor level. It should again be noted that due to the small size of the pavilion, in winter it is covered with snow almost to the roof, this significantly stabilizes the temperature inside.

3 Magnetometers and other measuring systems

The Geophysical Observatory Paratunka is quite typical division of the KMIS network and has passed the standard way of its development in the USSR and then in the Russian Federation. The same applies to the evolution of equipment used in magnetic monitoring: from variometers based on Bobrov's quartz sensors with analog recording and QHM quartz magnetometers for absolute observations to digital variation systems and absolute instruments based on nonmagnetic theodolites and Overhauser scalar sensors.

In the 1990s, the first component fluxgate magnetometers such as DCR, MAGDAS and FRG-601 were installed at observatory under of international agreements of IKIR with scientific organizations of Japan. At the same time, the Russian system on the basis of the digital magnetic variation station "Quartz-6" (IZMIRAN) with Bobrov's quartz sensors was developed. MMP-203 proton magnetometers and quantum M-61 magnetometers were used to measure the total field intensity. Significant progress was made at the beginning of the 21st century: the development of the observatory was focused on achieving the standards of the INTERMAGNET network [5]. Thanks to such trends the GFZ German Research Centre for Geosciences (Potsdam, Germany) in 2009 provided a full set of equipment for the variation magnetic measurements in accordance with new standards (fluxgate magnetometer FGE-DMI, scalar magnetometer GSM-90F1, recorder Magdalog) and modern absolute DIflux magnetometer LEMI-203 was received under INTAS project CRENEGON in 2004 [6]. Vector magnetometers dIdD GSM-9FD and scalar Overhauser magnetometer GSM-19W were purchased. The update of the measuring devices together with organizational arrangements and the development of software led to success - in 2013 the Paratunka was certified as an observatory of the INTERMAGNET. However, the modernization of experimental base is continued, as one of the ways to increase the reliability and quality of magnetic measurements, and as tendency of transition to domestic developments. So, Overhauser magnetometers were purchased under support of the Russian Science Foundation grant (a scalar POS-1 and a vector POS-4), the software package that provides for the joint processing of a significantly increased amount of data was developed with support of Russian Foundation for Basic Research (RFBR).

At present, the system for monitoring the magnetic field at the Observatory Paratunka includes a set of magnetometers operating on different principles and in different modes, a software system and a developed system of auxiliary equipment. The general diagram of these devices in pavilions, recording and communication systems and other devices is shown in Fig. 5. The list of magnetometers used in 2017 on the Observatory Paratunka and their main specifications are presented in Table 1 and 2. The diagram in Fig. 5 is rather simplified, for example, there are not shown magnetometer power supplies (main and emergency ones), communication lines and recorded computers of the digital temperature sensors, DIflux magnetometers in the absolute pavilion, etc. The arrangement of instruments is shown schematically, e.g., without their height separation or real spatial scales. Nevertheless, Fig. 5 makes it possible to assess the structural complexity of the system as a whole, which, in general, can be a source of noise and interference and greatly complicates the maintenance of the equipment, its tuning or repair.

We note the following features of the measuring system:

DOI: 10.1051/e3sconf/20172002002

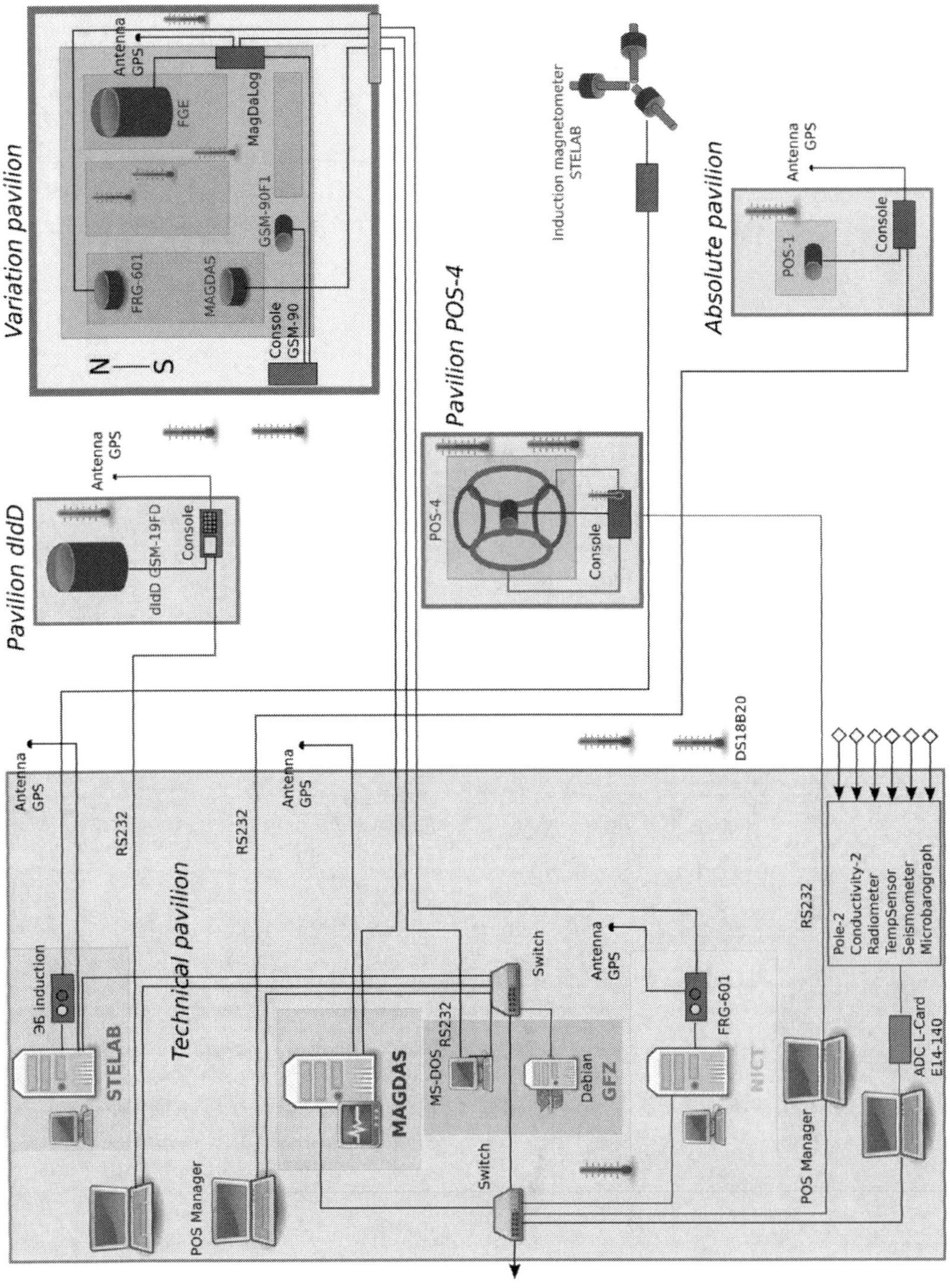

Figure 5. The diagram of the location of main devices of system of the magnetic field monitoring.

 DOI: 10.1051/e3sconf/20172002002

Table 1. Variation magnetometers at Observatory Paratunka, 2017

Name	Type	Specifications	Manufacturer
FGE	fluxgate	- variations dH, dD, dZ - resolution 0.1 nT - dynamic range ±10000 nT (regulated) - sampling 2 Hz - temperature dependence 0.25 nT/^{o}C - longtime trends less 3 nT/year - suspended system ±0.5^{o}C - recorder Magdalog (Germany) - GPS-synchronization	DMI, Denmark GFZ, Germany
MAGDAS	fluxgate	- variations dH, dD, dZ - resolution 0.061 nT/bin ADC - variation range ±2000 nT - sampling 1 Hz - tilts NS and EW - resolution 0.2" in range ±1°, sampling 1min - temperature of sensor - resolution 0.01^{o}C in range ±60°C, sampling 1 min - GPS-synchronization	ICSWSE, Kyushu Univ., Japan [7]
FRG-601	fluxgate	- variations dH, dD, dZ - dynamic range ±65000 nT - resolution 0.5 nT - sampling 1 Hz - GPS-synchronization	NICT, Japan
dIdD GSM-19FD	Overhauser scalar sensor in coil system	- variations dD, dI, total field intesity F - sensitivity of F 0.18 nT, resolution 0.1 nT - uncertainty of dI less than 1", of dD less than 4" - dynamic range 20000 — 120000 nT(F), ±180^o(D), ±90^o(I) - sampling 0.2-1.0 Hz - GPS-synchronization	GEM Systems, Canada [8]
POS-4	Overhauser scalar sensor in coil system	- total field components F, Z, Y(X) - dynamic range F 20000-100000 nT, components ±20000 nT - sensitivity F less than 0.05 nT, components less than 0.3 nT - absolute error F less than 1 nT, Z less than 3-10 nT, horizontal component — no specify - rate of readings 1 sec and more	Ural Federal University, Quantum Magnetometry Laboratory, Ekaterinburg, Russia
STELAB	induction	- variations (derivatives) dH, dD, dZ - sensitivity 0.45 V/nT at 1.8 Hz, 0.02 V/nT at 0.1 Hz - sampling 64 Hz, turnover frequency 1.8 Hz - GPS-synchronization	ISEE, Nagoya Univ., Japan

Table 2. Scalar and absolute magnetometers at Observatory Paratunka, 2017

Name	Type	Specifications	Manufacturer
POS-1	Overhauser scalar sensor	- total field intensity F - absolute error less than 1 nT - uncertainty less then 0.03 nT - dynamic range 20000 — 100000 nT - rate of readings from 1 sec and more - GPS-synchronization	Ural Federal University, Quantum Magnetometry Laboratory, Ekaterinburg, Russia [9]
GSM-90F1	Overhauser scalar sensor	- total field intensity F - resolution 0.01 nT - absolute error 0.2 nT - dynamic range 20000 — 120000 nT - rate of readings from 1 sec and more - Magdalog recorder (Germany) - GPS-synchronization	GEM Systems, Canada [10]
GSM-19W	Overhauser scalar sensor	- total field intensity F - resolution 0.01 nT - absolute error 0.1 nT - dynamic range 20000 — 120000 nT - rate of readings from 0.2 sec and more - GPS-synchronization	GEM Systems, Canada [11]
LEMI-203	Fluxgate sensor on nonmagnetic theodolite	- declination D and inclination I - range of fluxgate sensor 2000 — 100000 nT - resolution of sensor 0.01 nT (for range 2000 nT) - theodolite 3T2KP (UOMZ, Ekaterinburg) - scaling division 1" - accuracy - 5"	LC ISR, Ukraine
Mag-01H	Fluxgate sensor on nonmagnetic theodolite	- declination D and inclination I - range of fluxgate sensor ±200000 nT - resolution of sensor 0.1 nT - theodolite Wild-T1 (WILD Heerbrugg, Switzerland) - scaling division 6" - accuracy 3"	Bartington Instruments, GB [12]

1) all fluxgate variation magnetometers are installed with the orientation of the sensors in the HDZ system (with the horizontal axis X toward the magnetic north at the time of installation). This greatly simplifies the setting, but creates methodical difficulties during the processing of data, for example, when the variations of the horizontal components recorded by different instruments are compared;

2) all magnetometers, with the exception of POS-4, have an automatic synchronization of the internal timer with the UTC, using built-in GPS receivers. This ensures long-term stability of the timer, however, in some cases, when GPS synchronization failures (for example, in the MAGDAS and the recorder Magdalog), cause problems with timestamps of the data;

3) the used vector magnetometers for continuous recording are based on various physical principles of measurements: FGE, MAGDAS and FRG-601 are fluxgate devices, STELAB is induction magnetometer and dIdD GSM-9FD and POS-4 use the scalar sensors in ring system. This makes it possible to significantly improve the efficiency of comparing the data obtained by them, including the identification of noise, the malfunctioning equipment or the detection of the effects like temperature ones. In addition, in some cases there are methodical advantages. For example, the baseline values of the dIdD magnetometer are estimated by direct comparison of the variations dF, dD and dI with the absolute values of F, D, I obtained by absolute observations. Unlike dIdD the calculation of baseline values for other variometers requires the conversion of the absolute values of F, D, I to orientation system of variometer, which is the source of the errors;

4) magnetometers are installed in various places (on different pillars, in different pavilions), forming a gradiometric system with bases of various lengths and orientations. Magnetic measurements by such system are very effective for identification of the sources of noise in the observatory environment. In addition, the location of magnetometers in various pavilions allows us to have undisturbed magnetic data during the repair or adjustment of equipment;

5) standard processing is performed practically for all variation magnetometers and include the baseline calculation and estimation of the total field components. With this approach, there is a high interchangeability of data and the filling of gaps in records of one instrument is accomplished by a simple "substitution" of data obtained by another device;

6) some magnetometers uses the suspended system of sensors (FGE, dIdD) to compensate the slow inclinations of the pillar or sensors. Such systems increase the long-term stability of measurements, but they are sensitive to mechanical motion of the base: for example, after earthquakes the seismic waves moved near pillar produce the oscillations of suspended system with sensors, which causes the fictitious field variations up to hundreds of nT. Other magnetometers have rigid connection of the sensors with pillar and do not give similar effects (MAGDAS, FRG-601, POS-4). The MAGDAS magnetometer contains sensitive electronic levels that record the slopes of the device in the direction "north-south" and "east-west". In the POS-4 magnetometer the ordinary exact liquid levels are used, however they allow visually to fix only significant and slow slopes.

In the summer-autumn season at the station Karymshina, located near 15 km to the south-west of the observatory, measurements are performed using a magnetometer dIdD GSM-19FD. In other seasons the measurements at Karymshina are impossible due to a significant amount of precipitation (snow level at the end of spring may reach 6 m and more). Therefore, the risk of destruction of the pavilion increases and maintenance of the magnetometer is significantly complicated. The pavilion is a log house made of boards 50 mm thick, 2x1 m in size, 1 m high, with a flat roof. The pillar is made of glass blocks, the total height is 0.7 m, the height of the above-ground part is about 0.1 m, the cross section is 0.4x0.4 m. 220 V power supply is supplied to the pavilion from the station's service room, the control and data collection is performed using RS232 protocol, the communication line length is about 20 m The measurements are performed in the standard mode, the data is available through the institute's local network.

In addition to the equipment for measuring the magnetic field, there are many devices at the Observatory Paratunka that are implicitly included in the magnetic monitoring system. First of all, these are temperature sensors, since many magnetometers are sensitive to changes of the temperature. The basis of the temperature control network is the digital sensors DS18B20, using parasitic power and communication with the recording computer on a single-wire line. The sensitivity of these sensors is 0.01°C, the accuracy in the range of seasonal and diurnal temperature variations is 0.5°C. These sensors are joint very easy to network. Currently, about 15 sensors are located in magnetic pavilions at different altitude levels (for more details, see section 1) and outdoors at altitudes from 0.5 to 30 m. In addition to the DS18B20 sensor network, temperature sensors built-in in magnetometers (MAG-

DAS, FGE) and electric field meter CS110 and analog sensors based on thermoresistors are used. Meteorological parameters are monitored using Davis Vantage Pro II and WS2000 digital weather stations with remote sensors located at 10 and 20 m altitude and using the meteorological sensors of the CS110.

Observatory Paratunka was created as KMIS, that is, as a system for various geophysical measurements. At present, the initial structure of the observational system has changed significantly, but the principle of complexity remains the same: continuous observations of the ionosphere (vertical sounding by a digital ionosonde Parus), atmospheric electricity (electric field meters Pole-2 and CS110 for measurements of the vertical gradient of the potential Ez and the system Electrical conductivity-2 for measurements of the electrical conductivity of the surface air layer) are performed. Also the measurements by the vertical seismic sensor and the microbarograph are made.

4 Software and databases

A large set of magnetometers and other measuring systems at the observatory requires developed software, which allows efficient processing and analysis of the recorded data. Such software on the Observatory Paratunka has been developed and continues to be modernized in accordance with changing tasks (see, for example, [13], [14]). The basic principles of the software are described in [15] and reflect the following features of use:

1) the software has a modular structure, which allows it to be easily adapted to a specific situation, for example, when input and output formats or equipment structure are changed, or when software is installed at other IKIR FEB RAS observatories;

2) the software is designed to be used by data processing by a magnetologist directly at the observatory, so it uses the maximal visualization of data during various steps of processing and provides the functions of interactive work with data;

3) the software reflects the basic requirements for the magnetic data processing at the INTERMAGNET observatories.

The structure of the software provides the following main functions:

- reading of raw data from files obtained during measurements by various magnetometers, the possibility of analyzing them, including the identification of noise and their correction;
- processing of the results of absolute magnetic observations and calculation of the baseline values of the variation magnetometers;
- daily processing of data, the calculation of total field components and minute values in accordance with INTERMAGNET standards, including minute values with the Quasi-definitive and Definitive status.

The most important function of the software used is the comparison of data obtained by various magnetometers. The comparison is performed at various stages of the processing, including the analysis of the raw data and the preparation of the resulting files. The functional diagram of the software for processing the data of the magnetometers FGE and GSM-90 is shown in Fig. 6. The software is developed as a package of scripts and functions for the Matlab system (http://www.mathworks.com/) and completely compatible with the Octave system (GNU GENERAL PUBLIC LICENSE, http://www.gnu.org/software/Octave/). The data is processed in the Matlab command window, in interactive mode. The choice of Matlab and Octave packages as a working environment for processing is due to their high efficiency for working with data series, the advanced graphical system and the support for calculations in command interpretation mode, when scripts and functions are represented as text files and loaded into memory after calling. The text format of scripts and functions provides the magnetologist the opportunity to easily modify them, if necessary, taking into account the real conditions arisen during processing.

DOI: 10.1051/e3sconf/20172002002

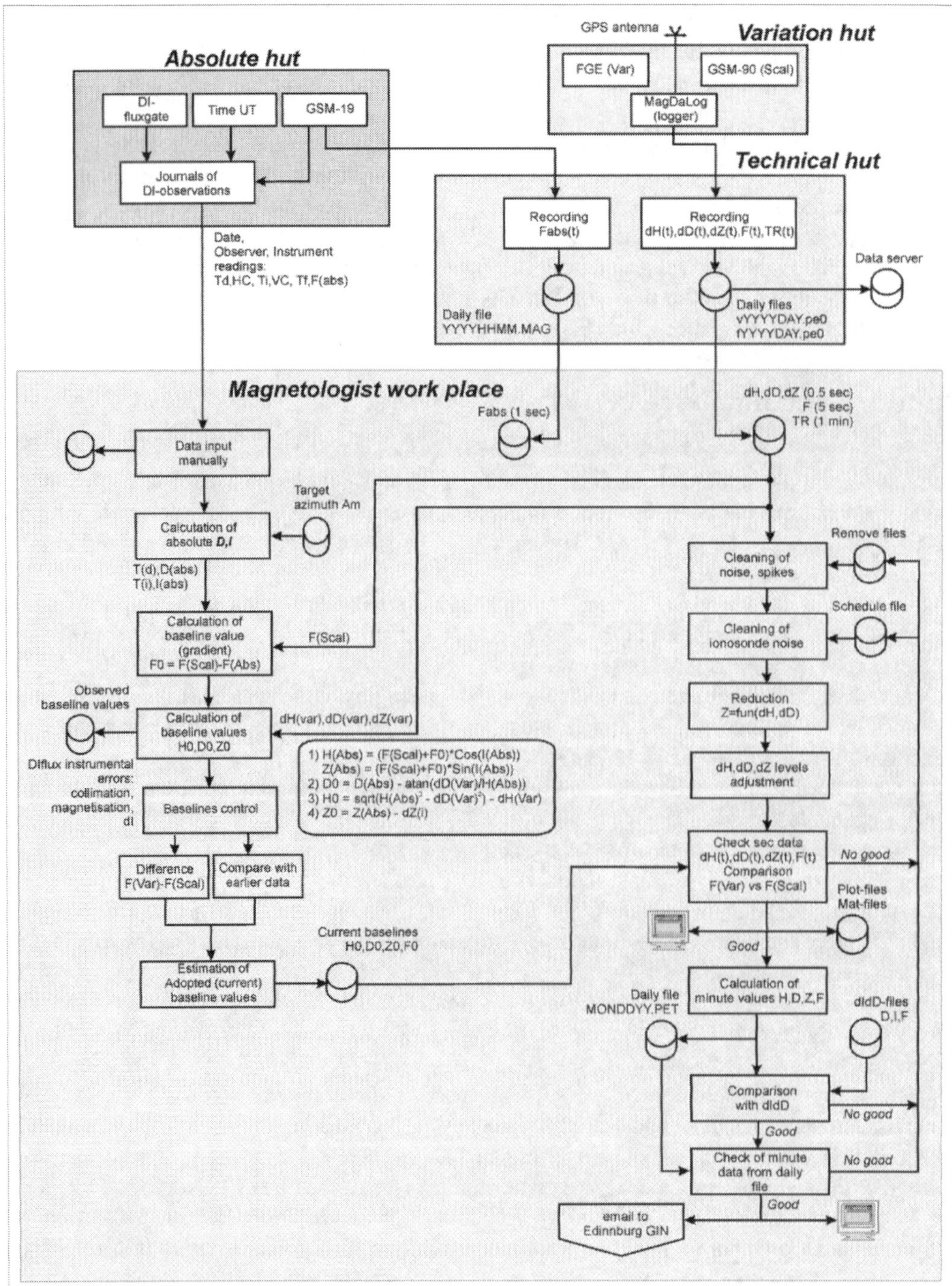

Figure 6. Diagram of full processing of the absolute observations (magnetometers LEMI-203 and GSM-19 or POS-1) and variation measurements (magnetometers FGE and GSM-90). The minute values in accordance with INTERMAGNET Standards with Reported and Adjusted status are the results of processing.

The results of current measurements on the Observatory Paratunka are saved in databases (DB) at institute's data server and also locally on the some workstations of the observatory. DBs on the server collect new data automatically: every day after 24:00UT the archiving program (scripts in Shell envelope) copies new data from the recording computers of magnetometers and other measuring systems and performs service procedures, for example, logging and informing the server administrator. Local databases at the workstation are filled up manually, if it need to have the results of measurements in real time.

In addition to the database with the results of current measurements, the historical data are available at the Observatory Paratunka and are presented in international databases:

1) the hourly values of the H, D, Z, F of the total field vector (IAGA standard) were obtained by the Bobrov's quartz sensors with analog recording on photographic paper during 1969-2004 and are presented in the World Data Centres (http://www.wdc.bgs.ac.uk/catalog/master.html) before 1997;

2) the minute values X,Y,Z,F of total field vector (INTERMAGNET standard) are obtained by digital vector and scalar magnetometers during 2007-2016 (http://intermagnet.org/data-donnee/data-eng.php);

3) the digital images of analogue magnetograms for 1967-2006 are obtained under support of grant "The creation of the database of images of old analogue magnetograms of Geophysical Observatory "Paratunka", Kamchatka, Russia, 1967-2006", VarSITI project [16]. The database is accessible from WDC (http://www.wdcb.ru/stp/magnetogr_list.en.html).

5 Some applications of magnetic monitoring results

Complete set of the equipment, advanced software, extensive archives and databases show that the Observatory Paratunka is a magnetic observatory that provides measurement of field variations at times from decades to seconds in accordance with IAGA and INTERMAGNET standards and provides an experimental basis for solving a wide range of fundamental and applied problems.

Near 50-year series of hourly data on the total field vector makes it possible to study slow variations of the magnetic field (secular variations), such as pole drift or jerks. Figure 6 shows the annual values of H, D, Z obtained at the Observatory Paratunka (data not presented in the WDC and required the verification, is not shown). Magnetic field changes calculated by the international model IGRF12 (www.ngdc.noaa.gov/IAGA/vmod/igrf.html) is also presented at Fig. 7. The inset shows a map (polar projection) with Russian magnetic observatories, which obtain the annual mean data during recent decades, and the position of the north magnetic pole during 20 years. Figure 7 shows that the Observatory Paratunka (PET) is the most eastern Russian observatory, on which regular observations of the total field vector are performed and provide a global representation for magnetic research. It is also clear that the IGRF12 model is good presented the slow field changes in Kamchatka, but there are significant constant differences, possibly related to an inadequate model description of the field in the region. These features are important for the fundamental basis of modeling the main Earth's magnetic field and its secular variations. In addition, the questions of correct modeling of the regional field are important, including the field of crust sources which necessary for inclined drilling [17].

Minute and second magnetic data obtained at the observatory in accordance with the standards of INTERMAGNET, are basic for many scientific studies and in applied problems. This is, for example, the study of processes in the magnetosphere and the ionosphere, including the processes with regional scales or related to Space Weather [18], [19], the study of the Earth's interior [20], the correlation of geophysical processes in the near-Earth atmosphere [21] and many others. An important applied aspect of magnetic measurements is the possible connection between variations of geophysical fields, for example, between magnetic field and seismicity of Kamchatka [22], [23]. High-frequency magnetic

 DOI: 10.1051/e3sconf/20172002002

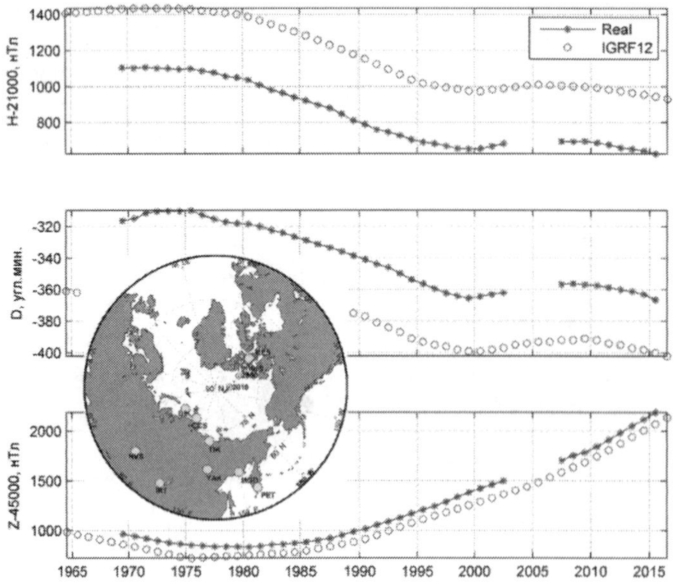

Figure 7. Slow changes of magnetic field at the Observatory Paratunka (PET) from real magnetic observations and by global model IGRF12 (annual mean values). The inset shows magnetic observatories at eastern region of Russia and the location of North magnetic pole.

measurements by induction magnetometer at the observatory provide information on fast processes in the magnetosphere and ionosphere [24], [25], [26].

Observatory Paratunka with a large set of various magnetometers and developed infrastructure makes it an important element in applied works:

- during magnetic surveys of various levels the Observatory can be the center of the reference data about field variations in the region. In addition, before and after field season the calibration and verification of field equipment can be carried out at Observatory;
- the Observatory can be the base point for magnetic measurements at the Repeat Stations (RS) network (the observation program on RSs was stopped in the 90's, but its relevance is now increasing);
- calibration and reconciliation of magnetometers and compasses used by various organizations for stationary and field observations can be performed on the Observatory (comparison of DIflux and scalar magnetometers of the Observatory Magadan was performed at the Observatory Paratunka in October 2012, similar comparison of Geophysical Center of RAS was performed in December 2014);
- Observatory Paratunka has a full set of field instruments for magnetic measuring the total field vector, including the absolute magnetometers Mag-01 and GSM-19W and a variation magnetometer dIdD GSM-19FD. In addition, a new vector magnetometer POS-4 is being tested on the Observatory and can be used in the field surveys.

6 Conclusions

The Geophysical observatory Paratunka was organized fifty years ago. Today this is modern observatory, where the Earth's magnetic field is observed in accordance with high international requirements and standards. The modern digital equipment, the advanced software and the presence of a developed infrastructure allow to obtain the high quality results necessary for solving of fundamental scientific and applied tasks. The Observatory Paratunka is certified as a magnetic observatory of the INTERMAGNET and is the coordinating center for magnetic measurements at other IKIR FEB RAS observatories (Magadan, Khabarovsk and Cape Schmidt). The observatory carries out measurements in cooperation with scientific organizations of Japan, Germany and India.

Further development of the Observatory supposes its participation in regional magnetic survey, including the Repeat Stations program, in studies of the regional features of magnetospheric-ionospheric processes and also in studies of the seismic activity of Kamchatka.

The author is grateful to the staff of the Observatory Paratunka, who for 50 years have provided and continue to provide high-quality measurements of the magnetic field in Kamchatka. This work was partial supported by grants of RFBR No.16-55-45007 and RSF No.14-11-00194.

References

[1] DS18B20 Programmable Resolution 1-Wire Digital Thermometer, https://datasheets.maximintegrated.com/en/ds/DS18B20.pdf, 20 (2015)

[2] I.Y. Babakhanov, M.Y. Basalaev, Z.F. Dumbrava, I.N. Poddelsky, S.Y. Khomutov, VI Int. Conf. *Solar-Terrestrial Relations and Physics of Earthquakes Precursors*, 234-238 (2013), in Russian

[3] S.Y. Khomutov, I.Y. Babakhanov, M.L. Basalaev, Z.F. Dumbrava, I.N. Poddelsky, *Materials of the Partnership conference "Geophysical observatories, multifunctional GIS and data mining", Geoinf. Res. Papers, BS1004, doi:10.2205/2013BS012_Kaluga, http://ebooks.wdcb.ru/doi/kaluga_conf.html* (2013)

[4] S. Khomutov, V. Sapunov, A.Denisov, D. Savelyev, I. Babakhanov, E3S Web Conf., VII International Conference "Solar-Terrestrial Relations and Physics of Earthquakes Precursors", **11**, 7 (2016)

[5] *INTERMAGNET technical reference manual. Version 4.6, Edited by Benoit St-Louis*, 92 (2012)

[6] J.L. Rasson, A.S. Potapov, V.E. Korepanov, S.Y. Khomutov, V.M. Krasnov, J. Bitterly, Earth Planets Space, **58**, 717-722 (2006)

[7] *MAGDAS-A Installation Manual, SERC, Kyushu Univ., Editor George N. Maeda*, 35 (2005)

[8] *dIdD v.7.0 Insrtuction Manual. GEM Systems, Canada*, 50 pp.(2010)

[9] *POS-1 Processor Overhauser Sensor – Magnetometer. User Manual, Ural State Technical University*, 21 (2004)

[10] *Observatory / Volcanology Magnetometer (GSM-90 v7.0), GEM Systems, Inc.* (2008)

[11] *GSM-19 v7.0 Instruction Manual. GEM Systems, Canada*, 149 pp. (2008)

[12] *Operation Manual for Mag-01H Fluxgate Declinometer/Inclinometer with non-magnetic WILD T1 Theodolite (OM2020 Issue 4), Bartington Instruments Ltd.*, 27 (2012)

[13] S.Y. Khomutov, I.Y. Babakhanov, M.L. Basalaev, Z.F. Dumbrava, I.N. Poddelsky, *Modern Information Technologies in Earth Sciences*, 42-43 (2014), in Russian

[14] S.Y. Khomutov, *Software for processing of results of magnetic measurements and preparing of data in accordance with INTERMAGNET Standards. Certificate of Rospatent No2016610829.* (2016), in Russian

[15] S.Y. Khomutov, Ind. Geophys. Union (Special Volume), **2**, 54-61 (2016)

[16] S.Y. Khomutov, VarSITI Newletter, **3**, 7-8 (2014)

[17] I.M. Demina, A.A. Petrova, Bulletin of KRAESC. Earth Sci., **1(15)**, 206-215 (2010), in Russian

[18] D.G. Baishev, A.V. Moiseyev, R.N. Boroyev, S.E. Kobyakova, A.E. Stepanov, O.V. Mandrikova, I.S. Solovev, S.Yu. Khomutov, Yu.A. Polozov, A. Yoshikawa, K. Yumoto, Sun and Geosphere, **10**, 133-140 (2015)

[19] O.V. Mandrikova, T.L. Zalyaev, I.S. Soloviev, S.Y. Khomutov, Bulletin of KRAESC. Phys.-Math.Sci., **3(10)**, 40-46 (2016), in Russian

[20] Y.F. Moroz, T.A. Moroz, S.E. Smirnov, Izvestiya, Physics of the Solid Earth, **8**, 49-61 (2011), in Russian

[21] S.E. Smirnov, G.A. Mikhaylova, O.V. Kapustina, Geomagnetism and aeronomy, **53**, 532-545 (2013)

[22] A.V. Buzevich, S.E. Smirnov, *The problems of geodynamics and earthquake prediction*, 285 (2001), in Russian

[23] *The method of detection of earthquake precursors*, http://www.emsd.ru/265-methods-eq (2017), in Russian

[24] K. Shiokawa, R. Nomura, K. Sakaguchi, Y. Otsuka, Y. Hamaguchi, M. Satoh, Y. Katoh, Y. Yamamoto, B.M. Shevtsov, S. Smirnov, I. Poddelsky, M. Connors, Earth Planets Space, **62**, 517-524 (2010)

[25] R.A. Rakhmatulin, A.Y. Pashinin, Y.V. Lipko, S.Y. Khomutov, I.Y. Babakhanov, Solar-Terr. Phys., **24**, 64-69 (2013)

[26] V.A. Parkhomov, N.L. Borodkova, A.V. Yakhnin, B. Tsegmed, V.E. Chilikin, S.Y. Khomutov, A.V. Pashinin, B.V. Dovbnya, *Physics of auroral phenomena. The 40th annual seminar, 13-17 March, 2017. Abstract Book*, 47 (2017)

 DOI: 10.1051/e3sconf/20172002003

Parallel adaptive sparse approximation methods for analysis of geoacoustic pulses

Alina Kim[1,2], *Olga* Lukovenkova[1,2,*], *Yuri* Marapulets[1], and *Alexander* Tristanov[1,3]

[1]Institute of Cosmophysical Researches and Radio Wave Propagation (IKIR), Far East Division, Russian Academy of Sciences, Acoustic Researches Laboratory, 684034 Kamchatka region, Elizovskiy district, Paratunka, Mirnaya str., 7, Russia
[2]Vitus Bering Kamchatka State University, 683032 Petropavlovsk-Kamchtsky, Pogranichnaya str., 4, Russia
[3]Kaliningrad State Technical University, 236022, Kaliningrad, Sovetskiy pr., 1, Russia

Abstract. The article is devoted to a new approach in the analysis of geoacoustic pulses. The authors proposed a mathematical model based on a sparse representation of the signal. An adaptive matching pursuit method has been developed to identify model parameters. A parallel implementation of this algorithm is proposed on the CUDA platform. This allows real-time processing and modeling of signals.

1 Introduction

The passive acoustic emission method is widely used to study of the strength of materials, in geophysics and to study of the seismic process. It is based on the study of the characteristics of acoustic emission pulses that arise as a result of plastic deformation of solid media. Researches in Kamchatka shows that the acoustic emission method is effective in monitoring acoustic disturbances occurring in the sound range during activation of sedimentary rocks deformation at the final stage in the earthquake preparation. Such emission signals are called geoacoustic emission (GAE), and their disturbances are often used as operational earthquake precursors.

Typical GAE signal represents the succession of relaxation pulses of different form, amplitude and frequency. The characteristics of pulsed radiation depend on the properties of the studied plastic processes. Geoacoustic pulses are characterized by a complex internal structure, a wide variety of time-domain waveforms, large amount of natural and technogeneous noise, short duration. These features significantly complicate the estimation of characteristics and analysis of signal structure. The classical time-frequency methods used to analyze similar signals in other research fields do not give good results. A short duration leads to a poor time-frequency resolution and the redundancy of the resulting decomposition does not agree with the additive character of the pulse generation process because the pulse is the sum of elementary pulses corresponding to separate deformations [1].

It was necessary to develop a model that makes possible qualitative analysis of geoacoustic signals. The term "model of an isolated pulse" should be regarded as a linear combination of functions that satisfies the following properties:

* Corresponding author: o.o.lukovenkova@yandex.ru

1) functions (model components) should describe elementary pulses;
2) model must have high accuracy and do not contain a redundant number of components.

Due to the complex structure of the signals the methods used to identify the model parameters have the property of adaptability and adapt to the local features of the signals and allow for real-time processing.

2 Sparse approximation and matching pursuit algorithm

To a first approximation the geoacoustic signal model may be described as the sum of elementary GAE pulses and interference

$$\mathbf{Signal} = \sum_{m=1}^{N} c_m \mathbf{g}_m + \mathbf{Noise}. \quad (1)$$

The model identification method is based on the sparse approximation ideas. At first, a sparse representation is constructed over a large, in general case, linearly dependent set of different functions. It's substantially increases the adaptability of the methods. At second, sparse representations are compact and not redundant. Finally, the sparse approximation methods were effective in solving similar problems in related fields of science.

The sparse approximation problem is reduced to the signal representation in the linear combination form of the minimum possible number of functions for a given accuracy from some predefined set (hereinafter, the dictionary).

$$\mathbf{Signal} = \sum_{m=1}^{N} c_m \mathbf{g}_m, \text{ where } \|\mathbf{c}\|_0 \to \min. \quad (2)$$

There is no algorithm that can solve this problem in polynomial time, but there are algorithms that can get an approximate solution. The two popular approaches to solving the sparse approximation problem are to replace the minimization of the l_0-norm by minimization the l_1-norm (Basis Pursuit, [2]) and replacing the minimization of the l_0-norm by minimization the l_2-norm (Matching Pursuit, [3]). Each of these approaches gives an estimate of the signal sparse representation on some dictionary $\mathbf{Dictionary} = \{\mathbf{g}_i\}_{1 \le i \le M}$

A series of experiments was performed on real geoacoustic signals (the experiment is described in) to select the most suitable algorithm. The best results for the processing time and the accuracy of the obtained sparse representations were demonstrated by the Matching Pursuit (MP) algorithm proposed by S.Mallat and Z.Zhang.

The algorithm is reduced to an iterative process of searching for dictionary elements that minimize the difference between the original function and the approximation at each step.

$$\begin{cases} \mathbf{Signal} = \sum_{m=1}^{N} c_m \mathbf{g}_m + \mathbf{R}_N, \\ \|\mathbf{R}_N\| \to \min, \\ \|\mathbf{c}\|_0 \to \min. \end{cases} \quad (3)$$

The classic MP algorithm allows to construct linear combinations as (1), satisfying the definition of the model, but has several disadvantages. The only way to improve the accuracy of the constructed approximations is the addition to the dictionary of new functions (atoms) that increases temporal and spatial costs (computing complexity of the MP is $O(M^3)$). The need to use a dictionary of fixed size leads to "coarse" discretization in space of model parameters and reduces the adaptive property of the algorithm.

The authors proposed an Adaptive Matching Pursuit algorithm which allow under conditions of limited computing resources to obtain qualitative signal approximation and adapt to the particular features of the signals in [4]. This algorithm is chosen as a method for identifying model (1) parameters.

3 Adaptive Matching Pursuit

The iterative refinement procedure was added to the classical MP algorithm to improve the quality of the approximation for the dictionary of fixed size. It lies in applying numerical method of alternating-variable descent type for searching atom with the highest value of the scalar product c_{max} with the signal (hereinafter, the "best" atom).

1) to determine the scalar product of each atom and all its shifts with the signal, to remember the parameters of the "best" atom, $\mathbf{p}^0$, and the value of the scalar product c^0_{max};

2) to choose a vector $\boldsymbol{\lambda}$ containing the steps lengths for each of the vector parameters $\mathbf{p}^0$ (it's recommended to select half of the parameter sampling interval as a step λ_i for the parameter p_i);

3) k=1;

4) to calculate the scalar products of each atom obtained by a combination of parameters from vectors $\mathbf{p}^{k-1} - \boldsymbol{\lambda}$, $\mathbf{p}^{k-1}$, $\mathbf{p}^{k-1} + \boldsymbol{\lambda}$ and all its shifts with the signal, to determine the "best" atom, to remember its parameters $\mathbf{p}^k$ and the value of the scalar product c^k_{max};

5) if $\mathbf{p}^k = \mathbf{p}^{k-1}$ then $\boldsymbol{\lambda} = 0.5\,\boldsymbol{\lambda}$;

6) if the required accuracy is not achieved $\left|c^k_{max} - c^{k-1}_{max}\right| \geq \varepsilon$, then k=k+1 and go to step 4 else STOP.

A stopping criterion was added to the algorithm together with the refining procedure. This criterion allows to reduce the experiment temporal cost because the algorithm "decides" itself when to stop instead of calculating the maximum number of iterations for each test signal. It reduces spatial costs. Significant atoms are remembered for each decomposition instead of storing a fixed number of atoms. And also the criterion improves the adaptive property of the algorithm. The number of atoms included in the decomposition directly depends on the complexity of the signal structure.

$$ERR_N \leq level\%, \qquad ERR_N = \frac{\|\mathbf{R}_N\|}{\|\mathbf{Signal}\|} \cdot 100\%. \tag{4}$$

The value 5-10% of error level was chosen empirically. The required accuracy of approximation for geoacoustic signals containing separate pulse is achieved.

The developed algorithm better adapts to local signal features and is more resistant to noise compared with the classical MP algorithm. Figure 1 shows the dependence graphs of approximation error on l_0-norm for MP and AMP algorithms.

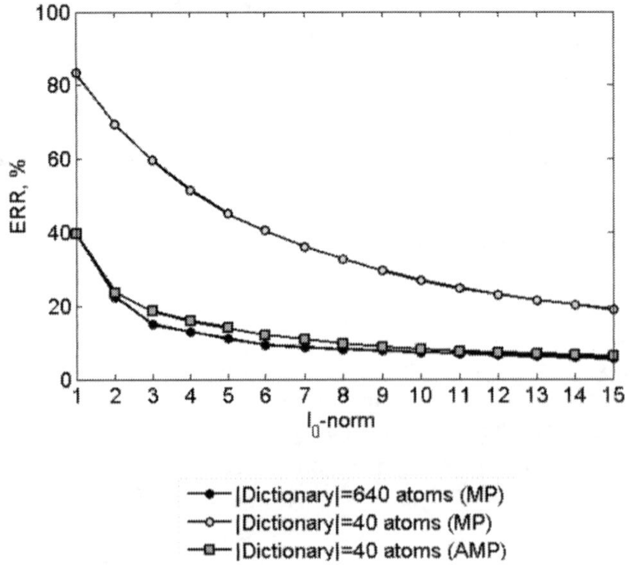

Fig. 1. Approximation error for MP and AMP algorithms.

4 Choice of dictionary

The second important task in modeling geoacoustic signals is the choice of a basic dictionary for research. The authors decided to use a combined dictionary consisting of different functions. The choice of the dictionary is devoted to the works [5,6,7].

First of all, Gaussian modulated functions having the less area of Heisenberg box and widely used in audio signal description models were included in the dictionary. This function can be described in the form

$$g(t) = A \cdot \exp\left(-\left(B_{\lim}(T_{end}) \cdot \Delta\right) \cdot t^2\right) \cdot \cos(2\pi f t), \qquad -T_{end}/2 \le t \le T_{end}/2, \tag{5}$$

where A is the amplitude chosen in such a way that the atom is normalized $\|g(t)\|=1$, $A>0$; T_{end} is the atom length; $B = B_{\lim} \cdot \Delta$ is the parameter affecting the attenuation of the envelope curve, $B>0$; $B_{\lim}$ is the critical value of parameter B, calculated in respect of T_{end} so that the pulse amplitude at the domain boundaries is not more than 5% of the maximum value

$$B_{\lim} = -\frac{4 \cdot \ln 0.05}{T_{end}^{\;2}}; \tag{6}$$

f is fill frequency in Hz; Δ is the gain of parameter B in respect of the critical value (Figure 2).

The model should be composed of signals describing typical geoacoustic impulses by the above definition. Berlage pulse used in the modeling of seismic processes well describes the temporal form of the geoacoustic pulse from the different parametric equations (Fig. 2).

Berlage pulse may be described as following analytic expression

$$g(t) = A \cdot t^{n_{\lim}(T_{end}, t_{\max}) \cdot \Delta} \cdot \exp\left(-\left(n_{\lim}(T_{end}, t_{\max}) \cdot \Delta\right)/t_{\max} \cdot t\right) \cdot \sin(2\pi f t), \quad 0 \le t \le T_{end}. \tag{7}$$

where A is the amplitude chosen in such a way that the atom is normalized $\|g(t)\|=1$, $A>0$; T_{end} – is the atom length; $t_{\max}$ is position of the envelope maximum value; $n = n_{\lim} \cdot \Delta$ is the parameter affecting the attenuation of the envelope curve, $n>0$; $n_{\lim}$ is the critical value of parameter n, calculated in respect of T_{end} and $t_{\max}$ so that the pulse amplitude at the domain boundaries is not more than 5% of the maximum value

$$n_{\lim} = \frac{\ln 0.05}{\ln \frac{T_{end}}{t_{\max}} - \frac{T_{end}}{t_{\max}} + 1}; \tag{8}$$

f is fill frequency in Hz; Δ is the gain of parameter n in respect of the critical value (Figure 2).

The most probable ranges of values for the parameters p_i (T_{end}, $t_{\max}$, Δ и f) for the proposed functions $[p_i^0; p_i^{end}]$ were chosen empirically.

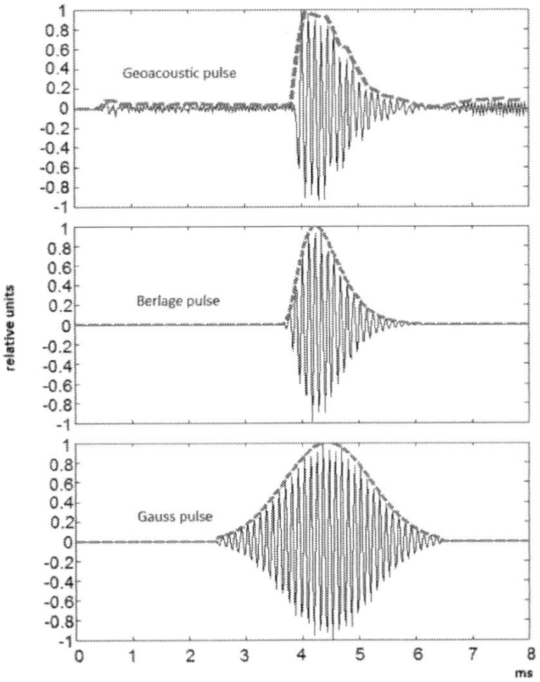

Fig. 2. Berlage and Gabor pulses

5 Model of geoacoustic signal

On the basis of the above, detailed requirements regarding GAE signal model (1) were formulated by authors:

1) functions $\mathbf{g}_i$ belong to the linear normalized space $L_2(R)$;

2) the system of functions $\mathbf{Dictionary} = \{\mathbf{g}_i\}_{1 \le i \le M}$ is redundant;

3) $\mathbf{g}_i$ are time shifted, modulated Gauss and Berlage pulses. Every atom is uniquely identified by and set of parameters $\mathbf{p}$: frequency f; parameters influencing the pulse envelope shape: the type of function, n, B;

4) the atoms are normalized $\|\mathbf{g}_i\| = 1$.

$$\begin{cases} \mathbf{Signal} = \sum_{m=1}^{N_1} \alpha_m \mathbf{g}_m + \sum_{n=N_1+1}^{N_1+N_2} \beta_n \mathbf{g}_n + \mathbf{R}_N, \\ \|\mathbf{R}_N\| \to \min, \\ N = N_1 + N_2 \to \min, \end{cases} \tag{9}$$

where $\mathbf{g}_m$ are the atoms approximating the pulse, $\mathbf{g}_n$ are the atoms approximating the parasitic component of pulse. Value N_1 характеризует characterizes the complexity of the pulse structure, N_2 is the noise level of the pulse. The difference between the original function and the approximation error $\mathbf{R}_N$ determines the goodness of decomposition to the signal.

The coefficients $\boldsymbol{\alpha}$, $\boldsymbol{\beta}$ and the parameters of the functions $\mathbf{p}$ are identified by the Adaptive Matching Pursuit algorithm. Figure 3 shows the modeling of the real geoacoustic pulse.

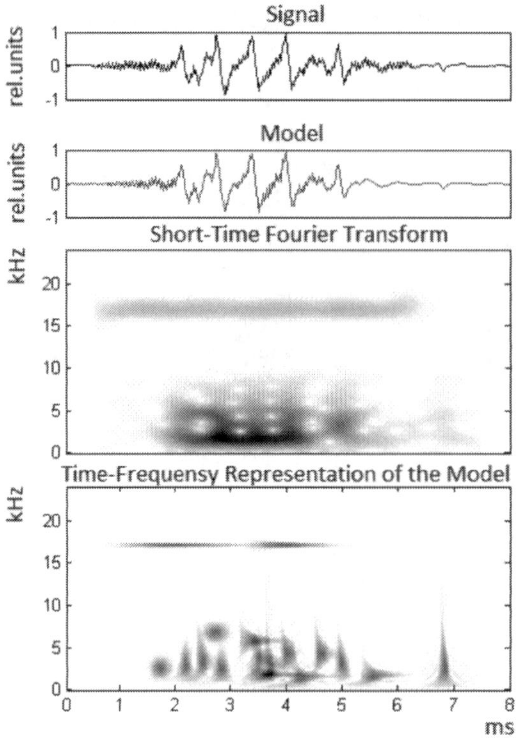

Fig. 3. Model of the real geoacoustic pulse

6 Software development

It was shown that the proposed algorithm is effective in modeling real signals. But high temporal cost of algorithm and a huge data amount continuously coming from registration points don't allow real time processing and modeling of signals.

The registration system records geoacoustic signals with a frequency 48 kHz in real time and stores them in 15-minute WAVE files. A large amount of computing resources is required to analyze the anomalies of a single file. At the first stage, the researcher must detect signal sections containing pulses, but it's complicated by the short length and high noise level of pulses. Next, each detected segment is analyzed using the AMP algorithm.

The automated system for processing GAE signals was developed to automate and accelerate the analysis process. It consists of two parts

1) pulse detection and elimination of not informative signal segments;
2) parallel realization of the algorithm AMP.

6.1 Pulse detection

It was decided to apply the threshold method for automatic pulse detection in a signal. The pulse is registered when the signal exceeds some predetermined threshold value.

Most often, the detection threshold is a constant value and it is determined empirically. However, the high risk of losing informative data in this case: impulses below the specified threshold are not registered. But with a low thresholds and a strong noisy signal, a large amount of non-informative data can be extracted.

An effective solution to this problem is using of an adaptive threshold, which automatically adjusts to the current noise level in the signal. The mean square deviation is used as the pulse detection parameter. Its calculation is carried out in non-intersecting windows of fixed length.

$$s_{th_k} = A \cdot \sqrt{\frac{1}{n} \cdot \sum_{i=(k-1)\cdot n}^{k\cdot n-1} (x_i - \mu_k)^2} \quad (10)$$

where s_{th_k} – threshold value in signal segment from $k{\cdot}n$ to $k{\cdot}n + n - 1$, x_i – discrete signal value at the i sample, μ_k – average signal in the window from $k{\cdot}(n\text{-}1)$ to $k{\cdot}n$ - 1, n – window length, A – experimentally determined parameter.

The experiments proved that using of the adaptive threshold function makes it possible to detect single pulses in signals with different degrees of noisiness. However, when pulses are frequently followed at the signal segment, the threshold values can increase sharply, resulting in a loss of informative data (Figure 4a).

Therefore, it is suggested to not include signal sections containing a pulse in the process of calculating the threshold function (Figure 4b).

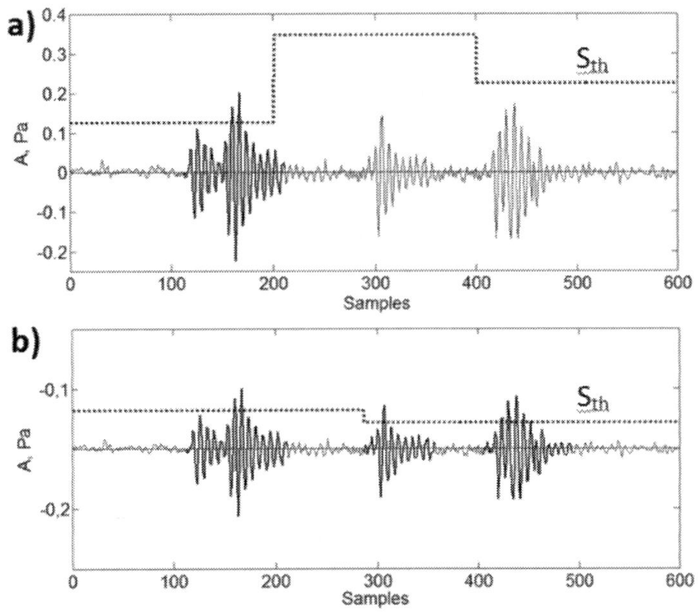

Fig. 4. Using an adaptive threshold to pulse detection in a signal

6.2 Parallel realization of adaptive matching pursuit algorithm

At the next stage, the system analyzes each detected pulse using the AMP algorithm. The most practical and cost-effective way to increase computing capability is the organization of parallel computing [8].

Let there be a dictionary consisting of M atoms with length of L_g samples, and a signal with length of L samples. The scalar product between **Signal** and each **Dictionary** line is calculated and written into a **C** matrix of dimension $M\times(L+L_g-1)$ to determine the "best" dictionary atom (Figure 5):

$$c_{i,j} = \sum_{k=\max(1,j+1-L_g)}^{\min(j,1)} \mathbf{Dictionary}_{i,k} \cdot \mathbf{Signal}_{L_g-j+k} \qquad (11)$$

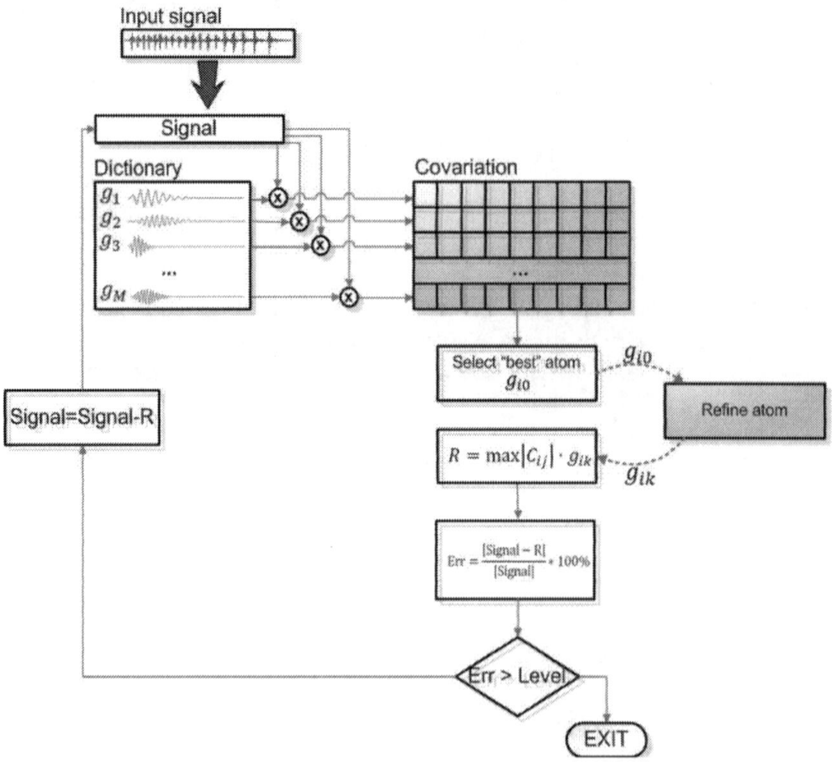

Fig. 5. Serial realization of AMP algorithm

The "best" atom is refined k times and stored in the output file. Figure 6 shows the scheme of the refinement procedure. The costliest part of the algorithm is the calculation of scalar products of dictionary atoms with the signal: matrix calculation time is more than 90% of the total algorithm execution time.

The same type processing of a large amount of information makes it possible to apply data parallelism to this part of algorithm. Each subtask calculates one element of the covariance matrix, depending on the input parameters i and j. The proposed algorithm decomposition is suitable for efficient execution in the style of the architecture SIMD (Single Instruction stream / Multiple Data stream), that allows to perform the same arithmetic operation on multiple data points simultaneously. One of the most popular technologies based on the SIMD concept is the software and hardware platform CUDA, used to organize parallel computing on graphics processors (GPU) [8]. The main part of the algorithm is executed on the central processor (CPU), but the costly process of calculating the covariance matrix is sent to GPU (Figure 7).

 DOI: 10.1051/e3sconf/20172002003

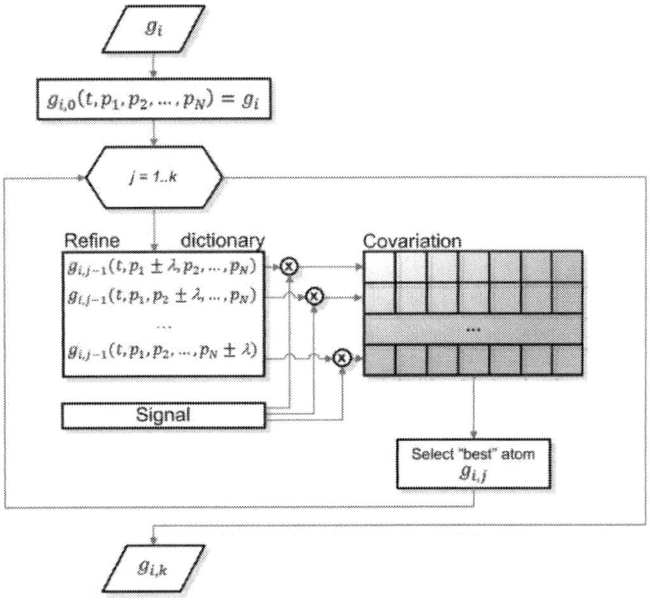

Fig. 6. Refinement procedure algorithm

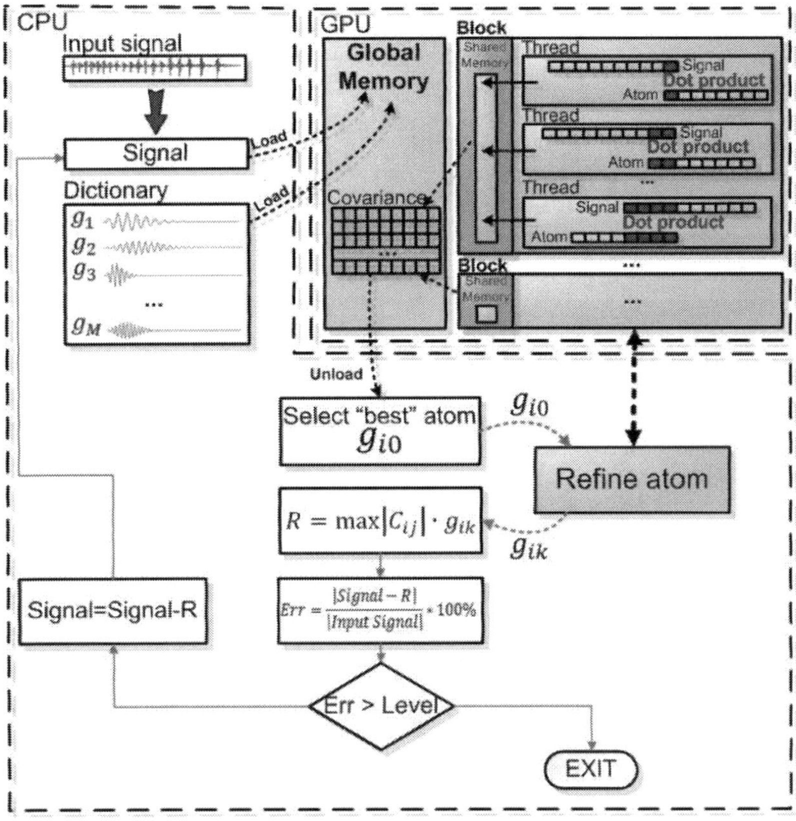

Fig. 7. Parallel realization of AMP algorithm

The signal and the atom dictionary are copied into the global memory of GPU. Next, a grid consisting of n_b blocks of n_t threads is run for execution. One thread computes one element of the covariance matrix and puts it in the shared memory of the block. After all threads of the block are executed, the shared memory contains an array of values that is copied to the resulting matrix **C** in the global memory. So each block fills the matrix **C** with n_t elements.

After the completion of all blocks, the resulting covariance matrix is unloaded to the CPU memory, and the algorithm execution returns to the central processor. The atom refinement algorithm also uses the GPU resources to calculate covariance matrices.

7 Results

The developed system for processing GAE signals is implemented in the Acoustic Research Laboratory IKIR FEB RAS. It runs on a computer that has Intel(R) Pentium(R) CPU G2120 (3.10 GHz) and NVIDIA GeForce GTX 760 graphics card (1152 CUDA cores, 2258 Gflops). The system was tested on real geoacoustic signals with a frequency 48 kHz. The runtime of the serial and parallel AMP realization was measured for different lengths signal sections (Table 1).

Table 1. The runtime of the serial and parallel AMP realization

Serial algorithm (ms)				**Parallel algorithm (ms)**			
Number of atoms / Pulse length (samples)	**1**	**5**	**10**	**Number of atoms / Pulse length (samples)**	**1**	**5**	**10**
50	35	155	317	**50**	10	22	86
100	110	542	1132	**100**	12	54	102
200	420	2087	4252	**200**	16	86	166
500	2532	12715	25297	**500**	46	268	496

So using the parallel realization of adaptive matching pursuit algorithm allowed us to reduce the time costs by an average of 8 times.

References

1. Yu.V. Marapulets, B.M. Shevtsov, *Mezomasshtabnaya akusticheskaya emissiya: monografiya* (2012)
2. S.S.Chen, D.L. Donoho, M.A. Saunders, SIAM J. Sci. Comput., **20**(1). P.129-159 (1998)
3. S. Mallat, Z. Zhang, IEEE Trans. Signal Process., **41**(12). P.3397–3415 (1993)
4. O.O. Lukovenkova, Vestnik KRAUNTS. Fiziko-matematicheskiye nauki, **2**(9). P. 59-64 (2014)
5. O.O. Lukovenkova, A.B. Tristanov, Tsifrovaya obrabotka signalov, **2**. P. 54-57 (2014)
6. A.B. Tristanov, O.O Lukovenkova. Tsifrovaya obrabotka signalov, **2**. P. 35-38 (2015)

7. A.B.Tristanov, Yu.V. Marapulets, O.O. Lukovenkova, A.A. Kim, Pattern Recognition and Image Analysis, **26**(1), P. 34–44 (2016)
8. J. Sanders, E. Kandrot, *CUDA by Example: An Introduction to General-Purpose GPU Programming* (2013).

E3S Web of Conferences **20**, 02004 (2017) DOI: 10.1051/e3sconf/20172002004
Solar-Terrestrial Relations and Physics of Earthquake Precursors

Choice of optimum heights for registration of ionospheric response onto earthquakes

Valerii Krasnov*, *Ivan* Gotur, *Yurii Kuleshov* and *Sergei* Cherny

Mozhaisky Military Space Academy, 197198, Saint Petersburg, Zhdanovskaya,13, Russia

Abstract To investigate the dependence of ionospheric disturbances on height we used model calculations, and the data of seismic and ionospheric observations during the Tohoku-Oki earthquake. High-altitude dependences of "portraits" of ionospheric disturbances are calculated for a case of influence of a seismic P-wave onto the ionosphere. We compared the "portraits" of ionospheric disturbances with the "portraits" of the seismic recording. The correlation coefficient of the recordings for the height of 100 km was about 0.81, for 130 km - 0.85, for 160 km - 0.77, for 180 km - 0.76, for 200 km - 0.7, for 230 km - 0.54 and for 250 km - 0.41. At the same time the maximum of F2-layer was at the height about 250 km. Thus, the height of a maximum of F2-layer was not optimum for registration of ionospheric disturbances due to the earthquake. It was preferable to carry out measurements of the ionospheric disturbances at the heights below 200 km. The profile of amplitude of the ionospheric disturbance had no sharply expressed maximum at the height of a maximum of F2-layer. Therefore it is problematic to use the approach of the thin layer for interpretation of TEC disturbances.

1 Introduction

Development of Global Navigation Satellite Systems has opened new opportunities to investigate ionospheric disturbances based on the space-time variations of total electron content (TEC). For example, the paper ensemble appears in the literature after strong earthquake, which describe results of influence of movement of a terrestrial surface or a tsunami upon the ionosphere. It is offered to use this kind of observations in tsunami warning systems, and also during monitoring of earth (sea) surface motion in region, which is difficult of access (for example, [1, 2]). The principal disadvantage of this method is integrated character of a measurable value of TEC along trajectory of a radio wave (for example, [3]). However, the assumption is accepted that the main contribution to TEC variations appears around the height of the maximum of ionosphere ionization (F2 layer). Integrated character of TEC does not allow defining precisely height of observable ionospheric disturbances and their coordinate. There are attempts to define the conditions at which ionospheric disturbances can be most precisely located and to the greatest degree are similar to a seismic signal. For example [4], it is recommended to carry out the measurements at a satellite elevation angle lower than 40°. The purpose of our work, by means of model calculations and the observation data of the Tohoku-Oki earthquake, to investigate dependence of the form of ionospheric disturbances on height and to define the optimum heights for observation of the ionospheric response onto the earthquake.

2 Models used for the purposes of calculations

To do investigation we used the data of seismic stations of the Czech Republic, the data of ionosonde, located in Pruhonice, and the data of ionospheric Doppler radio sounding. The detailed description of the network of observation points is presented in [5], and Fig.1 shows location of the network.

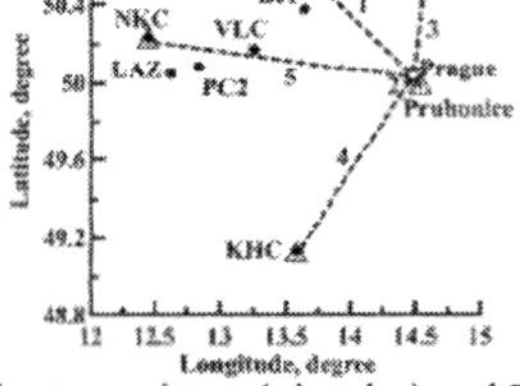

Fig. 1. Position of seismic stations (circles), Doppler transmitters (triangles) and Doppler shift receiver (Prague). Dashed lines show the five paths of Doppler radio sounding of the ionosphere.

*Corresponding author: vmkrasnov@yandex.ru

The epicentre of a strong earthquake (M=9.0) was at a distances of ~9000 km from a point of observation. Seismographs have registered arrival P, S, SS and Rayleigh waves. As an example, in this work we consider a case of influence of a P wave onto the ionosphere. The model of vertical motion of earth surface is developed on the basis of seismic data and in detail described in [6]. The basic oscillations of the earth surface motion $S(t)$ under action of P wave occurred in the frequencies range of 0.0195 Hz (the period - 51.2 s) and 0.039 Hz (the period 25.6 s, Fig.2).

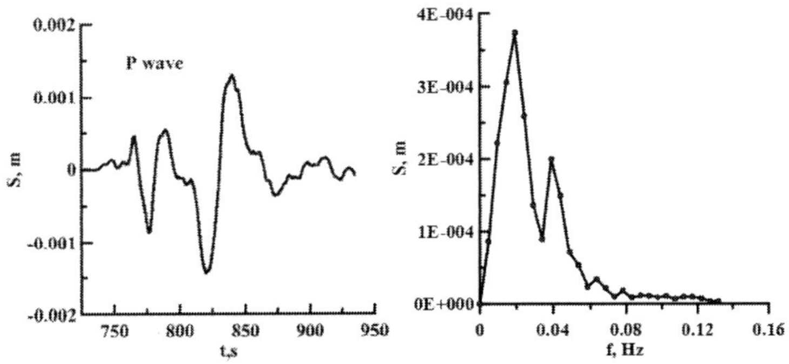

Fig. 2. "Portrait" of motion of the earth surface as a function of time for P wave (a) and its spectrum (b).

The model of radiation of an infrasound by ground motion is developed on the basis of the Rayleigh integral

$$p(t) = \frac{\rho_0}{2\pi}\int \frac{a(t-L/c)}{L}\,dS, \tag{1}$$

where $p(t)$ is the derived pressure history at any arbitrary point in space, t is time, ρ_0 is the air density at the surface, L is the slant range from the surface element to the position (observation point), $a(t-L/c)$ is the retarded acceleration normal to the earth surface, and dS is the surface element of the emitting source.

The knowledge of the height of the observation points allows us to resolve task of choice of the circle radius size to calculate the Rayleigh integral. For this purpose it is necessary to define maximal horizontal distance from the observation point up to the surface element, from which the acoustic ray can propagate to the observation points. For example, the earth's surface with radius of 240 km forms an acoustic field at a height of 215 km during the Tohoku-Oki earthquake [6].

The Rayleigh integral (1) assumes linear propagation of acoustic wave through a homogeneous medium without attenuation. However, the Rayleigh integral may be used and for calculation of infrasound field in the inhomogeneous atmosphere. In particular, there is the model for a point source, which allows calculation of the acoustic field taking into account atmospheric inhomogeneity, nonlinear effects, absorption and divergence of the wave front because of long-range acoustic wave propagation [7]. Krasnov et al. [6] have developed an algorithm, which can be used to adapt this model and the Rayleigh integral for the calculation of infrasonic fields caused by an earthquake at heights of the ionosphere.

To describe the evolution of an acoustic signal along acoustic ray we used the formula [7]

$$V = \frac{b}{\varepsilon c \rho_0}\frac{\partial}{\partial t}(\ln U), \tag{2}$$

where V, ρ_0, are the velocity and density of the fluid, respectively; c is a sound speed,

$$U(z,t_p) = \frac{1}{2q\sqrt{\pi}}\int_{-\infty}^{+\infty} U_{in}(0,t')\exp\left\{\frac{-(t-t')^2}{4q^2}\right\}dt', \tag{3}$$

where $\varepsilon = \dfrac{\gamma+1}{2}$; $b = \left(\zeta + \dfrac{4}{3}\eta\right) + \left(\dfrac{1}{c_v} - \dfrac{1}{c_p}\right)\chi$, ζ and η bulk and shear viscosities, χ is the thermal conductivity, γ is the ratio of specific heats cp/cv, $t_p = t + \int \dfrac{f(z)}{c}dz$, t is running time, $f(z) = \left(1 - \dfrac{3a+d}{2\rho_0 c^2}\dfrac{\partial c}{\partial z} - \dfrac{3a}{2\rho_0^2 c}\dfrac{\partial \rho_0}{\partial z}\right)$,

$a = \zeta + \frac{4}{3}\eta$, $d = \left(\dfrac{1}{c_v} - \dfrac{1}{c_p}\right)\chi$, $q^2 = \int \dfrac{b}{2\rho_0 c^3}dz$ is response time of "atmospheric filter".

DOI: 10.1051/e3sconf/20172002004

Within altitudes below ~100 km (region D and the lower part of region E of the ionosphere), both ions and electrons move together with neutral particles; therefore, it is possible to assume [8]

$$N(t) \approx N_0\left(1 \pm \frac{\gamma-1}{2}\frac{V(t)}{c}\right)^{2/(\gamma-1)}, \qquad (4)$$

where N_0 is the background concentration of electrons in the ionosphere and N is the concentration of electrons under the effect of the acoustic wave.

Above 130 km, the ionospheric plasma is considered magnetized, and the following formula [9] is required to estimate the influence of the acoustic perturbation on the formation of the disturbance of electron concentration:

$$\frac{\partial N}{\partial t} = -\left[N_0\left(\frac{v}{L}\sin^2\Theta + \frac{\partial v}{\partial l}\cos^2\Theta\right) + v\frac{\partial N_0}{dz}\cos\varphi\cos\Theta\right], \qquad (5)$$

where Θ is the angle between direction of the geomagnetic field and the acoustic ray trajectory, and φ is the angle between direction of the geomagnetic field and z axis, l is coordinate along acoustic ray, L is the slant range from the surface element to the position (observation point).

We calculate the Doppler shift of the radio wave as the time derivative of the transmitted phase which is given by

$$f_d = -\frac{f}{c_L}\frac{d}{dt}\int_s n\cos\alpha ds, \qquad (6)$$

where f is the carrier frequency of the radio sounder, c_L is the speed of light, α is the angle between trajectory of radio ray and z axis, s is the path of radio ray from transmitter to receiver, n is the reflection coefficient. We did not neglect by the magnetic field.

Validity of the models was confirmed by comparison of the calculations and data of the Doppler radio sounding of the ionosphere for five paths (Fig.1). The correlation coefficient between calculated and experimental records for seismic waves: P, S, SS and the first Rayleigh waves was at the range from 0.68 to 0.9. As an example, Fig.3 shows the calculated and experimental Doppler shifts for the P wave and cross-correlation coefficients (k) between the calculated and experimental curves.

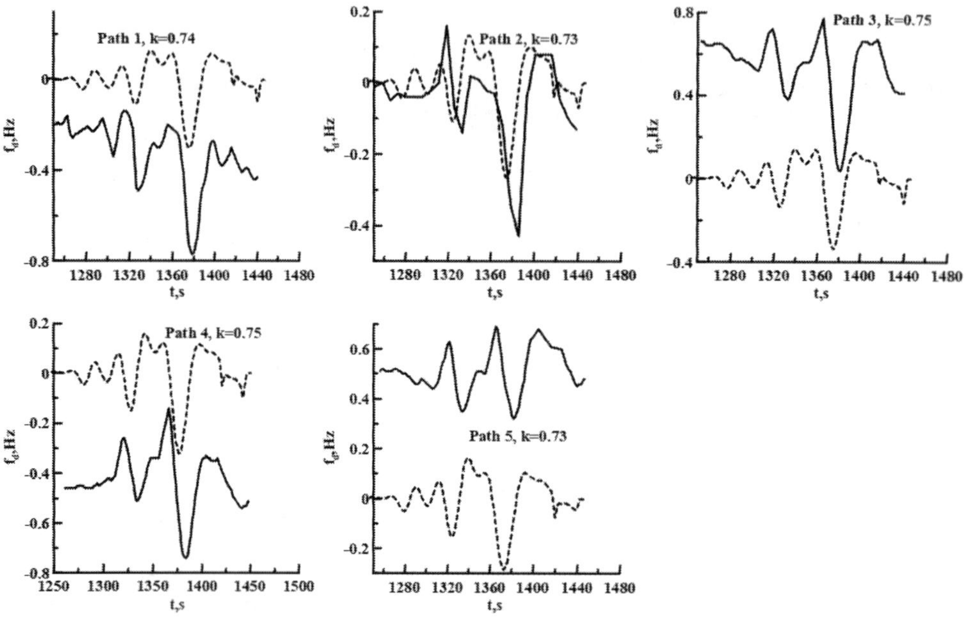

Fig.3. Experimental (solid lines) and calculated (dotted lines) forms of Doppler frequency for five radio traces of the ionospheric sounding during the P seismic wave propagation. Zero time is time of the earthquake.

3 Dependence of the ionospheric disturbances on height

The ionospheric electron density profile (Fig. 4) was estimated from the ionogram measured by the digital ionosonde located in Pruhonice (Fig.1)

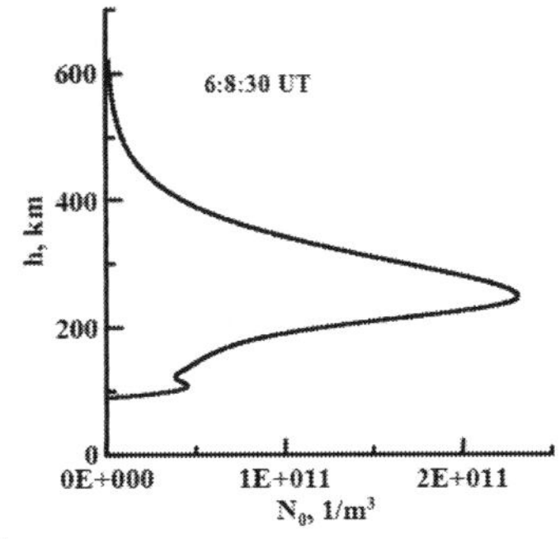

Fig.4. The ionospheric electron density profile.

The profile N(z) begins with height of ~90 km, and the maximum is located at the height of 250 km.

First of all it is interesting to compare amplitudes and forms of the earth surface motion and the motion of hydrodynamical particles at different heights of the atmosphere. Fig. 5 shows the calculated motions of hydrodynamical particles under action of the acoustic wave at different altitudes.

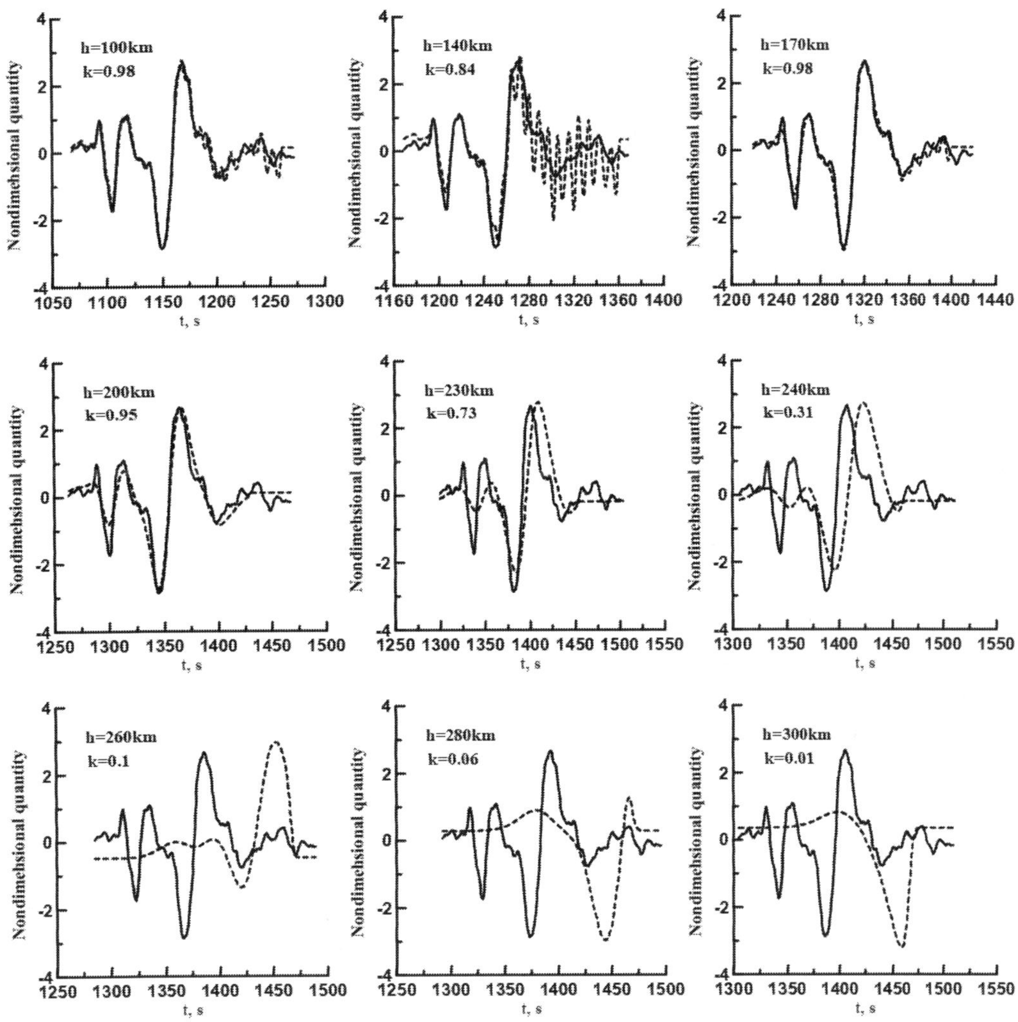

Fig. 5. The comparison of forms of the earth surface motion (solid line) and motions of hydrodynamical particles (dotted line) at different heights. K is a correlation coefficient.

Additional high-frequency oscillations of motions of hydrodynamical particles at heights of 100 and 140 km have arisen because of an interference of the acoustic waves radiated by different areas of the earth surface. These high-frequency oscillations are destroyed by the nonlinear processes during its propagation upward to the altitude of 170 km. As a whole, good similarity of motion of the earth surface and hydrodynamical particles is observed up to height of 230 km (k≥0.73). Then nonlinear processes strongly change the form of motion of hydrodynamical particles. In practice, seismographs observe a vertical speed ($V_z = \frac{dS}{dt}$) of an earth surface motion, and radiophysical methods observe disturbances of electron density. Fig. 6 shows the results of comparison of a seismogram for P waves and calculated disturbances of electron density at different heights.

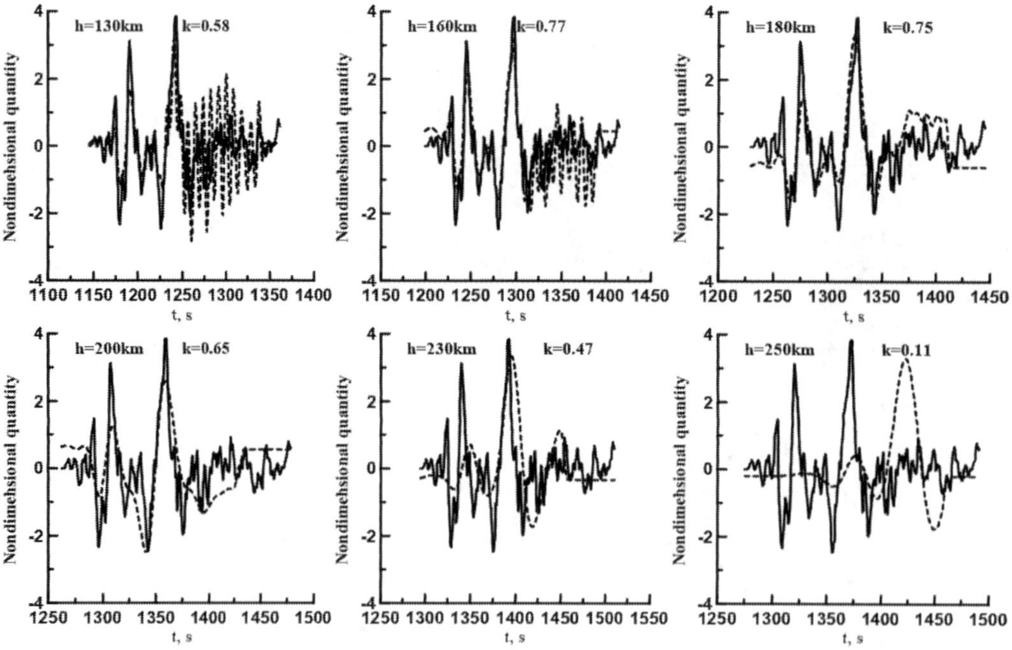

Fig. 6. The comparison of forms of P wave (solid line) and disturbances of electron density at different heights (dotted line). K is a correlation coefficient.

Good similarity of oscillations of vertical velocity of the earth surface and disturbances of electron density is observed up to height of 200 km (k≥0.7). It means that heights of the ionosphere below 200 km are most preferable to observe the ionospheric response to the influence of P-wave. It is obvious, that for registration of these disturbances it is necessary to use oblique trajectory of radio wave propagation from satellites to a receiver. This conclusion is in agreed with the results of experiment by Rolland et al. [4].

Disturbances of electron density arise in the ionosphere due to propagation upwards of an infrasonic wave. As a result, the disturbances of electron density are shifted from each other in time (Fig. 7).

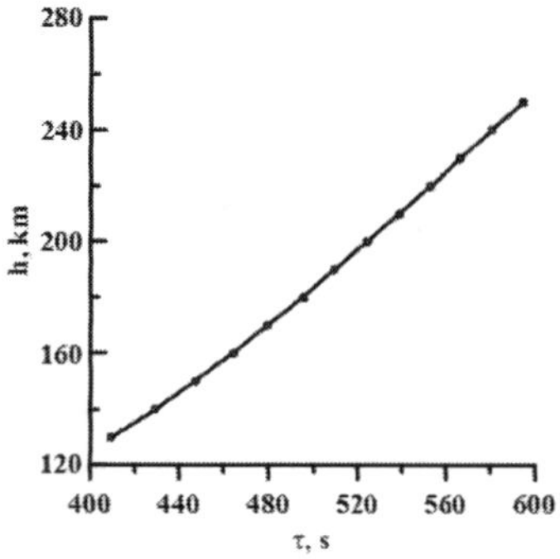

Fig. 7. Time delay of influence of the P-wave on formation of ionospheric disturbances at different height.

It leads to summation of the disturbances in different phases of motion, when we measure TEC along a trajectory of radio wave, and it leads to additional distinction of a seismogram and TEC. It is obvious, that the distinction of forms of oscillations will increase with decrease of wave-length of the disturbances.

It is assumed that the main contribution to TEC variations appears around the height of the maximum of ionosphere ionization (F2 layer). This allows one to consider the ionosphere as a thin layer located at the height H. TEC then represents a point of intersection of a line of sight with this thin layer. Let us consider dependence of deviation amplitude of the ionospheric disturbances on height for our case. Fig. 8 shows the calculation results of dependence of root-mean-square deviations (σ_N) of ionospheric disturbance on height.

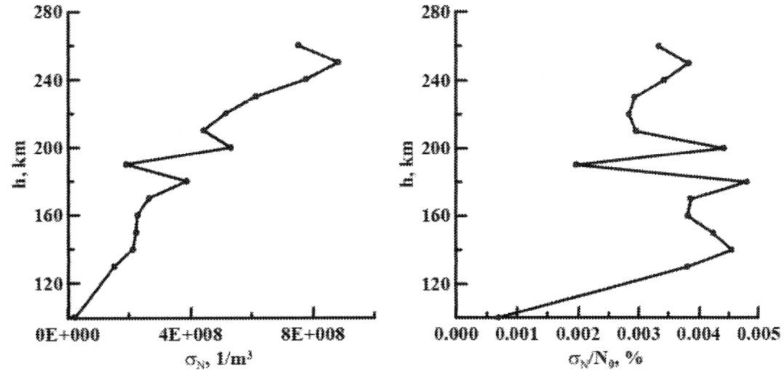

Fig. 8. The calculation results of dependence of σ_N on height: a) on absolute value; b) in percentage in relation to background value of electron density.

Small amplitude of the disturbances at the fixed height explains small values of TEC during earthquakes. There is a small maximum of disturbance around the maximum of the ionospheric F2 layer. Thus, it is a problem to use the approach of a thin layer for interpretation of the TEC.

Spectra of S, SS and Rayleigh waves of the Tohoku earthquake contain more low-frequency oscillations. Therefore nonlinear effects will essentially influence them at greater heights. Accordingly the upper border of optimum registration of the ionospheric response to these waves will be at higher height, than for P-wave.

4 Conclusion

We have considered influence of a P-wave on the ionosphere during the Tohoku-Oki earthquake. To do research we used the model, which allows calculation of the acoustic field taking into account atmospheric inhomogeneity, nonlinear effects, absorption and divergence of the wave front because of long-range acoustic wave propagation. High-altitude variations of "portraits" of ionospheric disturbances were determined. We compared the "portraits" of ionospheric disturbances with the "portraits" of the seismic recording. The correlation coefficient of the recordings for the height of 100 km was about 0.81, for 130 km - 0.85, for 160 km - 0.77, for 180 km - 0.76, for 200 km - 0.7, for 230 km - 0.54 and for 250 km - 0.41. At the same time the maximum of F2-layer was at the height about 250 km. Thus, the height of a maximum of F2-layer was not optimum for registration of ionospheric disturbances due to the earthquake. As a result we defined the border of heights (<200km) up to which the temporal oscillation of ionospheric disturbances are similar to the form of seismic records. The knowledge of this border and the characteristics of a directional radiation pattern of acoustic wave can help to choose the optimum trajectory of radiowave propagation from a satellite to a receiver to measure TEC. The profile of amplitude of the ionospheric disturbance had no explicitly expressed maximum at the height of a maximum of F2-layer. Therefore it is problematic to use the approach of the thin layer for interpretation of TEC disturbances.

References

1. P. Coïsson, P. Lognonné, D. Walwer, L. M. Rolland, Earth Space Sci. **2 (5)**, 125 (2015).
2. A. Komjathy, D. A. Galvan, P. Stephens, M. D. Butala, V. Akopian, B. Wilson, O. Verkhoglyadova, A. J. Mannucci, and M. Hickey, Earth Planets Space, **64**, 1287 (2012).
3. E. Astafyeva, L. Rolland, P. Lognonné, K. Khelfi, T. Yahagi, J. Geophys. Res.: Space Phys. **118 (9)**, 5942 (2013).
4. L. M. Rolland, P. Lognonné and H. Munekane, J. Geophys. Res. **116**, A05320 (2011).
5. J. Chum, F. Hruska, J. Zednik and J. Lastovicka, J. Geophys. Res. **117**, A08319 (2012).
6. V. M. Krasnov, Ya.V. Drobzheva, J.,F. Chum, J. Atmos.Sol.-Terr.Phys. **135**, 12 (2015).
7. Ya. Drobzheva, V. Krasnov, J. Atmos. Sol.-Terr.Phys. **65 (3)**, 369 (2003).

 DOI: 10.1051/e3sconf/20172002004

8. V. M. Krasnov and Yu. V. Kuleshov, Acous. Phys. **60 (1)**, 19 (2014).
9. V.M. Krasnov, G.M. Pelenitcin, O.E. Ryaskov, N.M. Salikhov, Vestn. Acad. Sci. Kazakh USR, **8**,44 (1991).

E3S Web of Conferences **20**, 02005 (2017) DOI: 10.1051/e3sconf/20172002005
Solar-Terrestrial Relations and Physics of Earthquake Precursors

Geophysical field disturbances and quantum mechanics

Vladimir Kuznetsov[1,*]

[1] *Institute of Cosmophysical Research and Radio Wave Propagation FEB RAS, Paratunka, Russia*

Abstract. Quantum processes impact into physics of geophysical field disturbances is discussed here in examples of phenomena such as an earthquake with processes preceding and accompanying it, volcanoes eruptions and diamond exploding pipes. Physics of shock waves generation in ionosphere and atmosphere, mechanism of atmosphere phenomena in supercooled clouds recorded by a stormglass is considered. The report treats of physics of ball and dark lightning, of generating in atmosphere the high-energy particles involved in sprites occurrence, and so on.
Geophysical phenomena considered here have no clear and consistent interpretation in the context of classical physics. We attempt to involve the recent achievements of quantum physics namely the quantum entanglement between elementary particles implicated in considered phenomena.

1 Introduction

It is well known that geophysics describes nature phenomena occurring in solid Earth (earthquakes, volcano eruptions, heat flow), in oceans (tsunami), in atmosphere (clouds formation, climate, atmosphere electricity, thunderstorms, sprites). Geophysics is a complex of sciences, which explore the structure of the Earth by the physical methods. Each phenomenon implies the formation of "its own "fields unique to it, they are thermal field of the Earth, gravity field, the Earth's magnetic field and the electromagnetic field. The division into fields determines the division of geophysics into relevant disciplines, such as meteorology, oceanography, etc. Geophysics phenomena such as earthquakes, volcano eruptions, lightning and sprites, climatology, etc. appear not to fit the only one section. This is particularly noticeable when talking about possible methods of prediction. Let us cite, for example, some of the ways of earthquakes forecast. They are acoustic monitoring, and monitoring of electric and magnetic fields, changes of gravity and stresses in lithosphere, monitoring of heat flow, change in groundwater level, etc. As many of these fields are experiencing a perturbation on the eve of an earthquake they must be considered jointly. Sometimes it turns out that it is impossible to find an "agent" which would be manifested simultaneously in these fields. The fact is that physicists use to solve the problem in the frames of one method inside the so-called classical physics.

Currently the situation in science began to change radically. To explain some phenomena the physicists appeal to quantum mechanics assuming the so-called classical physics in which frames the phenomena have no apparent explanation to be a component of quantum mechanics. Here physicists has called attention to a particular matter state, named quantum entanglement (QE) (quantum entanglement, nonlocality). Quantum properties were considered to manifest themselves only at very low

*Corresponding author: vvkuz38@mail.ru

temperatures close to the absolute zero. Recently it was shown that it is not entirely true. Secondly, entangled substances have a property effecting as expressed by Einstein, a "supernormal long-range action". Otherwise, particles of substances in QE state behave like a single substance, despite the fact that there are no interaction between its parts and particles can be at any large distance from each other. In quantum mechanics it means that all linked particles of a substance are described by a single wave function and they have a specific coherence. It is not appropriate to object to this approach from the point of GTR. Thirdly, talking about the role of QE in nature it is worth to refer to the recently published paper stating that the essence of gravity consists in quarks and antiquarks QE [1]. The development of this approach seems to allow to solve the problem of gravity quantization.

Is QE common at the Earth? The answer to this question is given at a recently published article alleging that "everything is linked" in the world [2]. According to the authors QE originates from the creation of the Universe i.e. from the Big Bang. In fact, a typical particle is entangled with many particles far outside our horizon. However, as QE is evenly propagating by a volume the two randomly chosen particles are unlikely to be associated through reduced density matrix which applies to any pairs that are separable [2].

Here we will focus on the fast perturbations of the geophysical environment, called quakes. Quake is the environment shaking against the background of its initially calm conditions. Where and how do quakes occur? In the monograph [3] these phenomena refer to such ones (not including a well known earthquake) like: ice quake, wood quake, protein quake, etc. Author of this monograph rightly believes that in such cases a violation of the fluctuation-dissipative theorem has occurred. We refer lightning (ball and dark), sprites, etc. to quantum phenomena. Obviously, the broader is the range of quake, the more difficult is to select an environment with a universal QE. In the subsequent discussion we will restrict ourselves to QE of hydrogen bonds protons (HB).

Hydrogen bond - a special type of bonding due to the fact that the hydrogen atom associated with a strongly electronegative element (nitrogen, oxygen, fluorine, etc.) lacks of electrons and it therefore is able to interact with the lone electron pair of another electronegative atom of the same or another molecule (the total proton). The presence of HB and their cooperative properties, particularly in water, lead to the fact that water properties change depending on the number of HB. For example, in ice there is a lot of HB and their number is decreasing with ice melting. The number of HB small in melt is less in hot water and there is almost no HB in boiling water. If water completely has lost its ability to form HB, it would turn into vapor, condensing into liquid at 100 °C. Present not only in water but also in the lithosphere hydrogen bonds are determining its special properties. The mechanism of HB is that the electron density on the line H...O (N, F, etc.) is shifted to the electronegative atom generating a nearing of electrons of electronegative atoms in the neighboring molecules and the distance O-H, O - N, etc. decreases.

Data on QE. "A surprising feature of the quantum world is that one particle instantly affects "entangled" with it another one, no matter how far apart they are. The term "entanglement" to describe this phenomenon came up with Schrödinger who like Einstein was not a fan of what Einstein called a "supernormal long-range action". But despite their skepticism, the quantum "entanglement" proved in many experiments is one of the most fundamental ideas in quantum mechanics..." ([4], p. 235). Quoting the book of S. I. Doronin, "Quantum magic" [5]: "The matter, that is the matter and all known physical fields are not environment basis but represent only a small part of the total Quantum Reality."

Some basic definitions from quantum mechanics:

Decoherence is a physical process in which the locality breaks and quantum entanglement between the constituent parts of the system reduces in its interaction with the environment. Her the subsystems "manifest" from their nonlocal states as separate independent elements of reality, they separate from each other, acquiring visible local forms.

Quantum tunneling is a quantum mechanical effect of particles passage through the energy barrier. Similar to the passage through the tunnel this process is called a tunneling. There is no analogue in classical mechanics.

Nonlocality is a feature of entangled states which cannot be related to the local elements of reality. It is not relevant to waves, fields and to the classical energy of any kind and type. Quantum nonlocality have no classical analog and it cannot be explained within classical physics.

Nonlocal correlations (quantum correlations) – a specific effect of nonseparability (quantum entanglement) which consists in concurrent behavior of separate parts of the system. It is a "telepathic" association of objects when each of them feels the other one "as itself." This "supernatural" coupling between distant objects is not explained by classical physics. Unlike regular interactions restricted for example by the light speed, the nonlocal correlations are instantaneous, that is, changes at each part of the system affects its other parts at the same the time regardless of the distance between them. Quantum physics has revealed the mechanism of this coupling, it got into a quantitative description of coupling laws and gradually begins to use in technical devices.

Nonseparability (quantum entanglement) - the inability to divide the system into separate and completely independent parts.

The principle of superposition of states - if the system can be in different states, it may be in a state which is a simultaneous "superposition" of two or more states from this set.

Recoherence - reverse to decoherence the process is regenerating a quantum entanglement between the system components.

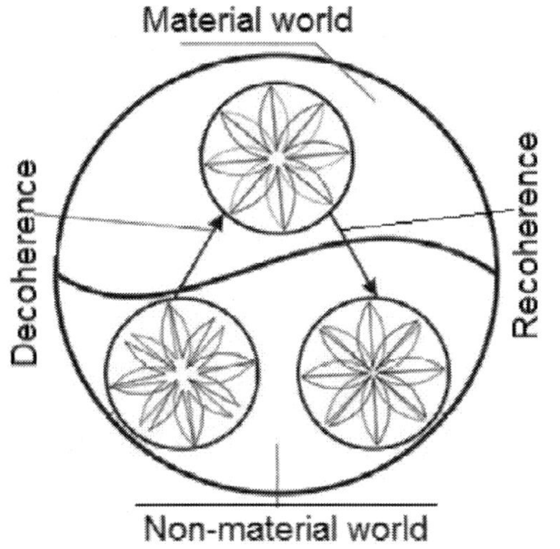

Fig. 1. Quantum jumps on the border between the material world and the non-material one [6]

2 Physics of the earthquake

As an unsolved problem physics of the earthquake challenges humanity for over 100 years and still remains unclear. As was recently announced by experts of the popular journal LiveScience, one of the ten mysteries of the Universe is formulated approximately as: "What is happening in the "heart" of the earthquake?"

A specialist in quantum mechanics Professor Jim al-Khalili formulates the 10 tasks of modern physics in his book "Paradox. Nine great mysteries of physics" [7]. Among them, the second position is for dark matter, the third one for dark energy, and question "will we learn one day to predict earthquakes?" is at the ninth position. To answer this question I stated about 25 years ago: "already begun the earthquake does not know what happens further." In other words there is no hope for reliable prediction. Intuitively I tried to find quantum "roots" of an earthquake and then I confirmed my statement.

Here the purpose is an attempt to understand physics. In previous papers the author showed that the earthquake can be presented as the output of a shock wave to the Earth surface, causing the emergence of the so-called strong motions of ground [8]. Let's follow this conception on data of quantum mechanics.

3 Experiments data

Our model is based on fundamentally new approaches to the explanation of the results obtained in research of rock samples under their compression by powerful presses. The authors have been repeatedly recorded spontaneous amplification of the intensity of acoustic emission followed by its spontaneous determination [8–10]. The author [8] attempted to explain this phenomenon from the point of view of self-organization in coherent structures by interaction of sound waves with opening cracks. The effects of acoustic emissions amplification was suggested to be similar to optical superradiance of Dicke [11].

The effects of acoustic superradiance are noted to occur not in all types of rocks. For example, the acoustic emission of granites (granodiorites) of two different types: Oshima (fine-grained) and Inada (coarse-grained) under their loading was revealed to be different though the samples had a formal resemblance [9]. Whereas samples of granites from Oshima deposit showed a sharp increase in intensity of acoustic emission, there was no effect at samples of granites from Inada. This result repeatedly confirmed suggests that Earth interior there may be geological bodies which rheology enables self-organization and generating of an earthquake inside them, while in other bodies at first glance identical to the first ones such phenomena do not occur.

The results of laboratory experiments and field observations (Fig. 2*A*) show that constant acoustic background has a fluctuating character ($\sim \sqrt{N}$), where N is a complete, the maximum possible number of sound impulses arriving at the receiver from the loaded sample. If impulses are generated coherently N appears. For comparison (Fig. 2*B*) shows the result of the experiment on the effect of quantum entanglement. Pay attention to the fact that in these figures the form of signals is identical. In our model we accept that the acoustic impulses are generated due to changes of hydrogen bonds length (breaks) at the rock samples.

4 The formation of SW at the lithosphere

Let's discuss the experiments on high-pressured kalicinite (KHCO3), the mineral in which hydrogen bonds show the capacity for quantum entanglement. On compression the structure of hydrogen bonds changes simultaneously over the whole volume of the sample (Fig. 3) resulting in the change of its unit-cell volume (or density) and correspondingly (according to our model) in the SW generation.

Similar results were obtained for other minerals containing hydrogen and oxygen (nitrogen, fluorine). By data of many authors such structural phase transition of hydrogen bonds happens instantly over a large volume and this phenomenon is called a cooperativity.

The parts of our model: 1) a shock wave formation at the breakup of the medium characteristics; 2) evaluation of an earthquake energy; 3) output of the shock wave at the surface and physics of

DOI: 10.1051/e3sconf/20172002005

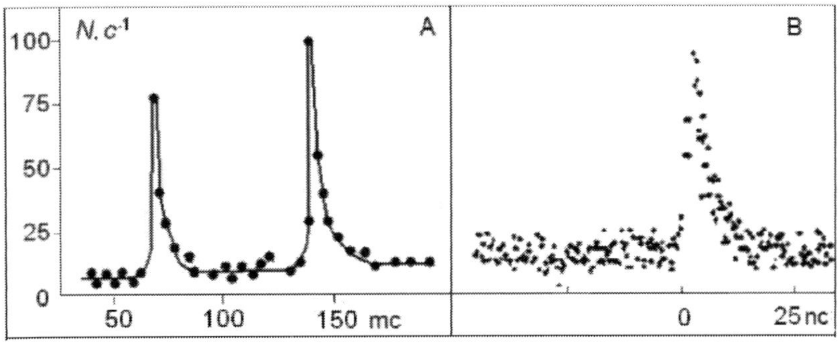

Fig. 2. A) Formation rate of sound impulses in diabase under a constant uniaxial compressive stress [10]. B) Number of photons pairs registered as a function of delay between the moments of two photons detection. Flat plot corresponds to random coincidences between uncorrelated photons, the peak corresponds to quantum entanglement [12]

accompanying phenomena; 4) model of an earthquake source – are discussed in [13]. As these parts are not associated with quantum mechanics as well as with material presented further they are not treated here. Model of aftershocks as quantum phenomena follows.

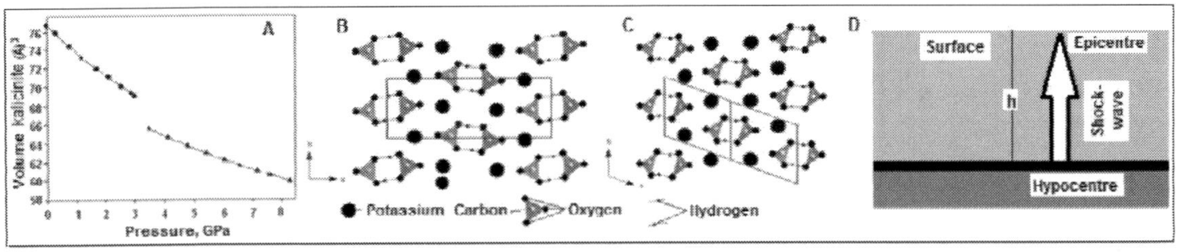

Fig. 3. The change of the unit-cell volume ($\triangle V = V = 4\%$) for kalicinite at the pressure of 3.2 GPa - *A*, the change of kalicinite crystal structure - *B*, *C* [12]. Shock wave generation inside the layer of lithosphere containing hydrogen bonds – *D* [13]

5 Aftershocks

After a strong earthquake at the region of its source a swarm of recurrent earthquakes with the depth of their hypocenters as a rule higher than the plane of the main shock occur. After the main shock the decline of aftershocks number in time is defined by a hyperbolic law (Omori law $N(t) \sim 1/t\zeta$, where $\zeta > 1$). As usually aftershocks fill the space previously occupied by the source of the main shock its volume may be evaluated in such a way (Fig. 4-*C*). Almost all hypocenters of aftershocks, some of them are comparable in magnitude to the main shock, are located above its hypocenter.

Though after a deep earthquake aftershocks are virtually absent, however, this relation is known to be rarely violated. After the deep (564 km) earthquake of 9 March 1994 (with a magnitude of $M_W = 7.6$) near the Tonga the authors [14] observed the first series of aftershocks using eight broad-band seismographs. Aftershocks (82 events) have been fading after the main shock for 42 hours according to the power law with magnitudes from 3.8 to 6.0. The main shock and most of aftershocks were located along the vertical plane of the main shock with the maximum distance from it of not more than 5 km.

Seismically active medium including entangled elements has a number of unique properties among which we are interested in the ability to memory. To use this feature, physicists are trying to create a quantum computer. In seismology this property manifests itself when, along with the main earthquake, foreshocks and aftershocks occur. If medium have not such a property, it is not able to generate foreshocks and aftershocks.

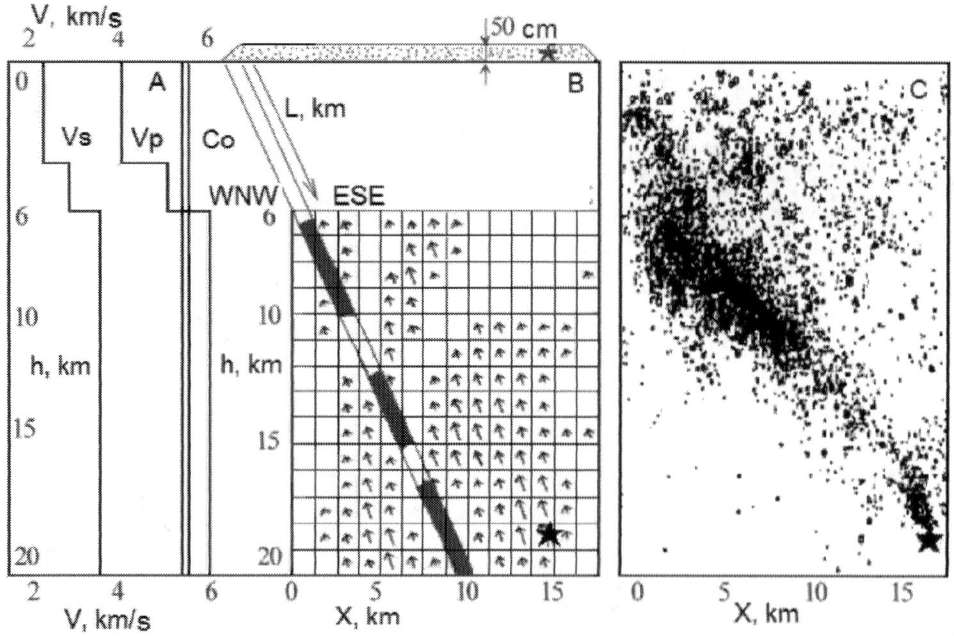

Fig. 4. Model of the Northridge earthquake ruptures (center) [14]. Left - depth distribution of *P*-, *S*-waves and shock waves velocities. Right - aftershocks [14]

6 Quantum effects in physics of earthquakes

Let's refer to Fig. 2, which shows that the rock sample loaded by press emits a sound impulse. It is generally accepted that the cause of these impulses is the cracks formation. We will not criticize this model [15] here. Here we suppose another mechanism of acoustic emission. Assuming that the rock sample includes HB as it follows from Fig. 3-*B* and 3-*C* we consider a model of acoustic emission generation due to the motion of a proton in the sun.

Let's use the model of HB [4], where the proton in the hydrogen bond linking the two pairs of DNA bases is regarded as being on two springs such that it can oscillate from side to side. It has two possible stable positions modeled here as a double energy well. The left-hand well (corresponding to the unmutated position) is slightly deeper than the right-hand well (the tautomeric position), and so the proton prefers to sit in the left one.

In Fig. 5 shows the component of the system [4]. Remember, in quantum mechanics a proton is an elementary particle which is simultaneously a wave. The movement of proton will be recorded as a phonon, i.e., the sound impulse. Moving along the hydrogen bond the proton meets an energy barrier and passes through it due to the effect of quantum tunneling. It was Per-Olof Lundin, Swedish physicist, who suggested that the protons in hydrogen bonds move by quantum tunneling resulting in formation of mutational forms of the nucleotides ([4] p. 277).

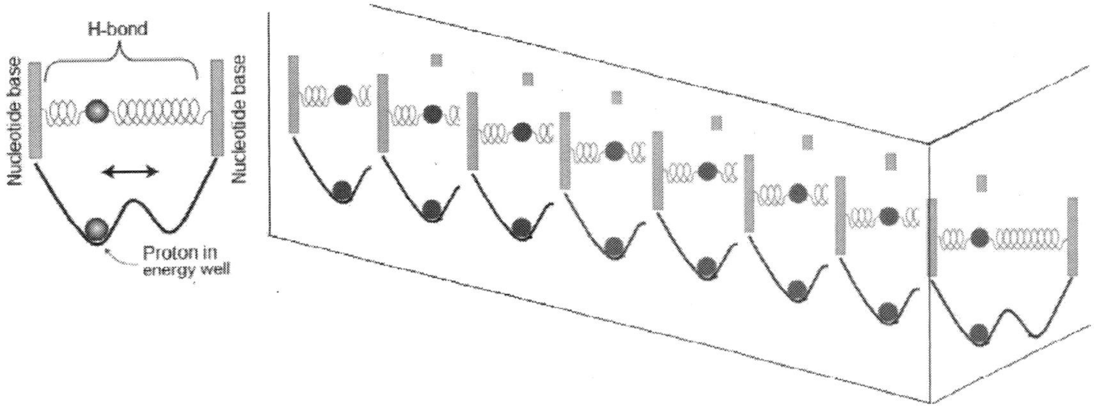

Fig. 5. Proton in hydrogen bonds linking the two pairs of DNA bases [4]. The range of HB – an element of an earthquake source

Why there is a sharp increase in the number of sound impulses, which is then reduced to the previous background values, and soon the process is repeated. In quantum mechanics the background acoustic emission is a random process of proton moving when protons become coherent. Upon reaching full coherence by protons of HB in a sample volume decoherence occurs (Fig. 1). The consequence of decoherence is a rapid increase in the number of sound inpulses N followed by recoherence, the system enters quantum domain. This stage is called a virtual quantum entanglement, here HB protons again "find" the same protons in the same quantum states. The process repeats.

Fig. 2-*B* shows the result of laboratory experiments on quantum entanglement of correlated photons [12]. Decoherence similar to that treated above is recorded: turning into classical domain the system turns back into quantum one in the process of recoherence.

In our model the reason for changing of HB potential is the pressure size of the press in laboratory experiments and the magnitude of gravity in nature. In [16] the dependence of rate constant for tunnel solid-phase reactions on temperature and pressure is considered. The temperature dependence is determined by intermolecular and weak intramolecular vibrations in reagents, leading to oscillations in the permeability of the potential barrier for the tunneling atom. The external pressure accounts for static reduction in the width and height of the potential barrier (dashed line in Fig. 5) under hydrostatic compression of matrix that leads to a higher transparency of the barrier.

Synchronous protons jump from top energy level down to the lower one as decoherence leads to coherent addition of sound impulses and shock wave (SW) initiation. Propagating from the hypocentre up to epicentre SW carries a pressure additional to lithostatic one at its front. This pressure, in turn, causes the same HB protons jump in upper layers of an earthquake source and an aftershock generation. An obvious condition for aftershock occurrence involves quantum entangling of protons over the total volume. If such entanglement occurs not only within the earthquake source, but also with HB protons of the external environment then SW can occur for these protons. As will be shown below, the aftershock can be generated between the protons of the lithosphere and supercooled water clusters of the cloud. It follows that decoherence in a quantum system is an earthquake.

7 The shock wave initiator and source of geophysical phenomena

After I made sure that the SW output to the earth surface very accurately correspond to the observed phenomena, named in seismology by "strong motions", I tried to test whether SW is a quake and whether it is possible to use this idea in other geophysics problems? It turned out that there are such problems and their solutions are given below in a brief overview. Their full discussion is presented in my articles and reports.

8 SW and volcano formation

My multiyear analysis of volcanological literature allows to make a few important observations: first, I noticed that volcanoes are in many seismically active regions and they are often close to epicenters of earthquakes. Second, the seismic pattern shows that beneath every volcano there is a magma chamber, and a volcano is a cone. Third, this chamber is not connected by a transport path of lava with another volume of the melt larger than the first one. Fourth, source of heat required for local melting of basaltic layer or mantle is an unsolved problem of volcanic activity manifestation. This melting should be localized, since a passage of seismic waves indicates that the crust and the upper mantle are solid. Moreover, thermal energy should be sufficient for melting of huge volumes of solid material. I could not come up with a source of heat, and sources mentioned in articles didn't suit. My idea is in the SW which, coming at the surface, produces strong ground motions accompanying earthquake and if medium is fractured then SW is absorbed by this environment, that is a local source of heating. There are well-known data on medium heating by SW. For example, [18] large-scale molecular modeling of SW propagation dynamics and quasi-isoentropic (adiabatic) compression of defective crystals have shown that so-called virtual melting is possible in distribution of SW over plastic environment. This effect is provided by reducing of melting point of the substance on 80%, for example, for copper $\Delta T \approx 4000K$, at the SW front. SW interacting with plastic environment, results in supercooled liquid.

This idea, if it appears correct, allows to explain why magmatic chambers do not cool for a long time being heat sources and they are used, for example, at thermal stations, etc. It would be possible to assess, for example, how many earthquakes (or SW) in the area of the volcanic source are necessary to provide heating of underground thermal source.

Volcano is usually conical and pouring out of its peak magma forms this cone. Compare a normal volcano with the structure of the volcanoes, for example, that on Io, Jupiter's moon, or with the famous supervolcano Yellowstone. Supervolcanoes have no cones and they are similar to plumes with magma pouring out of the outer core of the Earth by a channel [19].

9 SW and the formation of blow holes

Physical mechanism of diamonds elevating to the Earth surface is considered to be realized by forming blow holes or diatremes which resemble volcanoes, as diamonds are formed here at pT-conditions typical of the depths of about 100 km. How to transport a diamond with this depth? To this question the accepted answer by far is the volcano. Conception of diamonds elevating by volcano eruptions meets contradictions. A radically new idea of diamonds creating at small depths with their further deposition at the Earth surface is based on the model of the Hot Expanding Earth being developed by the author [20], his new physics of earthquakes, based on the SW formation in the lithosphere, its output at the surface and its interaction with a rarefaction wave initiated here is proposed. Model of Hot Earth accounts for changing of pT-parameters at the Earth's surface in its evolution, specifically, increasing of the Earth radius and, accordingly, reducing of pressure and temperature gradients [21].

Diamonds turned out to be formed about a milliard years ago at depth of 5 – 10 km and then they were delivered close to the earth surface through the environment moving after SW which acted as diatreme.

10 SW in ionosphere

New model of the formation of shock acoustic waves (SAW) in the atmosphere and the ionosphere is discussed [13].

Recording propagation of the shock acoustic waves (SAW) in ionosphere using a wide network of GPS stations modern research methods of phenomena in atmosphere and ionosphere don't solve the problem of mechanism of SAW generation. We consider attempts to assign occurrence of SAW to strong ground motions during an earthquake being incorrect, because the speed of these motions for SW initiation is too low.

Model of SAW generation as an aftershock of lithosphere earthquake is proposed here. Quantum entanglement in "earthquake source – supercooled water clouds" system postulated in our model may occur on HB protons in the source substance and, of course, on these present in the aquatic environment of such clouds. Our model removes all time and space inconsistencies observed between the earthquake source and SAW hypocenter. The same postulate is applied to aftershocks.

On the one hand the occurrence of SW in a cloud of supercooled water leads to the generation of a "hole" in the cloud, and on the other to the high-speed release of ice piece from clouds. It is a product of activitiy of rarefaction wave, which occurs simultaneously with SW. Probably, this mechanism leads to the appearance of UFOs and, at the same time, if ice speed is higher than that of sound in air there is a so-called "bombshell".

Not that various mechanisms of entanglement (many-particle quantum entanglement) except those adopted here (for simplicity) for HB protons are discussed in [22], but they will complicate our task greatly.

11 Shock-wave model of explosive hole

Model of explosive hole formation on the Yamal Peninsula and release of gas hydrates on Bennett island is suggested [23]. The reason for the such phenomena is the SW impact on the volume of gas hydrates deposits at the permafrost. Gas hydrates are destroyed through SW passing as an earthquake or an icequake which frequently occur in of gas hydrates deposits. Explosive hole formation or release of gas hydrates on Bennett island should need a series of earthquakes (shock waves).

12 Rock bump – SW output at a mine

Model in which a rock bump (an induced earthquake) as SW output at the inner wall a mine is discussed [24]. An associated rarefaction or stretching wave is moving in reverse (into the wall) direction. Ground motions in the direction of SW propagation with a velocity much smaller than that of SW occur. Phenomena on the inner surface of a mine result in the release of coal dust and methane. Spatial separation of charged particles causes an electrical breakdown, which initiates methane explosion.

13 Storm glass monitoring of quantum entanglement

English hydrographer admiral Robert Fitzroy was the first who described a strange barometer called then Fitzroy storm glass (SG) [25]. In 19-th century he used it for weather prediction namely for storms forecasting. Camphora crystallization within the liquid was believed to be related to the weather. Crystals form, height and amount were interpreted.

Numerous researchers of SG insist that SG has a little value in weather prediction and its operation principle is unclear. We were solid with these conclusions before we arrive at Kamchatka where the device began to record cyclones far before their coming.

Returning to Novosibirsk we made three similar SG. They turn to describe cyclones behavior at Kara Sea rather than Novosibirsk weather. Interested in quantum effects I develop now a quantum model of SG. As I propose SG quantum entangles with supercooled clouds of north cyclones rather than south cyclones. We called this device by an entanglemeter suggesting that it records a quantum entangling degree between clod supercooled water and camphora with its hydrogen bonds. Protons shifting in HB accounts for crystal growth at SG.

14 Lightning, MeV-particle fluxes, sprites

Ball lightning (BL) is unique a natural phenomenon [26]. First, this phenomenon is far ahead of all ones in the number publications. Second, despite a huge research number, it can not interpreted in the frames of classical physics. The only way to understand the BL nature we see is to consider its physics from quantum mechanics principles and achievements.

Here we propose a model in which the properties of SHM follow from the context of nonlocal multiparticle quantum entanglement (entanglements, the coherence - entanglement) between protons of hydrogen bonds (HB) in water clusters after the destruction of dark lightning. Dark zip. Discusses the problem of emergence in the clouds streams of high-energy particles. Describes several situations in which such discovered streams of dark lightning, preceding the ordinary and create additional ionisation, streams of fast electrons Mawei energy prior to the earthquake, lightning strikes, triggering the appearance of ball lightning, light effects, sprites, etc. All these phenomena seem to have a common origin based on the manifestation of the effects of quantum coherence of protons in hydrogen bonds of water clusters in the cloud [27].

15 Dark lightning

Emergence of high-energy particle fluxes in clouds is discussed here. Several cases in which they were discovered are dark lightning, preceding an ordinary one and creating additional ionisation, prior to an earthquake fluxes of fast electrons with MeV-energy, lightning strikes, triggering an appearance of ball lightning, light effects, sprites, etc. All phenomena listed seem to have a common origin based on manifestation of quantum coherence of HB protons in water clusters of a cloud [27]. Discovered as a phenomenon in 1991 dark lightning (DL) is a flux of fast electrons, moving with velocities close to that of light that sharply distinguishes them from ordinary lightning, which electrons are not so rapid. Ultrafast electrons of DL colliding with the air molecules give rise to gamma rays that generate electrons and their antiparticles, whose subsequent annihilation with air produces more gamma rays. Lightning familiar to us follows the track left by DL which electric current is less by several orders of magnitude than that of usual lighning. DM voltage is known to be much higher than the potential of a storm cloud.

16 Quantum effects in atmosphere electric field

"Crystal structure", similar to that which occurs in crystalline dusty plasma is a possible reason for the stability of water-jumper (bridge) between two water containers under the influence of a constant electric intensity. Perhaps a similar mechanism is responsible for tornado water column behavior like that of a rigid body [28]. An interesting experiment with a stable water bridge between two beakers containing water and platinum electrodes was described in 2007 (referenced in [28]). Each electrode had a constant voltage of 15-20 kV. Touching from the start the beakers were pulled apart after the voltage being applied and water in both beakers rose by contacting beakers walls up to connecting of both water flows moving towards each other. Water bridge ("bridge") of circular section with a diameter of 2-4 mm and a length of about 2-3 cm was observed.

In the 80-90 I was developing my hot model of the expanding Earth. At that time I was interested in research on so-called "plasma crystal" and its quantum properties. Analyzing the results of this research, I created my model of the Earth inner core [20]. The idea of a "plasma crystal" helped me find the most likely model of water bridge: under the electric field water of bridge behaved as a rigid body.

It is known that water column of tornado behaves as a rigid body. If we imagine that water column of tornado is a water bridge with a potential of a parent cloud the problem would be solved. I have not found here quantum entangled HB protons but it is revocable.

17 Conclusions

In February 2013 I put sited my Memorandum with justification for a new science of quantum Geophysics in www.vvkuz.ru . By that time a number of works on this subject I had written and published. Since that time, my scientific interest is associated exclusively with this science. Now I published a dozen articles and made some conference reports. My book called "Quantum Geophysics" is conceived. An excellent book al-Khalili and McFadden "Life on the edge. Your first book about quantum biology" [4] was a deciding factor. I used a picture from this book (Fig.5, p. 290) above and I turn to the next picture of this book (Fig. 6).

"The three strata of reality. The top layer is the visible world, filled with objects such as falling apples, cannonballs, steam trains and airplanes, whose motions are described by Newtonian mechanics. Lying beneath is the thermodynamic layer of billiard-ball-like particles whose motion is almost entirely random. This layer is responsible for generating the "order from disorder" laws that govern the behavior of objects such as steam engines. The next layer down is the layer of fundamental particles ruled by orderly quantum laws. The visible features of most of the objects that we see around us appear to be rooted in either the Newtonian or thermodynamic layers but living organisms have roots that penetrate right down to the quantum bedrock of reality [4]."

The earthquake source (Fig. 6) is a chain of HB protons which are in a coherent state. After a decoherence occurs, SW arises and then by the above mentioned script. HB protons of the source can be quantum linked with HB protons of with supercooled water in clouds ("roots of clouds" reach the field of quantum mechanics). Shock crystallization may occur in a cloud as an earthquake aftershock. Reaching ionosphere the resulting SW is converted to SW on the electrons, which is recorded then by GPS methods. Such a cloud is able to generate not only the lightning but also the dark lightning in which water clusters are coupled into a chain so that a voltage of several MeV instantly appears between the chain ends.

This article deals with some, not all quantum aspects of geophysics. Among the latter, for example, are quantum effects occurring at the boundary of the Earth inner core where a phase change leading to a reverse of geomagnetic field polarity proceeds, etc. [20]. Let us note the role of quantum linked particles which accumulation in atmosphere causes an abrupt climate change, behaving as the flicker-noise.

DOI: 10.1051/e3sconf/20172002005

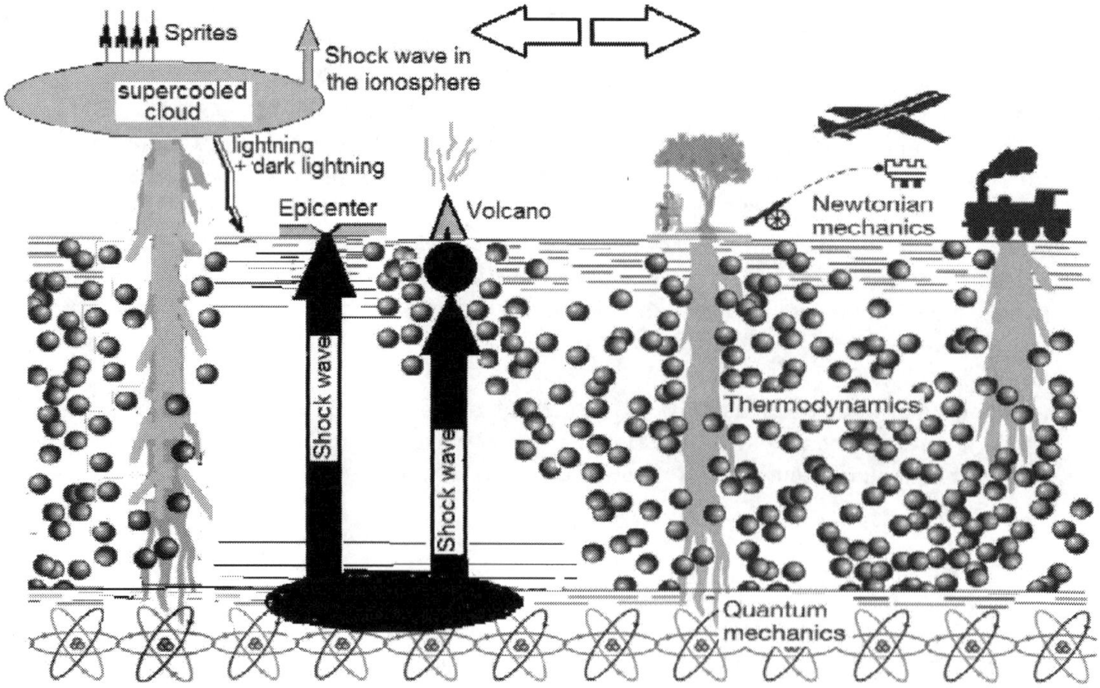

Fig. 6. The three strata of reality. Right part (→) – given at [4]. Left part (←) – my addition of phenomena and geologic processes discussed here

References

[1] J. Sonner, Phys. Rev. Lett. **111**, 211603 (2013)
[2] R.V. Bunity, S.D.H. Hsu, Phys. Lett. **718**(2), 233-236 (2012)
[3] P. Csermely, *Weak Links: The universal key to the stability of networks and complex systems* (Springer, 2009)
[4] J. Al-Khalili, J. McFadden, *Life on the Edge. The coming of quantum biology* (Bantam Press, 2014)
[5] S.I. Doronin, *Quantum magic* (2007), ISBN: 978-5-9573-0844-7
[6] V.D. Shkilev, About the numbers and fractals from the perspective of quantum mechanics, URL:http://www.gramota.net/materials/1/2012/1/21.html (2012)
[7] J. Al-Khalili, *Paradox. The nine greatest enigmas in science* (Bantam Press, 2012)
[8] V.V. Kuznetsov, Physical mesomechanics **12**(6), 87-96 (2009)
[9] K. Kusunose, X. Lei, O. Nishizawa, T. Satoh, PEPI **67**(1-2), 194-199 (1991)
[10] S.N. Zhurkov , V.S. Kuksenko, V.A. Petrov, et al., *Physical processes in earthquake foci: Sat. scientific* (Nauka, Moscow, 1980) 78-85
[11] R.H. Dicke, Phys. Rev. **93**, 99 (1954)
[12] R.A. Bertlmann, A. Zeilinger, *Aspect A. Bell's theorem: the naïve view of an experimentalist. Quantum [Un] speakables – From Bell to Quantum information* (Springer, 2002)
[13] V.V. Kuznetsov, Vestnik KRAUNC. Phys.-Mat. science **3**(14), 64-71 (2016)

[14] D.A. Wiens, J.J. McGuire, P.J.U. Shore, et al., Nature **372**, 540-543 (1994)
[15] V.V. Kuznetsov, PMTF **42**(4), 184-189 (2001)
[16] L.I. Trakhtenberg, A.A. Fokeev, V.L. Klochikhin, et al., Chemical physics **23**(6), 56-68 (2004)
[17] V.V. Kuznetsov, *New approaches to the physics of earthquakes and eruptions of volcanoes. Report at the conference. Institute of Volcanology and Seismology. DVO RAN* (2015)
[18] V.I. Levitas, R. Ravelo, Proc. Nat. Acad. Sci. USA. **109**(33):13204-7 (2012)
[19] V.V. Kuznetsov, Geophysical journal **30**(6), 76-92 (2008)
[20] V.V. Kuznetsov, *Introduction to the physics of the hot soil* (Kamgu, Petropavlovsk-Kamchatsky, 2008) 367
[21] V.V. Kuznetsov, Geology and MCP Siberia, (2017)
[22] V.V. Kuznetsov, Quantum magic **8**(1), 1101-1119 (2011)
[23] V.V. Kuznetsov, Volcanology and Seismology (2017)
[24] V.V. Kuznetsov, Vestnik KRAUNC. Phys.-Mat. science **1**(10), 34-43 (2015)
[25] N.D. Kuznetsova, V.V. Kuznetsov, Journal of Unconventional Science **3**(9), (2015)
[26] V.V. Kuznetsov, Quantum Magic **8**(3), 3101-3107 (2011)
[27] V.V. Kuznetsov, Herald kraunts. Phys.-Mat. science **4**(15), 63-73 (2016)
[28] V.V. Kuznetsov, Journal of Unconventional Science **11**(4), (2016)

DOI: 10.1051/e3sconf/20172002006

Complex atmospheric-lithospheric observations of acoustic emission at «Karymshina» site in Kamchatka

Igor Larionov[1,*]

[1] *Institute of Cosmophysical Research and Radio Wave Propagation FEB RAS, Paratunka, Russia*

Abstract. The results of complex observations of acoustic emission in the near surface rocks and in the atmosphere by the ground surface are described. The instrumentations for the observations are a laser strainmeter-interferometer and a microbarometer installed close to each other. It was shown that during the increase of deformation rate in the near surface rocks, increase of acoustic emission intensity in the atmosphere by the ground surface is registered. The effect of meteorological factors on the observation results is evaluated.

1 Introduction

At present time, the topical task in Acoustics is the joint investigation of acoustic wave processes at the lithosphere and atmosphere boundary. Investigation of acoustic waves, propagating in "solid body - gas" system, is necessary to built the models of lithosphere, atmosphere, ionosphere and magnetosphere coupling which is one of the main directions in Geophysics.

The lithospheric-atmospheric effect manifests the most in seismically active regions at the filal stage of earthquake preparation [1]. The high interest to the joint research of seismically active wave processes is determined by the fact that acoustic waves of infrasonic range can propagate at large distances and reach ionospheric heights under favorable conditions (weak attenuation and scattering, weak refraction in the atmosphere) [2]. As a result, they play a significant role in energy transfer between the geospheres.

2 Measurement technique

In 2016, a laser strainmeter-interferometer, constructed according to Maikelson interferometer scheme, was put into operation at the Laboratory of Acoustic Research of IKIR FEB RAS [3]. The principle of operation of a simplest interferometer is that the strainmeter base change causes additional phase increment in a laser radiation wave. The method of measurement is the following. Displacement of interferometer mirrors arranged at the base ends l by the value $\lambda/2$ results in the change of fringe pattern by one band where λ is the light wave length on which an interferometer is operating. The total relative displacement is $\Delta l = N(\lambda/2)$, where N is the fringe pattern band number. The capabilities of the interference methods are limited by the measurement accuracy of band fraction displacement ΔN,

*Corresponding author: igor@ikir.ru

 DOI: 10.1051/e3sconf/20172002006

which is determined by fringe pattern sharpness F_k and is characterized by the relation $F_k = \frac{\Delta\lambda}{\delta\lambda}$, i.e. it is a relation of the distance between the maxima to the maximum half width $\delta\lambda$.

The advantage of a laser strainmeter against the mechanical analogues is the absence of a mechanical sensitive element. Limitations on the registration accuracy are imposed on the measurements, carried out by an "open" strainmeter without a light guide. They are determined by the effect of meteorological parameter variations. The strongest effects are those from wind, precipitations, air temperature and pressure. To exclude the wind and precipitation effects, special cover is used. Air pressure and temperature changes are quite slow processes which occur within several hours that allows us to estimate faster deformation processes. On the basis of estimation data, the strainmeter, installed in such conditions, has the accuracy of relative deformation measurement not less than 10^{-8} [4, 5].

A differential microbarometer ISGM-03M (Fig. 1), produced by «Geofizichesie izmereniya» scientific-engineering center (Novosibirsk), was used to measure the atmospheric acoustic radiation. It is installed 4 meters from the strainmeter in a pit at the depth of 0.5 m from the ground surface and covered from precipitation. The microbarometer is intended for the transformation of atmospheric pressure variations into the electric channel in the operating frequency band from 0.02 to 4 Hz and in the amplitude range from 0.001 to 100 Pa.

Fig. 1. Overall view of the microbarometer ISGM-03M

The principle of the device operation is based on the measurement of elastic membrane displacement relatively a fixed electrode under the effect of atmospheric pressure by a capacitance sensor of displacement. The capacitive transducer of membrane displacement is powered from a reference frequency generator stabilized in amplitude and frequency to decrease instrumentation amplifier internal self-noise. To decrease the zero drift and the amplifier low-frequency in the operating temperature range radically, a signal isolating transformer was introduced into the scheme. The primary coil is also used as high frequency filter (HFF) element. When an amplifier is constructed in this way, zero temperature drift is produced by the final stage, having a minimum amplification factor, and does not exceed 0.1 mkV per Celsius degree. Thus, the amplifier self-noise level is determined by input stage noise on the reference frequency and reduced to the minimum by selection of corresponding elements.

The membrane assemblies of the microbarometer, including a membrane, backplate (the base of which is fused quartz), a tube for static pressure balancing and connected with the environment air via four acoustic inputs, provide self-noise level of the microbarometer and its high temperature stability as well as small size and low weight. A thin membrane made of a corrosion-resistant alloy with controllable tension is used in the membrane unit construction. Membrane tension force stability in operating temperature range is achieved by a temperature-compensated construction of the membrane unit that provides the conversion factor (sensitivity) consistency of the microbarometer and additional decrease of self-noise.

The important feature of the experiment is the fact that the devices, installed close to each other, are covered by a three-meter snow layer in winter-spring season that provides temperature stabilization. It is controlled by a temperature sensor mounted at the basis of the strainmeter optical system. To analyze the effect of other meteorological factors, we applied the data from a weather station located 20 meters from the strainmeter. We should note that atmospheric pressure changes slowly within a day and does not affect fast deformation processes and the microbarometer owing to the tube for static pressure relief. Wind velocity has a significant impact on the instrument readings which manifests as increases in signal amplitude oscillations.

3 Main results and discussion

The rates of deformation and acoustic pressure changes were under analysis. A method described in the papers [3,4,5] was used. Fig. 2 shows the graphs of deformation (Fig. 2*a*) and acoustic pressure rates (Fig. 2*b*) compared with wind velocity (Fig. 2*c*). It is clear from the graphs that wind puts variable pressure on the snow cover and thus affects the instrument readings in the form of increased rate of observable value changes.

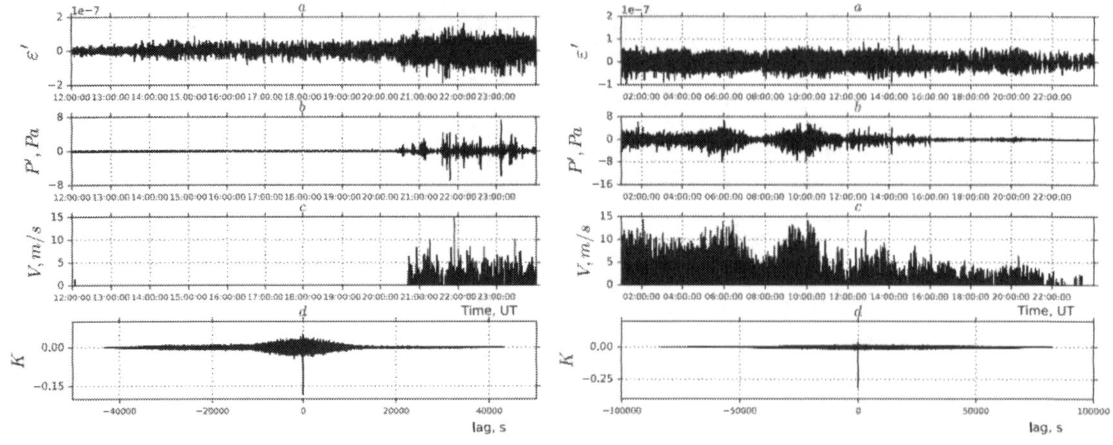

26.02.2017 18.04.2017

Fig. 2. Fragment of acoustic-deformation change record during strong wind effect: *a* - deformation change rate, *b* - pressure variation change rate, *c* - wind velocity, *d* - cross-correlation function of deformation and pressure rate series

To estimate the correlation of strainmeter and the microbarometer data, we constructed a cross-correlation function between the value series under investigation (Fig. 2(*d*)). It is clear that the correlation coefficient is small ($K < -0.15$), but the peak at the zero log with negative sign is clearly defined.

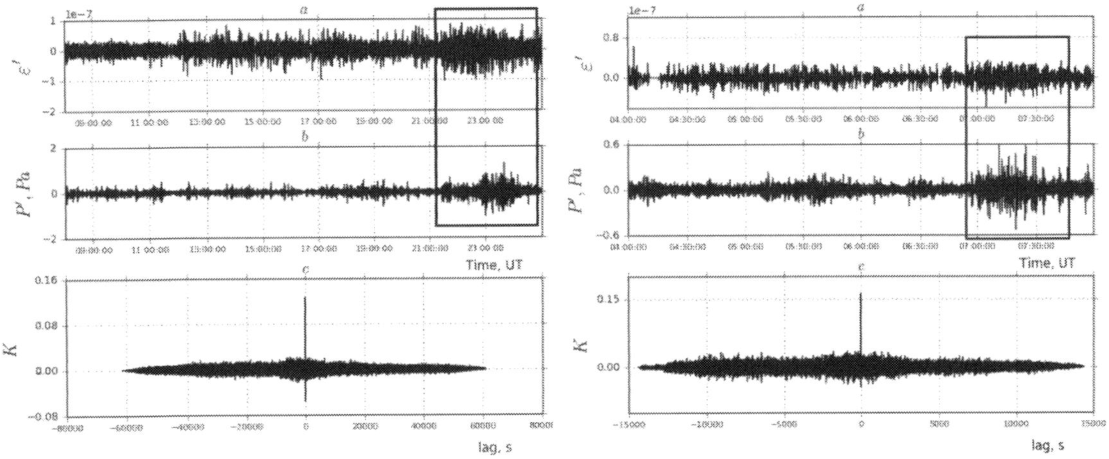

Fig. 3. Fragment of acoustic-deformation change record at the absence of wind effect: *a* - deformation change rate, *b* - pressure variation change rate, *c* - cross-correlation function of deformation and pressure rate series

We also chose the periods of absence of wind effect during the period under observations, they are shown in Fig. 3. It is clear from the presented fragments of estimation of deformation (Fig. 3*a*) and acoustic pressure (Fig. 3*b*) rates during the wind absence that their simultaneous increases are sometimes observed. The most vivid segments are marked by rectangles in Fig. 3. Thus, we may suppose that the occurring deformation disturbances cause acoustic vibration not only in solid medium [3–5] but also during the solid-gas transition.

For the data series in Fig. 3(*a-b*), cross-correlation function is also calculated (Fig. 3*c*). A peak at the zero log with a positive sign is clear on the graph. Thus, in case of wind effect, a reverse correlation is registered. It may be determined by the influence on the observation system from above. At the absence of wind, simultaneous increases of rates may be determined by deformation process dynamics in the lithosphere, i.e. the influence from below that causes positive correlation. It requires further investigations.

4 Conclusions

In the result of joint atmospheric-lithospheric observations it was shown that increase of deformation process rate causes the increase of acoustic pressure variations in the near ground layer of the atmosphere. Taking into account the fact that acoustic wave of infrared range can propagate at long distances, they can reach the ionospheric heights in favorable conditions. Thus, the possibility of propagation of deformation acoustic waves at high altitudes in the atmosphere is shown.

Differently directed effect of disturbing factors on strainmeter and microbarometer readings cause different phase of observable value rate change.

References

[1] V.V. Adushkin , A.A. Spivak , S.B. Kishkina et al., Izvestiya Physics of the Solid Earth, **7**, 34-51 (2006)

[2] V.V. Kuznetsov, V.V. Plotkin , S.Yu. Khomutov, Doklady Earth Sciences,**370**, 243-248 (2000).

[3] I.A. Larionov, Y.A. Nepomnyashchiy, E3S Web of Conferences, **11**, 00010 (2016)

[4] I.A. Larionov, Y.V. Marapulets, and B.M. Shevtsov, Solid Earth, **5**, 1293-1300 (2014)

[5] Yu.V. Marapulets, B.M. Shevtsov, I.A. Larionov, M.A. Mishchenko, A.O. Shcherbina, and A.A. Solodchuk, Russian Journal of Pacific Geology, **6**, 457-464 (2012)

InSAR data analysis at Kamchatka during 2016

Igor Larionov[1,*] and *Andrey* Perezhogin[1,2,**]

[1] *Institute of Cosmophysical Research and Radio Wave Propagation FEB RAS (IKIR FEB RAS), 684034, Mirnaya str., 7, Paratunka, Elizovskiy district, Kamchatka, Russia*

[2] *Vitus Bering Kamchatka State University, 683032, Pogranichnaya str.4, Petropavlovsk-Kamchatsky, Russia*

Abstract. Geophysical monitoring in seismically active areas depends on geodeformation processes in the earth's crust. Observations of earth's crust strain-stress using gps-measurements, laser interferometers give only an opportunity to analyze the dynamics in time without the possibility of extrapolation to adjacent areas. In this regard, it is useful to apply a radar interferometry technology to measure the displacements of the earth's surface. The report includes the results of processing the radar data of the Sentinel-1A satellite. Several qualitative interferometric pairs were obtained during the period from June to October 2016. A high coherence coefficient is observed in open areas in the vicinity of volcanic structures and adjacent territories, as well as on the west coast of Kamchatka, where there is no high vegetation. The main factor that significantly reduces the coherence of images is the forest cover. Possibility of estimating the surface displacement at regions with a high coherence coefficient is discussed.
Keywords: InSAR, strain-stress, Sentinel-1A

1 Introduction

There are several instrumental geophysical stations in Kamchatka region. The aim is a monitoring of geophysical variations of seismic and volcano's activity and its influence on geophysical processes. The main impact on geophysical fields is a stress-strain state of the earth's crust [1–4]. In this paper we analyze Sentinel-1A satellite SAR data in order to compare modeling results with InSAR data. We investigate pairs of radar data in Southern Kamchatka because of the location of varios geophysical stations near Petropavlovsk-Kamchatsky.

2 InSAR data and instruments

InSAR is an effective technique to investigate strain-stress of Earth's surface due to subduction zone displacement. There are several applications in monitoring of seismic and volcano activities. For example, InSAR is used to determine stress of the San Andreas fault in California [5]. Uzon caldera strain-stress at Kamchatka was observed by satellite [6].

*e-mail: igor@ikir.ru
**e-mail: d72156@gmail.com

DOI: 10.1051/e3sconf/20172002007

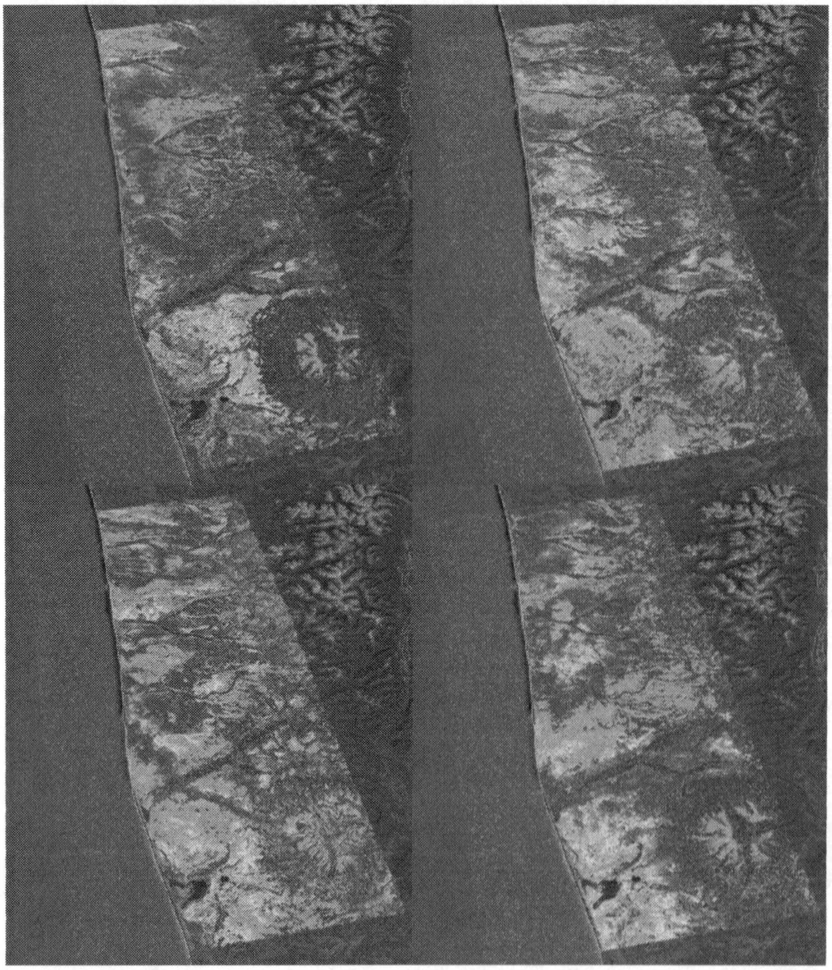

Figure 1. InSAR data for pair 2016.06.28-07.22 (upper left), 2016.07.22-08.15 (upper right), 2016.08.15-09.08 (bottom left), 2016.09.08-10.02 (bottom right)

In paper we consider region of Southern part of Kamchatka. We use Copernicus Open Access Hub for IW data of Sentinel-1A. Processing of Sentinel-1A data was performed using GMTSAR software [7].

During period from January 2016 to December 2016 we find out a high coherence only in summer and autumn months. Difference of phases is represented at fig. 2. There are many continuous regions with stable phase differences. Indeed this territory has no forests. But territory near Petropavlovsk-Kamchatsky is surrounded by forests. So we have a high coherence only at mountains and rocks.

There is no high coherence for winter period data because of unregulated character of snow surface. We analyze images in January and February 2016. If we discuss spring period there are the same problems with coherence. Most suitable periods are from June to October when the ground is free of snow.

So for investigation of strain-stress at Kamchatka we need a representative collection of InSAR data for summer and autumn periods.

3 Conclusion

We get InSAR pairs of data for analysis of a stress-strain field at Kamchatka region. There is sufficiently high coherence for pairs of images during summer period near volcanoes without forests.

References

[1] A. Alekseev, A. Belonosov, B. Petrenko, Problems in the dynamics of the lithosphere and seismicity. Computational Seismology pp. 81–97 (2001)

[2] A. Perezhogin, B. Shevtsov, Computational technologies **14** (2009)

[3] M. Bobrova, A. Perezhogin, Herald of KRAUNTS. Physics and mathematics **2** (2011)

[4] V. Saltykov, Y. Kugaenko, Pacific Geology **31**, 96 (2012)

[5] Y. Fialko, Nature **441**, 968 (2006)

[6] P. Lundgren, Z. Lu, Geophysical Research Letters **33** (2006)

[7] D. Sandwell, R. Mellors, X. Tong, M. Wei, P. Wessel, Eos, Transactions American Geophysical Union **92**, 234 (2011)

Analysis of magnetic data of ground station network during strong magnetic storms

Oksana Mandrikova[1,2*], *Igor* Solovev[1], *Sergey* Khomutov[1], *Dmitry* Baishev[3], *Phani* Chandrasekhar[4]
[1]Institute of Cosmophysical Research and Radio Wave Propagation FEB RAS, 684034 Kamchatka region, Elizovskiy district, Paratunka, Mirnaya str., 7, Russia
[2]Kamchatka State Technical University, 683003 Kamchatka region, Petropavlovsk-Kamchatskiy, Kluchevskaya st., 35, Russia
[3]Yu. G. Shafer Institute of Cosmophysical Research and Aeronomy SB RAS, 677980 Yakutsk, Lenin Ave., 31, Russia
[4]CSIR-National Geophysical Research Institute, Telangana, Hyderabad -500007, India

Abstract. The paper suggests a new technique for magnetic data analysis to detect and to estimate short-time anomalous increases in geomagnetic disturbance intensity. Based on the technique and applying the data of meridionally arranged stations in the north-east of Russia («Kotel'nyi» KTN, «Tixi» TIX, «Chokurdakh» CHD, «Zyryanka» ZYK, «Mys Shmidta» CPS, «Yakutsk» YAK, «Magadan» MGD, «Paratunka» PET, «Khabarovsk» KHB) and the near equatorial Indian stations ("Hyderabad" HYB and "Choutuppal" CPL), we analyzed the geomagnetic field variations during strong magnetic storms. This approach allowed us to detect simultaneously occurring local increases in geomagnetic activity observed a few hours before events and during the main phases of magnetic storms.
The development of the method was supported by RSF Grant No.14-11-00194. Experimental investigations were supported by RFBR Grant No. 16-55-45007.

1 Introduction

The work is focused on the development of methods for the research of dynamic process in the magnetosphere during increased solar activity and magnetic storms. The influence of solar activity on the Earth's magnetosphere is very complex, many aspects of which have not been sufficiently studied so far [1]. The state of the magnetosphere is an important factor of cosmic weather, which affects many aspects of our life. Therefore, researches in this area have great scientific and practical interest [1-3].

* Corresponding author: oksanam1@mail.ru

The magnetic field variations have a complex structure, and they consist of non-smooth local features of varying amplitude and duration in disturbed periods [1-7]. Analysis of formed local structures allows us to receive the information about the intensity and dynamics of geomagnetic disturbance development in the detail [2-7]. The traditional approaches and methods of time series analysis are not sufficiently effective to solve this problem [8-10]. At the present time, modern mathematical methods and information technologies in this field are intensively developing [2-7, 11-17]. Algorithms to isolate the interferences during the primary processing of magnetic data [2, 13] and methods for the anomaly detection during magnetic storms [2, 13] were developed on the basis of DataMining approaches. It is possible to optimize the process of geophysical data registration and the global databases formation. Authors [12] proposed an algorithm for the interplanetary magnetic field processing and the Dst-index computation by neural networks. Authors of the present paper developed a method for magnetic data analysis based on the wavelet transform [4, 18]. The method allows them to receive the information about the dynamics of geomagnetic field disturbances during magnetic storms. At the present time, the wavelet transform is widely used in the tasks of magnetic data analysis [6, 15-17, 19, 20]. In the work [19], the wavelet transform is used to study the connection between short-period oscillations of the geomagnetic field, the parameters of the solar wind and the interplanetary field during magnetic storms. Problems of removing noise and eliminating the periodic component of the field caused by the Earth rotation are solved on the basis of the wavelet transform [16, 17]. Automatic algorithm to determine the periods of the initial phase of a magnetic storm based on the discrete wavelet transform was proposed (15). On the basis of the wavelet spectrum analysis of the geomagnetic field variations, a method for prediction of strong geoeffective solar flux was proposed in the paper [6]. In the present paper, the authors used the continuous wavelet transform and the wavelet packet to analyze the geomagnetic field variations during strong magnetic storms and to obtain a pattern of distribution and dynamics of disturbances in the analyzed areas. In the analysis we used data from the meridionally located stations in the northeast of Russia («Kotel'nyy» KTN, «Tixie» TIX, «Chokurdakh» CHD, «Zyryanka» ZYK, «Mys Schmidta» CPS, «Yakutsk» YAK, «Magadan» MGD, «Paratunka» PET, and «Khabarovsk» KHB) and data from the near-equatorial Indian stations. The paper shows the application of developed computational solutions, which allows us to study small-scale shot-period changes (from a few seconds to tens of minutes) in the geomagnetic field parameters and long-period variations (from several hours to several days) of the complex structure. During the data processing, features of disturbances, describing active large-scale processes in the magnetosphere during the storms were detected in dynamics. Weak increases of the geomagnetic activity occurring synchronously on the eve of the events and possibly associated with approaching storms were detected.

2 Methods of data analysis

2.1 Isolation and estimation of the geomagnetic disturbance intensity

Using the continuous wavelet transform, which is defined by equation [21, 22]:

$$(W_\Psi f)(b,a) := |a|^{-1/2} \int_{-\infty}^{\infty} f(t)\Psi\left(\frac{t-b}{a}\right)dt,$$

where Ψ is a wavelet, $f \in L^2(R), a, b \in R$, $a \neq 0$, we can obtain the detailed information about the properties of the function f. Wavelet Ψ is a window function and has a zero mean and, when the scale a tends to be zero, coefficients $(W_\Psi f)(b,a)$ describe the properties of the function f in the neighborhood of $t = b$.

If at time $t = b$ a local singularity arises in the function, the absolute values of the wavelet coefficients in the neighborhood $t = b$ increase. To isolate the local features, a threshold function can be applied [4, 5, 18, 23]:

$$P_{T_a}(W_\Psi f_{b,a}) = \begin{cases} W_\Psi f_{b,a}, \text{если } (W_\Psi f_{b,a}) \geq T_a \\ 0, \text{если } |W_\Psi f_{b,a}| < T_a \\ -W_\Psi f_{b,a}, \text{если } (W_\Psi f_{b,a}) < -T_a \end{cases}, \quad (1)$$

where $T_a = U * St^l{}_a$ is the threshold function where $St^l{}_a = \sqrt{\frac{1}{l-1}\sum_{k=1}^{b}\left(W_\Psi f_{b,a} - \overline{W_\Psi f_{b,a}}\right)^2}$, is the standard deviation calculated within the time window of the length l, $\overline{W_\Psi f_{b,a}}$ is the average value calculated within the time window of length l, U is the threshold coefficient.

If we consider the variations of the magnetic field as f, then the absolute values of wavelet coefficients

$$e_{b,a} = |(W_\Psi f)(b,a)|. \quad (2)$$

characterize deviation of the field variation from its characteristic level on the scale a at time $t = b$ [23]. Since the magnitude of the deviation of the magnetic field variation from its characteristic level serves as a measure of the degree of its perturbation [1, 1, 3, 5, 8-10, 17], application of operation (1) makes it possible to isolate the periods of increasing geomagnetic activity. Then the intensity of field perturbations at time $t = b$ can be estimated as

$$E_b = \sum_a P_{T_a}(W_\Psi f_{b,a}). \quad (3)$$

In the case of field positive disturbances (current variation increase relatively the characteristic level), the value

$$e^{\pm}_{b,a} = (W_\Psi f)(b,a) \quad (4)$$

is positive ($e^{+}_{b,a}$). In the case of field negative disturbances (variation decrease relatively the characteristic level), the value is negative ($e^{-}_{b,a}$). Then the intensity of the positive and negative disturbances of the field at time $t = b$ can be estimated, respectively, as

$$E^{+}_b = \sum_a P_{T_a}(e^{+}_{b,a}), \text{ и } E^{-}_b = \sum_a P_{T_a}(e^{-}_{b,a}). \quad (5)$$

DOI: 10.1051/e3sconf/20172002008

The parameters of function (1), the window length l and the threshold coefficient U are adaptive, and determine the size of a time window within which geomagnetic disturbances are estimated and determine the size of a time window within which we estimate the geomagnetic disturbances and the intensity of isolated geomagnetic disturbances. Application of adaptive parameters of the threshold function allows us to identify both short-term insignificant increases in geomagnetic activity that can be observed on the eve of magnetic storms (Fig. 1) and longer strong increases during magnetic storms (Fig. 2).

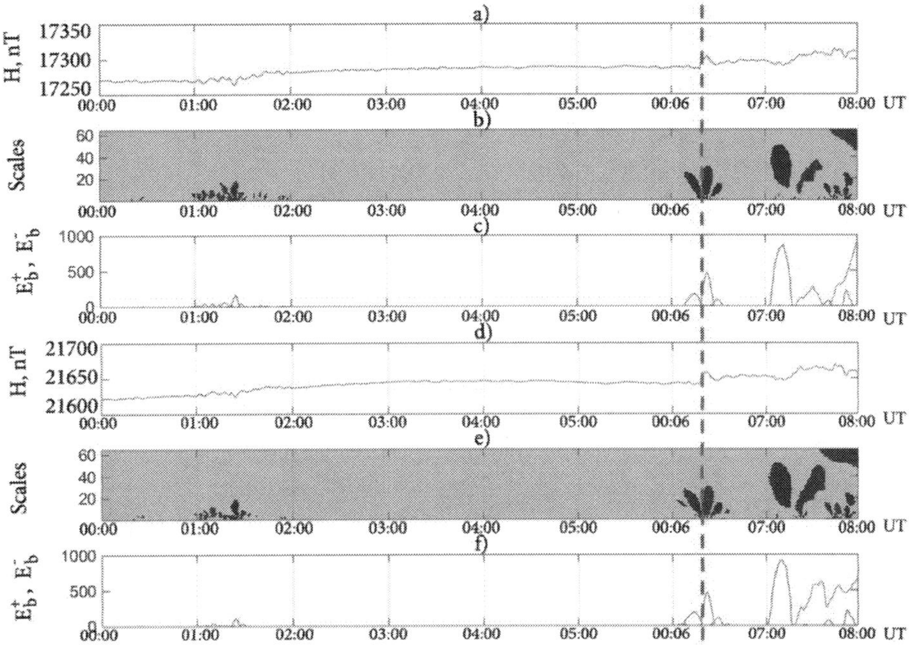

Fig. 1 Results of data processing for January 7, 2015 from 00:00 UT to 08:00 UT (the parameters of the threshold function l=720, U=3): a) H-components of the Earth's magnetic field (Magadan); b) result of application of operations (1) (Magadan); c) result of application of operations (1), (5) (Magadan) positive disturbances are indicated by red line, and negative disturbances are indicated by blue line; d) H-components of the Earth's magnetic field (Paratunka); e) result of application of operations (1) (Paratunka); f) result of application of operations (1), (5) (Paratunka) positive disturbances are indicated by red line, and negative disturbances are indicated by blue line; The dashed line indicates the beginning of the magnetic storm

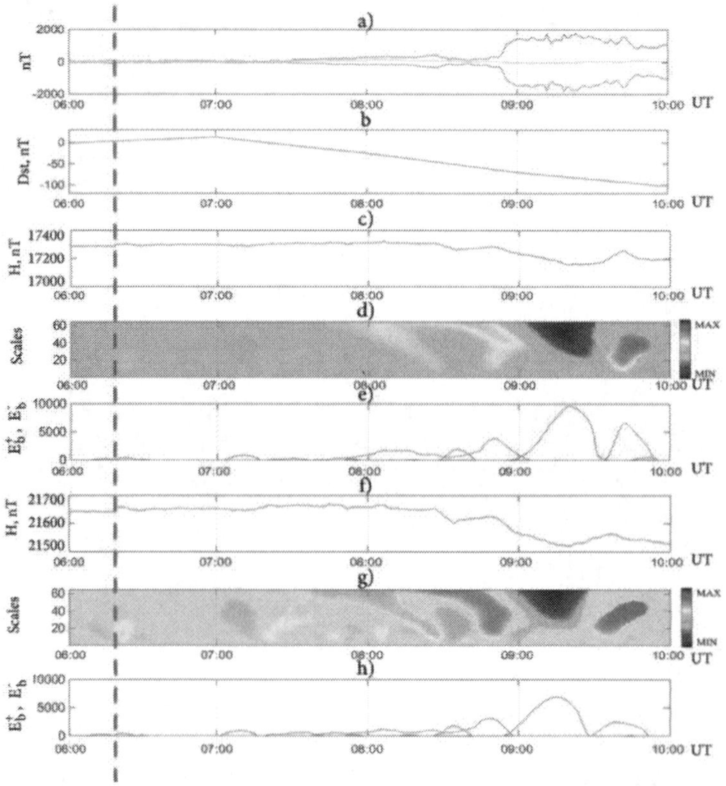

Fig. 2 Results of data processing for January 7, 2015 from 06:00 UT to 10:00 UT (the parameters of the threshold function l=720, U=3): a) H-components of the Earth's magnetic field (Magadan); b) result of application of operations (1) (Magadan); c) result of application of operations (1), (5) (Magadan) positive disturbances are indicated by red line, and negative disturbances are indicated by blue line; d) H-components of the Earth's magnetic field (Paratunka); e) result of application of operations (1) (Paratunka); f) result of application of operations (1), (5) (Paratunka) positive disturbances are indicated by red line, and negative disturbances are indicated by blue line; The dashed line indicates the beginning of the magnetic storm

2.2 Extraction of geomagnetic pulsations

To extract geomagnetic pulsations, following the results of [4, 18], we used the expansion of the geomagnetic field variation into wavelet packets:

$$
\begin{aligned}
f_{(0,0)}(t) &= \sum_{n} c_{j,k,n}\varphi_{j,k,n}(t) + \sum_{(j,k)\in D_i} \sum_{n} d_{j,k,n}\Psi_{j,k,n}(t) + \\
&+ \sum_{(j,k)\notin D_i} \sum_{n} d_{j,k,n}\Psi_{j,k,n}(t) = f_{trend}(t) + f_{dist}(t) + e(t)
\end{aligned}
\quad , \tag{6}
$$

where $\varphi_{j,k} = \{\varphi_{j,k,n}\}_{n\in Z}$ is basis in the node (j,k) of a wavelet packet tree obtained by a scaling function; $\Psi_{j,k} = \{\Psi_{j,k,n}\}_{n\in Z}$ is the wavelet-basis in the node (j,k) of the wavelet packets tree; coefficients $c_{j,k,n} = \langle f, \varphi_{j,k,n} \rangle$; $d_{j,k,n} = \langle f, \Psi_{j,k,n} \rangle$; D_i is s set of index pairs, (j,k) is a node of the wavelet packet tree, where j is the scale, k is the number of the component, the subscript (0,0) indicates the origin of the discrete data belonging to the node of the tree with indices (0,0) (it is assumed without loss of generality).

Component $f_{trend}(t) = \sum_n c_{j,k,n}\varphi_{j,k,n}(t)$ in quiet periods describes the trend of the geomagnetic field variation, component $f_{dist}(t) = \sum_{(j,k)\in D_i} g_{j,k}(t)$, where $g_{j,k}(t) = \sum_n d_{j,k,n}\Psi_{j,k,n}(t)$ in disturbed periods describes pulsations from a range determined by a set of indexes D_i. Component $e(t) = \sum_{j\notin D_i}\sum_n d_{j,k,n}\Psi_{j,k,n}(t)$ is taken as noise.

In this work, in order to extract the pulsations in the ranges Pi1, Pi2 and Pi3, we used the decomposition into wavelet packets according to the scheme shown in Gig. 3. In accordance with the scheme in Fig. 3, we obtain sets of index pairs $D_1 = \{(1,1);(2,1);(2,2);(3,1);(4,1);(5,1)\}$, $D_1 = \{(6,1);(7,1)\}$, $D_1 = \{(8,1);(9,1)\}$. Fig. 4 shows an example of extraction of pulsations Pi1 and Pi2 during the magnetic storm on January 7, 2015.

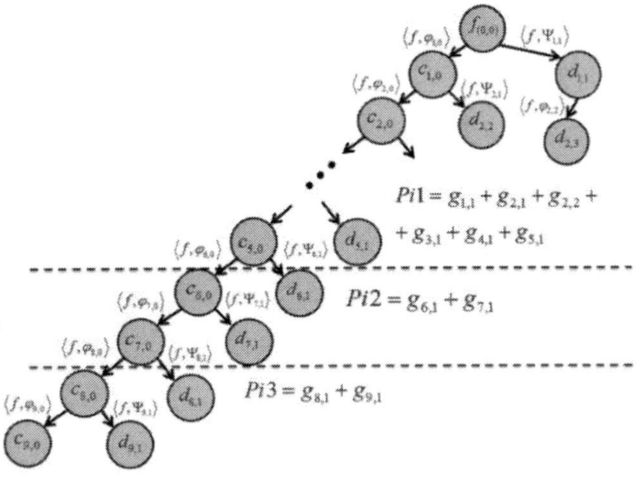

Fig. 3 Components of the wavelet packet containing disturbances in the ranges Pi1 - Pi3

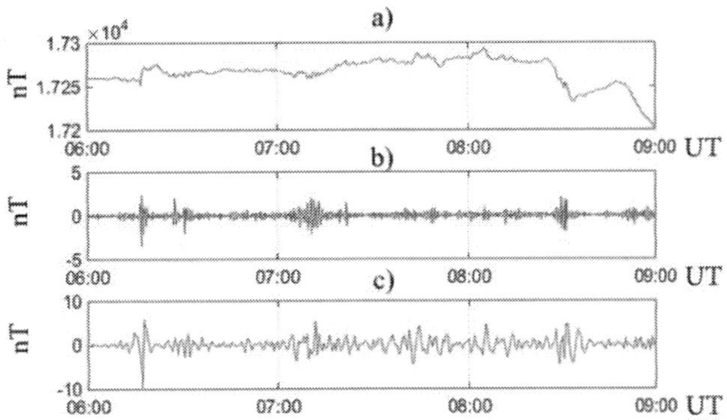

Fig. 4. Result of pulsation extraction on the basis of wavelet packets: a) H-component of the Earth's magnetic field; b) extracted disturbances in the range of pulsations Pi1; c) extracted disturbances in the range of pulsations Pi2

3 Results of data processing and analysis

Data processing and analysis for the station network in north-east of Russia (Table 1) were carried out using the developed method. To analyze the processes in the magnetosphere at equatorial latitudes, we used the data of Indian stations, HYB "Hyderabad" and CPL "Choutuppal" (Table 1). The results of data processing during the magnetic storm on January 7, 2015 are presented below in detail. The results of data processing during strong magnetic storms, occurred in 2015, are presented in Table 1.

Table 1. Observatories of the north-eastern segment of Russia.

Observatory	**IAGA code**	**Geo-graphic latitude**	**Geo-graphic longitude**	**Geo-magnetic latitude**	**Geo-magnetic longitude**	**Local Time (LT)**
Kotel'nyy (1)	KTN	76°00.0′	137°52.6′	66°44.4′	162°00′	UTC+09
Chokurdakh (1)	CHD	70°37.0′	147°52.6′	62°8.4′	152°36′	UTC+11
Tixie (1)	TIX	65°44.2′	150°53.0′	54°16.2′	146°14′	UTC+09
Yakutsk (1)	YAK	62°02.1′	129°42.1′	52°26.4′	163°13′	UTC+09
Magadan (2)	MGD	59°33.1′	150°48.3′	51°32.4′	146°2.4′	UTC+11
Paratunka (Petropavlovsk-Kamchatskiy) (2)	PET	52°58.3′	158°15.0′	45°51.6′	137°57′	UTC+12
Khabarovsk (2)	KHB	48°29.0′	135°04.0′	39°15′	15°48.6′	UTC+10
Hyderabad (3)	HYB	17°22′00′	78°28′00′	8°50′00′	151°78′00′	UTC+5:30
Choutuppal (3)	CPL	17°17.33′	78°55′	8°37.2′	152°34.8′	UTC+5:30

Note. Observatory affiliation is indicated in parentheses: (1) – IKFIA SB RAS, (2) – IKIR FEB RAS, (3) – CSIR-National Geophysical Research Institute.

The analyzed event on January 7, 2015 was caused by a coronal emission of solar matter (CME on January 4, http://ipg.geospace.ru/space-weather-review/). The storm had a classical nature with clearly expressed main phases in the Dst-variation (Fig. 5c). Fig. 5 shows the results of disturbance detection and estimation of their intensity from 06.01.2015 to 07.01.2015 (operations (1) and (5) were applied with the values of the threshold function parameters $l = 1440$, which corresponds to 24 hours, $U = 3$). The analysis showed a smooth increase of the geomagnetic disturbance intensity (Figure 5b) during the initial phase of the storm, and the Dst-index had positive values. The disturbance intensity reached the maximum during the period of a significant decrease in Dst-index and increase in the AE and AL indexes characterizing the occurrence of an intense substorm in the auroral zone.

The results of detection of Pi2 and Pi3 pulsations (see the diagram in Figure 3) showed that they occurred at the beginning of the magnetic storm at all the stations under analysis. The highest intensity of the pulsations in the initial phase of the storm was observed at the stations located to the south. At the end of the storm main phase, the intensity of Pi2 and Pi3 pulsations increased at high latitude stations and near the auroral zone, and the values of AE and AL indexes simultaneously increased (Fig. 6).

DOI: 10.1051/e3sconf/20172002008

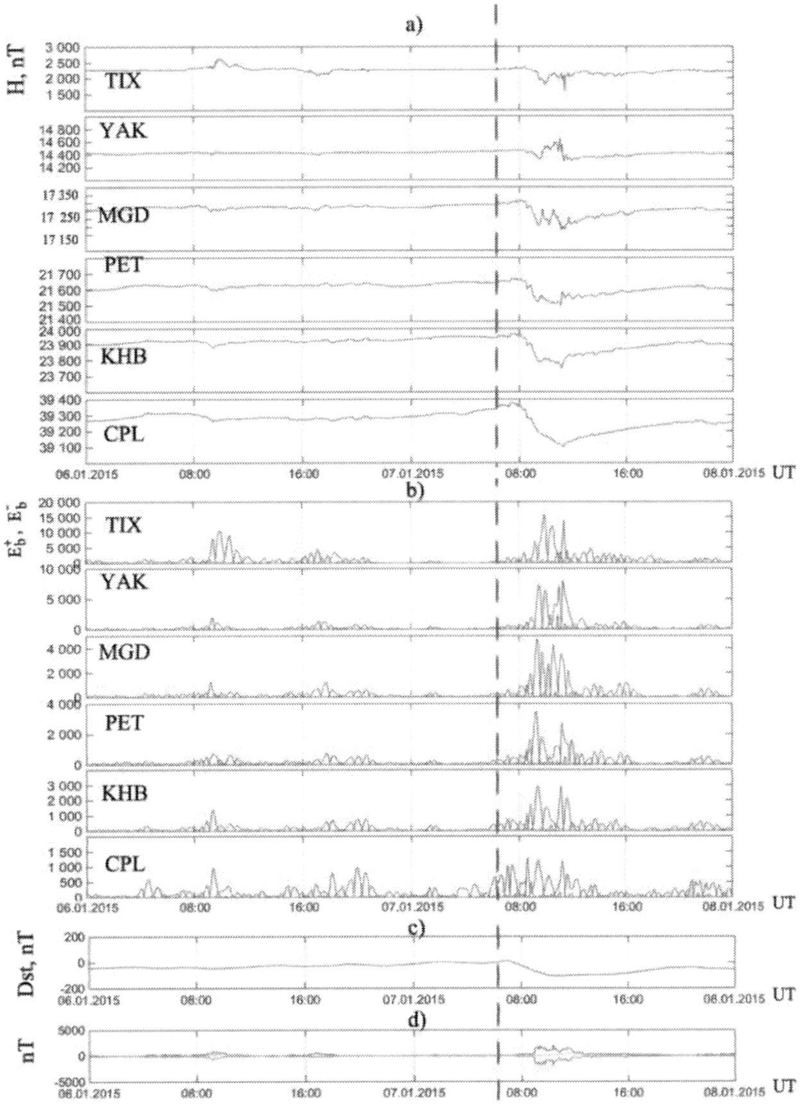

Fig. 5 Results of data processing for January 06-08, 2015: a) H-components of the Earth's magnetic field; b) geomagnetic disturbances intensity, based on the eq. (1), (5) (positive disturbances are indicated by red line, and negative disturbances are indicated by blue line); c) Dst-variation; and d) AE-indices (blue line), AU (yellow line), AL (red line). The dashed line indicates the beginning of the magnetic storm

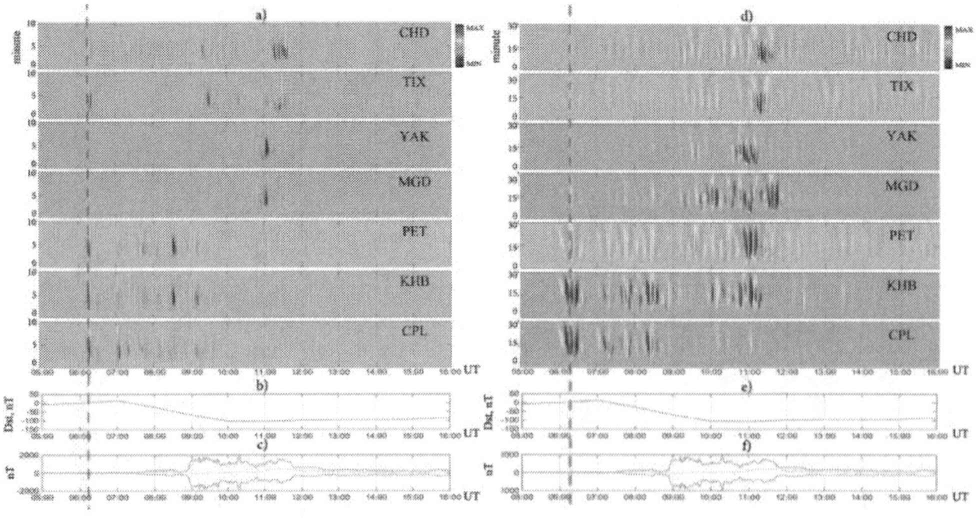

Fig. 6 Results of data processing for January 7, 2015 from 05:00 UT to 16:00 UT a) wavelet-spectrum of Pi2 pulsations; b) Dst-index; c) AE-indices (blue line), AU (yellow line), AL (red line); d) wavelet-spectrum of Pi3 pulsations; e) Dst-index; and f) AE-indices (blue line), AU (yellow line), AL (red line). The dashed line indicates the beginning of the magnetic storm

The wavelet spectrum of geomagnetic disturbance intensities presented in Fig. 7 (wavelet portrait of the storm) shows a complete pattern of the disturbance dynamics in the analyzed regions during the event period. Analysis shows that during the storm the disturbance dynamics had a general character, which probably characterizes large-scale processes in the magnetosphere (accumulation, redistribution and energy release). During the initial phase of the storm, the strongest disturbances in a wide frequency range (periods of oscillations from a few seconds to several hours) are observed at the equatorial station CPL. During the main phase of the storm, the most active three periods are clearly distinguished. Positive high-intensity disturbances (in Fig. 7a in red) are observed at the equatorial CPL station from 7:00 UT to 8:30 UT. They occur at mid-latitudes (YAK, MGD, PET, KHB stations) with a time shift of about half an hour, and then they occur at high latitudes (stations CHD and TIX) with a shift of another half an hour. The disturbance spectrum has a complex structure; an increase in the amplitudes, an expansion and a shift of the spectrum into the high-frequency region are observed, that it is probably due to the accumulation of energy in the magnetosphere and its discharge at the moments of extremes. Negative disturbances, the space-time distribution of which has a similar character, appeared at the equatorial station CPL (Fig. 7a, blue color) from 8:30 UT. The third active period is characterized by positive disturbances that occur approximately at 9:30 am at the stations near the auroral zone (PET, MGD, YAK stations). Further, they shift to the near equatorial latitude (with a delay of about 15 min) and high latitude (with a delay of about 30 minutes). Moreover, the following local features were detected in the wavelet portrait of the storm. Local active processes (Fig. 7a, dotted line), which coincide with variations in IMF small amplitude, are observed at the equatorial region (CLP station) from 7:00 to 9:00. Active regions (Fig. 7, dashed line), which are obviously related to the processes in the auroral zone, are also detected at MGD and YAK stations from 9:30 to 10:30. During the period of increasing values of AE and AL indexes, the strongest disturbances of a wide range of frequencies (oscillation periods from several seconds to several hours) occurred at high latitudes (CHD and TIX stations).

DOI: 10.1051/e3sconf/20172002008

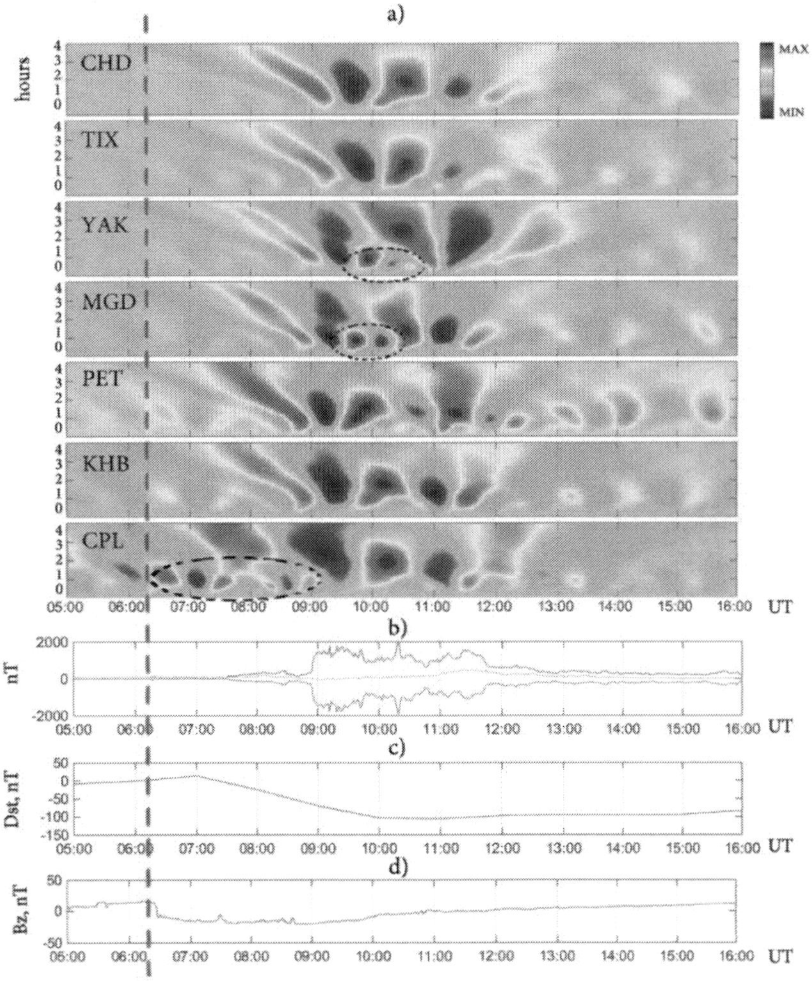

Fig. 7 Results of data processing for January 7, 2015 from 05:00 UT to 16:00 UT a) wavelet-spectrum of geomagnetic disturbances intensity; b) AE-indices (blue line), AU (yellow line), AL (red line); c) Dst-variation; and d) Bz-component of the interplanetary magnetic field (satellite ACE). The dashed line indicates the beginning of the magnetic storm

To analysis the small-scale short-period variations (from several seconds to tens of minutes) of the geomagnetic field, we used the operation (1) with the values for the parameters of the threshold function of $l = 1440$ samples, which corresponds to 24 hours, and $U = 7$. Fig. 8 shows the results of this operation on the eve and during the event on January 7, 2015. Analysis of the results shows that on the eve of the magnetic storm on January 6, 2015 from 09:20 to 18:10 UT, short-term increases in geomagnetic activity occurred simultaneously at the stations under analysis (Figure 8). According to space weather data (http://ipg.geospace.ru/space-weather-review/), several small leaps of the solar wind speed, caused by an accelerated stream of solar wind from the coronal hole, were recorded at the beginning of the day of January 5. The solar wind speed varied between 400 and 600 km / sec, and the Bz component was within ± 10 nT. According to the data of mid-latitude stations, the geomagnetic field was calm on the 5th and 6th of January (http://ipg.geospace.ru/space-weather-review/). The comparison of detected anomalous pre-increases with the data of auroral activity shows that the values of AE and AL indices were increased at these instants. Taking into account the data of space weather, the synchronous occurrence of this effect at various stations and the simultaneous increase in geomagnetic activity in the auroral zone, it is possible to assume with high probability the solar nature of this effect and the possible connection with the approaching magnetic storm. It can also be noted, that the intensity of the pre-increases grew up in the direction from high latitudes to the equator, and at the equatorial station CPL. The intensity of the pre-increases exceeds the intensity of geomagnetic disturbances in a given frequency range during the event. The results presented in Table 2 also confirm the possibility of anomalous increases in geomagnetic activity on the eve of magnetic storms.

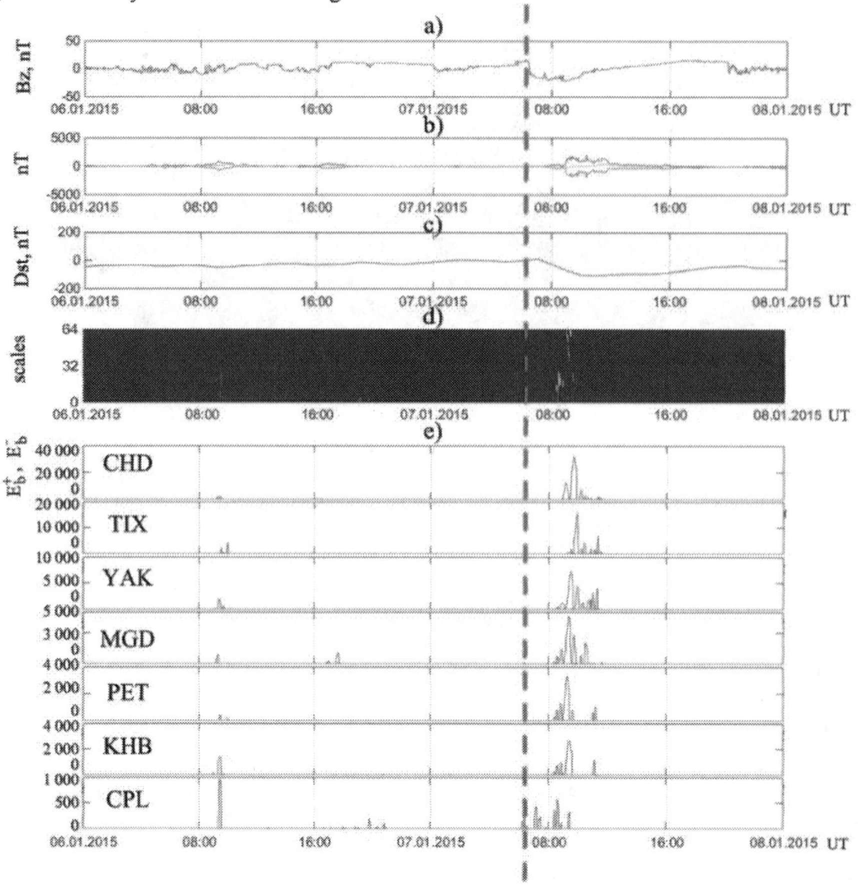

Fig. 8 Results of the data processing for January 06-08, 2015: a) Bz-component of the interplanetary magnetic field (satellite ACE); b) AE-indices (blue line), AU (yellow line), AL (red line); c) Dst-variation; d) detected synchronous increases in geomagnetic activity (using the operations (1)); and e) the intensity of increases in geomagnetic activity (using the operations (1), (5)). The dashed line indicates the beginning of the magnetic storm

Table 2. Results of the data processing during magnetic storms, which occurred in 2015

Date of a storm	Storm source	Time of storm beginning (UT)	Kp max	Dst max	Anomalies detected before a magnetic storm
					Time interval before a magnetic storm beginning
07.01.2015	CME	6:15	6	-103	21 hours
					12 hours 10 minutes

					4 hours 55 minutes
17.03.2015	CME	4:45	8	-233	16 hours 30 minutes
					12 hours 10 minutes
					11 hours 15 minutes
21.06.2015	CIR/CME	16:55	8	-111	10 hours
					7 hours
19.12.2015	CME	16:18	7	-170	25 hours 30 minutes
					15 hours 30 minutes
					7 hours 20 minutes
					1 hour

4 Conclusions

On the example of the event on January 7, 2015, the paper demonstrates that the proposed method makes it possible to perform a detailed spatio-temporal analysis of magnetic storms from geomagnetic data and to identify both large-scale processes in the magnetosphere and local dynamic features of events. The advantage of the proposed method is also its numerical implementation and the possibility of obtaining accurate quantitative estimates. The analysis of geomagnetic field variations, performed by the station network data during the magnetic storm on January 7, 2015, showed the presence of a general nature of the processes in the analyzed areas. At the initial phase of the storm, the strongest disturbances were observed at the stations located near the equator. During the main phase of the storm, geomagnetic disturbances reached their maximum values. During the periods of increasing values of AE and AL indexes, the strongest disturbances were observed at the stations of the auroral and high latitudinal zones. Moreover, local features, probably characterizing the processes in the auroral zone and in the equatorial region, were detected in the disturbance intensity spectrum.

In the opinion of the authors, the short-term increases in geomagnetic activity, detected on the eve of events and occurring synchronously at stations from high latitudes to the equator, are important. Similar pre-increases of geomagnetic activity were noted by the authors in the papers [6, 24]. Studies of this work confirm the possibility of anomalous increases in geomagnetic activity on the eve of magnetic storms.

Taking into account the limited amount of information on the state of the near-Earth space, the presence of noise, possible hardware failures, etc., application of geomagnetic data recorded on the earth's surface is important for the tasks of space weather forecast.

The development of the method for geomagnetic data analysis was supported by RNF Grant No. 14-11-00194. The primary analysis of the data was supported by RFBR Grant No. 16-55-45007. The authors are grateful to the institutions supporting magnetic observatories whose data were used in the study.

References

1. Yu.I. Yermolaev, M.Yu. Yermolaev, IZV ATMOS OCEAN PHY+, **46(7)**, 799–819 (2010)
2. S.R.P. Nayar, V.N. Radhika, P.T. Seena, ILWS2006, 328-331 (2006)
3. A.G. Hafez, E. Ghamry, H. Yayama, K. Yumoto, ASR, **51,** 39-49 (2013)

4. Z. Xu, L. Zhu, J. Sojka, P. Kokoszka, A. Jach, J. Atmos. Solar–Terr. Phys., **70,** 1579-1588 (2008)
5. Jach, P. Kokoszka, J. Sojka, L. Zhu, J. Geophys. Res., **111(A9)** doi:10.1029/2006JA011635 (2006)
6. P. Paschalis, C. Sarlanis, H. Mavromichalaki, Sol. Phys., **182(1),** 303-318 (2013)
7. K.P. Macpherson, A.J. Conway, J.C. Brown, J. Geophys. Res., **100**, 735-744 (2001)
8. J.W. Woolley, P.K. Agarwarl, J. Baker, International Journal for Numerical Methods in Fluids,**63,** doi:10.1002/fld.2117 (2010)
9. Soloviev, A. Chulliat, S. Bogoutdinov, A. Gvishiani, S. Agayan, A. Peltier, B. Heumez, EPS, **64(9),** 743-752, (2012)
10. N. Rotanova, T. Bondar, V. Ivanov, Geomagn. Aeron., **44,** 252-258 (2004)
11. J. Rybàk, A. Antalovà, M. Storini, Space Sci. Rev., **97** 359-362 (2001)
12. N. Zaourar, M. Hamoudi, M. Mandea, G. Balasis, M. Holschneider, EPS, **65(12),** 1525-1540 (2013)
13. A.S. Smirnova, S.D. Snegirev, O.A. Sheyner, Vestnik of Lobachevsky University of Nizhni Novgorod, **6(1)**, 88–93 (2013)
14. O.V. Mandrikova, I.S. Solovev, T.L. Zalyaev, EPS, **66.** doi:10.1186/s40623-014-0148-0 (2014)
15. V.P. Golovkov, V.O. Papitashvili, N.E. Papitashvili, Geomagn. Aeron. , **29**, 667–670. (1989)
16. K. Nowożyński, T. Ernst, J. Jankowski, Geophys. J. Int., **104,** 85-93 (1991)
17. M. Menvielle, N. Papitashvili, L. Hakkinen, C. Sucksdorff, Geophys. J. Int., **123**, 866-886. (1995)
18. O.V. Mandrikova, S.E. Smirnov, I.S. Solov'ev, Geomagn. Aeron., **52(1)**, 111–120 (2012)
19. S.R. Bogoutdinov, A.D. Gvishiani, S.M. Agayan, A.A. Solovyev, E. Kihn, Izvestiya, Physics of the Solid Earth, **46 (11),** 1004–1016 (2010)
20. R.V. Sidorov, A.A. Soloviev, Sh.R. Bogoutdinov, Izvestiya, Physics of the Solid Earth, **48(5),** 410-414 (2012)
21. R. I. Krasnoperov, A.A. Soloviev, Gornyi Zhurnal **10,** 89-93, doi: 10.17580/gzh.2015.10.16 (2015)
22. Soloviev, S. Bogoutdinov, A. Gvishiani, R. Kulchinskiy, J. Zlotnicki, Data Science Journal, **12,** doi:10.2481/dsj.WDS-019 (2013)
23. J. Uwamahoro, L.A. McKinnell, J.B. Habarulema, ANGEO, **30,** 963-972 (2012)
24. Mahrous, A. Radi, M. Youssef, A. Faheem, S. Ahmed, N. Gopalswamy, COSPAR, D23-0052-10 (2010)
25. G. Pallocchia, E. Amata, G. Consolini, M. F. Marcucci, I. Bertello, ANGEO, **24**, 989-999 (2006)
26. O.V Mandrikova, Problems of the evolution of open systems, **2(10)**, 161-172 (2008, in russian)
27. O.V. Mandrikova, Yu.A. Polozov, Information technologies, **7,** 61-65 (2014, in russian)
28. V.V. Geppener, O.V. Mandrikova, E.A. Zhizhikina, SCM'2015, 251-254, Doi: 10.1109/scm.2015.7190473 (2015)

29. O.V. Mandrikova, E.A. Zhizhikina, Computer Optics, **39**(3), 420-428 (2015, in russian)
30. O.V. Mandrikova, I. Solovjev, V.V. Geppenerc, A-KR. Taha, D. Klionskiy, DIGIT SIGNAL PROCESS, **23**, 329–339 (2013)
31. C.K. Chui *An introduction in wavelets (*Academic Press, 1992)
32. Daubechies *Ten Lectures on Wavelets* (CBMS-NSF Lecture Notes nr. 61, 1992)
33. O.A. Sheiner, V.M. Fridman, Radiophys Quantum El., **54(10)**, 655-666 (2012)
34. O.V. Mandrikova, V.V. Bogdanov, I.S. Solov'ev, Geomagn. Aeron., **53(2)**, 268-273. (2013)

Algorithms for changing the structure of geospace self-organizing question-answering sensor networks

Vladimir Mochalov[1,*], *Anastasia* Mochalova[1]

[1]Institute of Cosmophysical Research and Radio Wave Propagation FEB RAS, 684034 Kamchatka region, Elizovskiy district, Paratunka, Russia

Abstract. Optimization problems of construction, development and changing the structure of geospace self-organizing question-answering sensor networks (GSQASN) are considered. The task specifies the coordinates of various network nodes. It is required with the specified functional, structural, cost and spatial constraints to change the structure of the GSQASN by adding new nodes, moving to new positions or removing some existing nodes. After the formation of the GSQASN structure we solve the task of question-answer agents placement into GSQASN structure in order to be able to answer the given types of questions under the established limitations. The functional scheme of a given category nodes placement into GSQASN structure and approximate bio-inspired algorithms for solving the tasks are proposed. The results of the work can be used in the construction of specific GSQASN and in the GSQASN design support systems.

1 Introduction

New space technologies (nanosatellites, CubeSats, SmallSats and etc.), private space companies and the projects for launching thousands of small satellites to organize space networks with different purposes give principally new opportunities to monitor geospheres. We should note the increasing number of separate monitoring systems applying the data obtained from the geospace. The examples of such systems are the following: systems of estimation of magnetospheric parameters based on whistler distributed registration (AWDANet system); distributed systems for lightning activity monitoring (WWLLN, Blitzortung.org, Earth Networks and etc.); the Earth's ionosphere monitoring systems (ionospheric vertical sounding systems; GPS/GNSS ionospheric sounding allows to get the data about variations of the total electron content (TEC) in the ionosphere with high spatial and temporal resolution; and etc.), different monitoring systems applying satellite data (including global navigational systems to determine locations and time synchronization of measurements), space weather monitoring systems and so on. The term «geospace» is understood as the region of space that goes from the solar photosphere to the atmosphere of

* Corresponding author: vmochalov@ikir.ru

Earth. It includes the solar photosphere, chromosphere and corona, the solar wind, Earth's magnetosheath, magnetosphere, thermosphere, ionosphere and atmosphere.

In this paper on the basis of the personal results obtained before in the area of sensor networks designing [1-4], semantic analysis and question-answering systems [5-7] we introduce a new notion of «geospace self-organizing question-answering sensor networks» (GSQASN) that means a distributed network which monitors ambient environment parameters applying the data from geospace and allowing nodes not only to interact with each other owing to the self-organizing property, but to answer the defined types of natural language questions as well.

In the work we use a model of the GSQASN structure (Fig.1), where on the functional level the following types of GSQASN nodes can be defined: (1) functional nodes (F-nodes) that collect information in some neighborhood of their location; (2) transit nodes (T-nodes) that manage routing and retransmit the information collected by F-nodes to the information collection centers (ICC) to be utilized further; (3) ICCs that manage the GSQASN and process information collected by the GSQASN. In general case there can be multiple ICCs in the GSQASN, and the information that has arrived into each of them is available to one or multiple users for making decisions and performing certain actions. It means that information received by F-nodes should be retransmitted, with a required degree of reliability, to several ICCs by means of transit nodes allocated within the given object in a certain way. We think that ICC is capable of performing F-node and T-node functions. F-node can perform the T-node functions and information between nodes can be transmitted both via the wire and wireless networks.

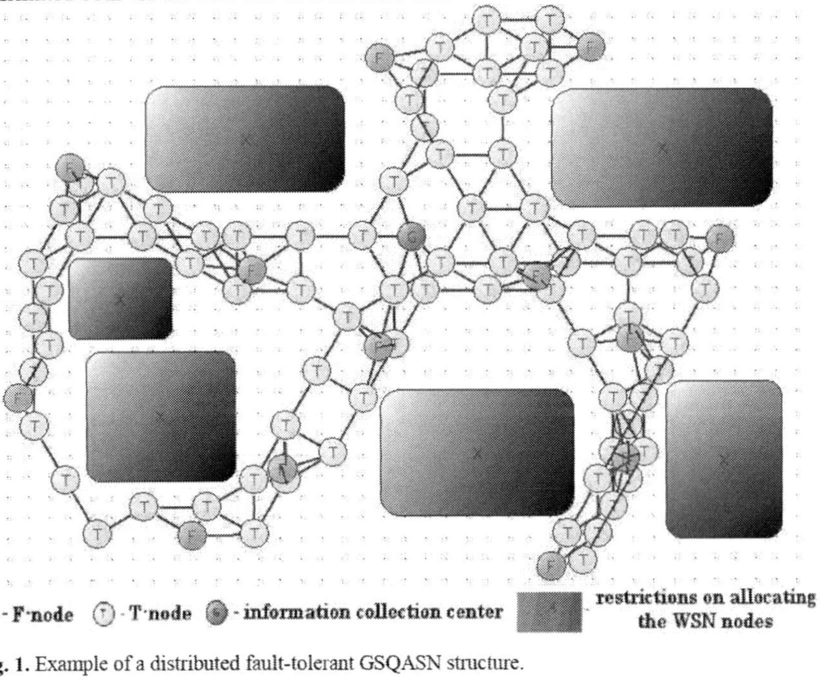

Fig. 1. Example of a distributed fault-tolerant GSQASN structure.

 DOI: 10.1051/e3sconf/20172002009

2 The functional scheme of a given category nodes placement into GSQASN structure

Designing of GSQASN requires the solution of many complicated problems referring to different areas of research, they are: projecting of network nodes (measurement stations, sensors and etc.); construction of different physical-mathematical models of monitoring processes of ambient environment parameters applying the geospace data; model selection for information collection from the GSQASN; development of methods and algorithms for GSQASN structure synthesis; estimate of measurement error and limitations, estimate of spatial and other limitations of network node placement; ensuring of the defined functional and structural parameters of the synthesized GSQASN; development of fitness functions of network nodes placement; development of algorithms for question-answering agents placement into the GSQASN structure; development of self-organization algorithms for different GSQASN levels (for example, the function level performed by the nodes, routing level, level of tasks distribution between nodes and question-answering agents).

Two types of GSQASN structures can be distinguished:

• distributed fixed network in which all the GSQASN nodes are not moved after the initial placement;

• decentralized mobile self-organizing network compound of fixed nodes (distributed fixed network segment) and mobile units (mobile network segment) which can be moved in different directions and, as a consequency, form different network structures in dynamics by break and establish network connections with other nodes, removal and installation of new nodes into the network structure. Mobile units can be installed both on satellites and on mobile robots (drones, above-water and under-water vehicles). Some functions of the mobile units are: formation of a self-organizing GSQASN structure; geographically-distributed acquisition of data from GSQASN nodes; organization of the interaction of a GSQASN mobile network segment with a fixed one. When applying the mobile robots, the following is possible: accurate nodes placement, distribution nodes over territory; moving, removing, reprogramming of nodes; charging and replacement of GSQASN nodes power sources; planning of cooperative behavior of mobile robots in the process of general aim solution and redesigning of GSQASN nodes placement based on the aims and current measurements of the whole GSQASN.

Types of node structures of a distributed fixed GSQASN are: star (Fig.2); cluster network structure (Fig.3); mesh self-organizing structure (Fig.1).

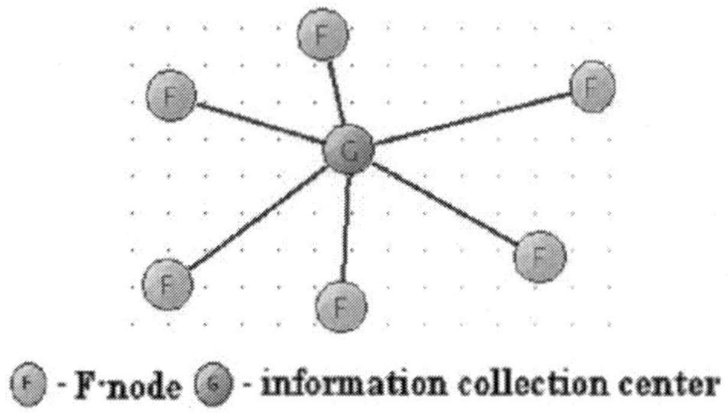

Fig. 2. Star network structure.

DOI: 10.1051/e3sconf/20172002009

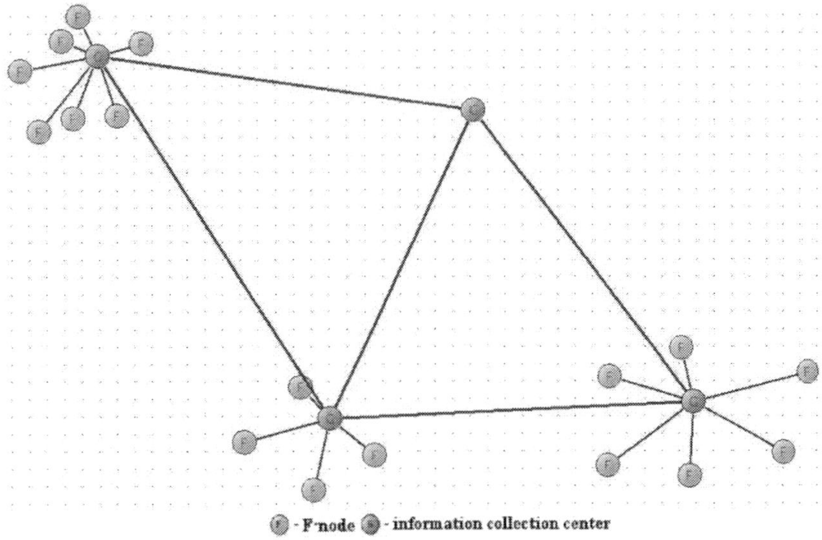

Fig. 3. Cluster network structure.

Possible ways of problem statement for GSQASN structure synthesis

1. Synthesis of ICCs allocation. In this task we know the spatial restrictions for allocating the ICCs. Also, we know the allocation of pre-installed ICCs. It is necessary to allocate ICCs in such way, that the designed GSQASN structure would have the «desired properties» assigned by a designer. During the synthesis of ICCs allocation it is possible to optimize (by their removal or moving) some pre-installed ICCs noted by the designer.
2. Synthesis of F-nodes allocation. We know the spatial restrictions for allocating the F-nodes. Also, we know the allocation of ICCs and allocation of pre installed F-nodes. It is necessary to allocate new F-nodes in such way, that the designed GSQASN structure would have the «desired properties» assigned by a designer. During the synthesis of F-nodes allocation it is possible to optimize (by their removal or moving) some pre-installed F-nodes noted by the designer.
3. Synthesis of T-nodes allocation. We know the description of the GSQASN allocation object, spatial restrictions for allocating the T-nodes. Also, we know the allocation of F-nodes, ICCs and pre installed T-nodes. It is necessary to allocate T-nodes in such way, that the designed GSQASN structure would have the «desired properties» assigned by a designer. During the synthesis of T-nodes allocation it is possible to optimize (by their removal or moving) some pre-installed T-nodes noted by the designer.
4. Complex sequential synthesis of ICCs, F-nodes and T-nodes allocation. This statement suggests a sequential allocation of ICCs first (statement 1), then F-nodes (statement 2) and then T-nodes (statement 3).

Fig. 4 illustrates a functional scheme of a given category nodes placement into GSQASN structure. This scheme can be used as the basis for synthesis of ICC, F-nodes and T-nodes placement.

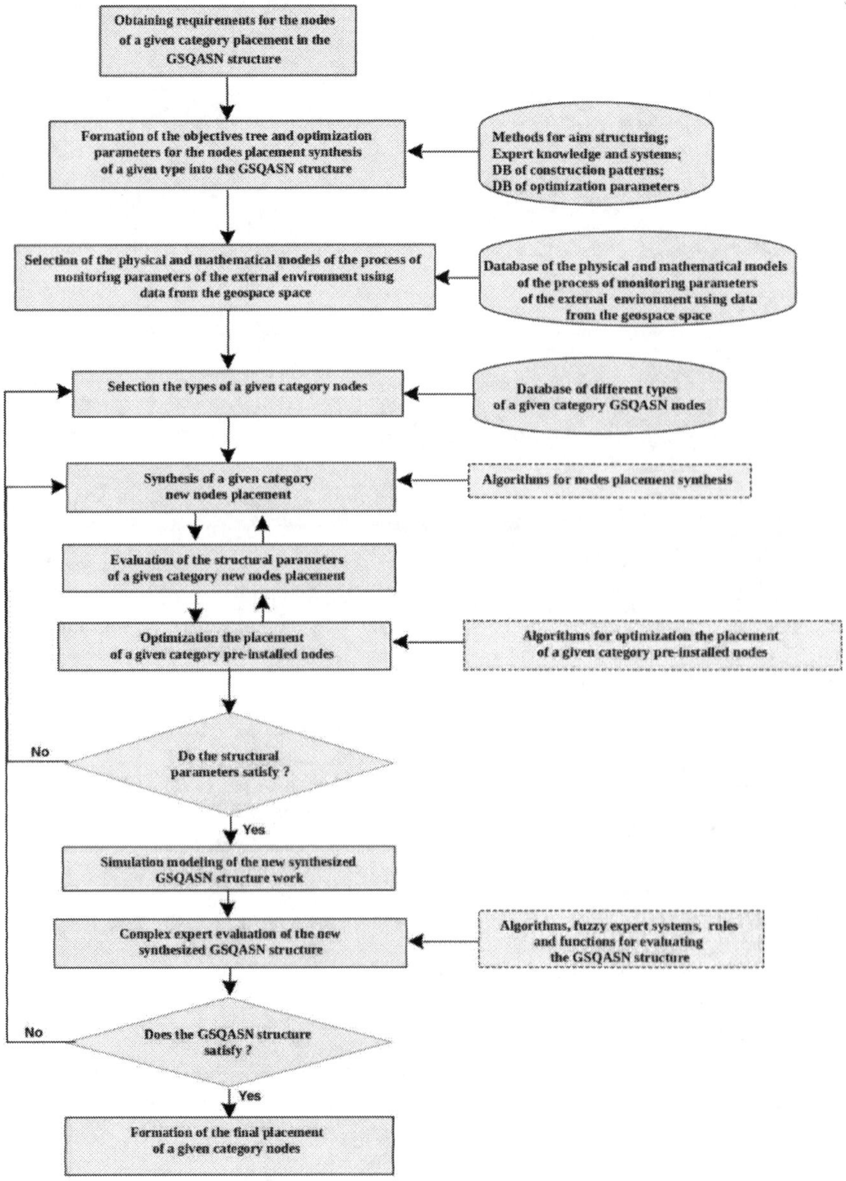

Fig. 4. The functional scheme of a given category nodes placement into GSQASN structure.

3 Bio-inspired algorithm for synthesis of a given category nodes placement into GSQASN structure

In recent years, the research area of *Natural Computing* is rapidly developing. It unites mathematical methods in which the principles of natural mechanisms of decision making are embedded [8]. Scientists have developed bio-inspired algorithms (BA) of modeling animals behavior ([9], [10], etc.]) for solving various optimization problems that either do not have exact solution, or the solutions search space is vary large and complex constraints of the objective function are presented, as well as NP-complete.

The described recommendations on applying BA and the proof in [11] that even the constrained variant of the problem of minimal coverage on plane is NP-complete allow us to conclude about the possibility to apply self-organizing bio-inspired algorithms for a self-organizing GSASN structure design.

Bio-inspired algorithms can be seen as multi-agent systems, each agent in which operates autonomously on very simple rules. The most frequently used bio-inspired agents (B-agents) include: ants, bees, termites, fireflies, birds, fish, bats, cats, wolves, etc.

The initial data of the algorithm for synthesis of a given category nodes $type_x$ placement are the following: allocation of pre-installed nodes of $type_x$ (set of nodes Ω_{pin}); allocation of installed nodes of $type \neq type_x$; description of the object that the GSQASN needs to be located at (its dimensions, scheme, spatial requirements for nodes allocation and etc.); characteristics of ready-to-use nodes of $type_x$; adopted self-organization and routing algorithms; information collection model; functional requirements; optimization parameters; fuzzy expert systems etc.

The algorithm below is based on the adaptation of the multi-agent bio-inspired algorithm for wireless sensor network design proposed in paper [1].

Step 1: Define a set M_{ALL} of all optimization parameters; the functions for calculating the parameters of M_{ALL}; a subset of optimization parameters $M_I \in M_{ALL}$,

Step 2: Determine the membership functions of fuzzy sets that characterize the optimization parameters of the M_{ALL}; fuzzy expert system to derive the confidence factor to meet the functional requirements of the designer. Create an empty set of the best solutions Ω_{BEST}. Determine the maximum number of solutions b_K that will be stored in Ω_{BEST},

Step 3: Create a set Ω_p of the possible placement points of type $type_x$ nodes (the set Ω_p can be formed with the help of: the algorithms for covering the object of placement with a mesh (based on an equilateral triangle or hexagon or square) or with circles with a given radius; covering algorithms in accordance with the choice and recommendations of the designer; other covering algorithms). Create an empty set Ω_T. Create node of type $type_x$ in each point of Ω_p and add this node to Ω_T set,

Step 4: Create a shared global memory of the stored pheromone (SGMSP) to share some "knowledge" between B-agents. The pheromone is stored on the edges of a fully connected undirected weighted graph (FCUWG), the nodes of which are the ones of $type_x$. To store the edges of the graph in computer memory it is required to create a two-dimensional array *feromoneNetwork* with $N(N-1)/2$ memory cells of type *float*, where $N=|\Omega_T|$ is the number of nodes. All values of *feromoneNetwork* must be initialized as zeros,

Step 5: Execute bio-inspired multi-agent algorithms.

5.1. Create a two-dimensional array *feromoneDif* to store changes in the pheromone using for example the following Java-code:

```
float feromoneDif [][] = new float[N-1][];
for(int i = 0; i < N; i++)
feromoneDif [i] = new float[N - (i + 1)];
```

All values of *feromoneDif* must be initialized as zeros (in the above code, the zero-initialization is done automatically). Define the number of different bio-inspired agents *m*, the strategy for choosing the initial location of the agent and other parameters needed for the agent to perform the work.

5.2. For each agent, perform the following steps (the code can be parallelized, i.e. to run in a separate thread for each agent):

5.2.1. Form, using the movement rules of the B-agent, SGMSP the route M_T (array) of agent moving on nodes Ω_T,

5.2.2. Create an empty extensible array of nodes M_{STR}, in which the nodes of the designed structure will be placed,

5.2.3. Select the design strategy:
(a) sequentially add nodes to the network -> Go to step 5.2.4;
(b) sequentially remove nodes from the network -> Add to M_{STR} all nodes from M_T array in the same sequence order. Go to step 5.2.11;

5.2.4. Create an empty set H_P, which will contain the caches of such internal parameters of the functions of computing estimates M_I, which will increase the speed of computing estimates M_I for the next iteration. Create a variable *i* to store the index of the current node from the M_T array and initialize its value to 0 (i=0). Set the node T_C (T_C=M_T[0]) as the current one,

5.2.5. Add to M_{STR} the node T_C. Form the network structure S_S of nodes M_{STR},

5.2.6. Calculate using the caches H_P the values estimations of the optimization parameters of the set M_I having structure S_S. Clear H_P. Save the caches of the internal parameters of the functions of computing estimates M_I of the current iteration to the set H_P,

5.2.7. Using a fuzzy expert system, calculate the reliability coefficient K_{DI} of meeting the requirements of the designer for parameters of a set M_I of structure S_S,

5.2.8. If $K_{DI} > p_I$, where p_I is a set threshold, go to step 5.2.11,

5.2.9. If $i<| M_T |$, put $i = i + 1$ and accept the next node $T_C = M_T[i]$ as the current one. Repeat steps 5.2.5. – 5.2.9. while *i* does not become equal to $| M_T |$,

5.2.10. Exit with notification of the failure from the function of agent design of the network structure,

5.2.11. Steps of eliminating optimization:

5.2.11.1. Select the strategy of eliminating optimization:
(a) step-by-step optimization with consideration of optimization parameters M_I. The fuzzy expert estimation of the structural parameters M_I is used;
(b) step-by-step optimization with consideration of all optimization options M_{ALL}. The unit of simulation modeling and complex assessment of the network is used;

5.2.11.2. Revert the M_{STR} array,

5.2.11.3. In a loop, temporarily exclude each node $T_O \in M_{STR}$ from M_{STR}, then compute the confidence factor K_D of meeting the requirements of the eliminating optimization strategy parameters. If in absence of node T_O evaluation of network structure stops meeting the designer requirements, put T_O back to M_{STR} into its place.

5.2.11.4. Revert the M_{STR} array,

5.2.12. Perform simulation modeling of the network. The results of the modeling and structural-parametric estimates of the various parameters are the input to the complex expert system for evaluation of network structure. Calculate with the latter the confidence factor K_{DALL} of meeting all the designer requirements,

5.2.13. If $K_{DALL} > 0$ then, in accordance with one of the following strategies, increase the pheromone amount in the array *feromoneDif*:
(a) consequent update – increase the amount of pheromone on the edges of the agent sequential traveling on nodes of M_{STR} by the value equal to $\Delta\tau_{ij,k}(t)=Q_{agent}*K_{DALL}$, where Q_{agent} is the amount of pheromone secreted by the agent on one edge;

(b) full-mesh update – increase the amount of pheromone on all edges of the fully connected graph constructed on the basis of nodes of M_{STR} by the value equal to $\Delta \tau_{ij,k}(t)=Q_{agent}*K_{DALL}$.
5.2.14. If K_{DALL} is greater that the estimate of the worst solution from Ω_{BEST}, or ($|\Omega_{BEST}|<b_K$ and $K_{DALL}>0$), then add into Ω_{BEST} the current solution. By the solution we mean the couple (M_{STR}, K_{DALL}). If $|\Omega_{BEST}| \geq b_K$ then leave in Ω_{BEST} only b_K best solutions.
5.3. After all agents have performed step **5.2**, update SGMSP (*feromoneNetwork* array) in accordance with the following well-known rule [9]: $\Delta \tau_{ij}(t+1)=(1-p)* \tau_{ij}(t)+ \Delta \tau_{ij}(t)$, where $\Delta \tau_{ij}(t)$ is the amount of pheromone on edge (i,j) in the array of pheromone changes *feromoneDif*, and $p \in [0,1]$ is the coefficient of pheromone evaporation. To enhance the intermediate best solutions, the amount of pheromone on the edges of the routes of the best solutions Ω_{BEST} should be increased (an example is using "elite" ants).
5.4. If the stopping criterion is not met, go to step **5.1.**
Step 6: If it is necessary to continue the search, then create new set Ω_T and add to it type $type_x$ nodes located in points of possible nodes placement (for example, to cover the object of nodes placement with a mesh more densely in comparison with the previous coverage). After it go to Step **4**.
Step 7: Return the best solution from Ω_{BEST}.

4 Modifications of bio-inspired algorithm for changing the GSQASN structure

Under the change of the GSQASN structure we assume adding new nodes, moving to new positions or removing some existing pre-installed nodes with the specified functional, structural, cost and spatial constraints. Moving the node to new position is supposed to remove it from the old position and add it to a new one. The algorithm for adding new nodes to the GSQASN structure was considered in Section 3. Consider the algorithms of moving to new positions or removing some existing pre-installed nodes based on the modification of the algorithm discussed in Section 3.

Algorithm 1 for changing the GSQASN structure based on the modification of the algorithm discussed in Section 3

• At the beginning of the Step 4 add to the set Ω_T those pre-installed nodes of $type_x$ of set Ω_{pin} that can be removed or moved to new positions.
• In the implementation functions of GSQASN structure evaluation take into account the cost of removing and moving some pre installed nodes, as well as changing the cost of maintenance these nodes.
• All other steps of the bio-inspired algorithm for synthesis of a given category nodes placement into GSQASN structure remain unchanged.

Algorithm 2 for changing the GSQASN structure based on the modification of the algorithm discussed in Section 3

• Perform all steps of the algorithm considered in Section 3 so that new nodes of $type_x$ will be added to the GSQASN structure, but optimization of some pre-installed nodes placement of $type_x$ will not be performed.
• To optimize the placement of some pre-installed nodes of $type_x$ perform steps 1-5 of the algorithm discussed in Section 3 with the following modifications:

 DOI: 10.1051/e3sconf/20172002009

a) On the step 3, add only those pre-installed nodes of $type_x$ from the set Ω_{pin} to the set Ω_T, which can be removed or moved to new positions (don't add other nodes to the set Ω_T).
b) In the implementation of functions of the GSQASN structure evaluation take into account the cost of removing and moving of some pre-installed nodes, as well as changing of the cost of the node maintenance.

5 The scheme for generating the answer using the GSQASN question-answer agents

The question-answer agent (QA-agent) performs the function of generating the answer from natural language questions by collection, aggregation and accumulation information from some F-nodes that are serviced by this QA-agent. After the accumulation of sufficient information from the group of F-nodes, the QA-agent generates an answer. The QA-agents can interact with each other through self-organization to be able to answer the given types of questions under the established limitations. Physically, the QA-agent is a software/hardware add-on that can upgrade any type of GSQASN nodes. The QA-agents can differ from each other by technical capabilities (due to various hardware and software add-ons) and functionality capabilities (the ability to answer different types of questions, performed functions, etc.).

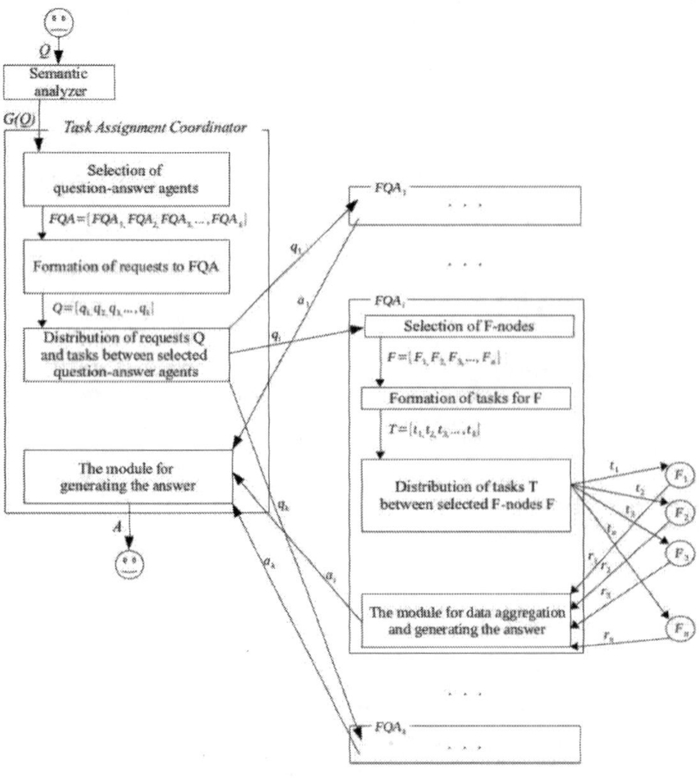

Fig. 5. The scheme for generating the answer using the GSQASN question-answer agents.

In general, formulated in the natural language question Q enters to the input of one of the QA-agents that perform the functions of task coordinator for other QA-agents. After it the question Q enters to the input of the semantic analyzer module, which create ontological-semantic graph $G(Q)$. The graph $G(Q)$ enters to the input of the module for selection of QA-agents, which are best suited for generating the answer to question Q. The information about selected QA-agents is placed into a set $FQA=\{FQA_1, FQA_2, FQA_3, ..., FQA_k\}$. The set FQA enters to the input of generating requests for QA-agents module. As a result, for each QA-agent $FQA_i \in FQA$ this module form a request q_i. Each q_i request is transmitted to the QA-agent FQA_i, which first tries to find the necessary information in the local database and if it is not found FQA_i select serviced F-nodes from which the necessary information should be collected. The information of selected at this stage F-nodes is placed into the set $F = \{F_1, F_2, F_3, ..., F_n\}$. Further, the task formation module generates a task t_j for each F-node $F_j \in F$. Each F-node $F_j \in F$ that receives the task t_j executes it (using information stored on this F-node or receive it from the environment with the help of a sensor installed on the F-node) and sends the response r_j back to the QA-agent which generated the task t_j. On the basis of all the responses $r_1, r_2, r_3, ..., r_n$ obtained from F-nodes $F_1, F_2, F_3, ..., F_n$ QA-agent FQA_i generates a_i answer and send it back to the task assignment coordinator. Thus, the task assignment coordinator receives all answers from QA-agents and on the basis of them makes up a general answer A, which is transmitted to the user as an answer to the question Q (Fig.5).

6 Question-answer agents placement into GSQASN structure

The selection of GSQASN nodes for assigning them additional QA-agents functions can be executed both at the stage of the GSQASN nodes placement synthesis (by adding the logic of QA-agents placement to the implementation functions of GSQASN structure evaluation) and after the GSQASN nodes have already been placed. In general, the task of QA-agents placement into the GSQASN structure is formulated as follows: with the well-known allocation of all GSQASN nodes (ICCs, F-nodes and T-nodes) we need to solve the optimization problem of selection some GSQASN nodes to assign them additional QA-agents functions to be able to answer the given types of questions under the established limitations.

Consider the algorithm for QA-agents placement into the GSQASN structure based on the modification of the algorithm discussed in Section 3. In this modification we synthesize the placement of QA-agents nodes (a given category nodes $type_x$=QA-agent).

Algorithm for QA-agents placement into the GSQASN based on the modification of the algorithm discussed in Section 3

- On the step 3 create a set Ω_p of all possible GSQASN nodes for assigning them additional QA-agents. Create an empty set Ω_T. Create node of $type_x$=QA-agent on each node of Ω_p and add this node to Ω_T set.
- To implement the functions of the GSQASN structure evaluation add logic for QA-agents placement evaluation.
- All other steps of the bio-inspired algorithm for synthesis of a given category nodes placement into GSQASN structure remain unchanged.

References

1. V.A. Mochalov, *Proceedings, The IEEE 17th ICACT*, 34-42 (2015)
2. V.A. Mochalov, *Lecture Notes in Computer Science* **9247**, 11-22 (2015)
3. V.A. Mochalov, A.P. Pschenichnikov, ICACT TACT **4**, 659-663 (2015)
4. V.A. Mochalov. Certificate of registration of computer software "Program for synthesis of monitoring networks by bio-inspired algorithms" No. 2016612039 (2016).
5. Mochalova A.V., Mochalov V.A, *Lecture Notes in Artificial Intelligence, Advances in Computational Intelligence*, **10061**, 53-66 (2017)
6. A.V. Mochalova, V.P. Zacharov, V.A. Mochalov, ICACT TACT **5**, 902-906 (2016)
7. V.A. Kuznetsov, V.A. Mochalov, A.V. Mochalova, ICACT TACT **5**, 651-658, (2016)
8. G. Rozenberg, T. Bäck, J.N. Kok, *Handbook of Natural Computing*, (2012).
9. F. Bonavear, M. Dorigo, G. Theraulaz, *Swarm Intelligence: from Natural to Artificial Systems*, 320 p, (1999)
10. D. Teodorovic, *Bee Colony Optimization (BCO)*, Available: http://www.sf.bg.ac.rs/downloads/katedre/oi/1.BCO-Book-Chapter.pdf .
11. R.J. Fowler, Inf. Process. Lett., **12(3)**, 133-137 (1981).

DOI: 10.1051/e3sconf/20172002010

Application of «Sensor signal analysis network» complex for distributed, time synchronized analysis of electromagnetic radiation

Vladimir Mochalov[1,*], *Anastasia* Mochalova[1]

[1]Institute of Cosmophysical Research and Radio Wave Propagation FEB RAS, 684034 Kamchatka region, Elizovskiy district, Paratunka, Russia

Abstract. The paper considers a developing software-hardware complex «Sensor signal analysis network» for distributed and time synchronized analysis of electromagnetic radiations. The areas of application and the main features of the complex are described. An example of application of the complex to monitor natural electromagnetic radiation sources is considered based on the data recorded in VLF range. A generalized functional scheme of stream analysis of signals by a complex functional node is suggested and its application for stream detection of atmospherics, whistlers and tweaks is considered.

1 Introduction

At the present time authors are developing a new software-hardware complex «Sensor signal analysis network» (SSAN) for construction and functioning of sensor networks for distributed analysis of time synchronized signals. In this paper we accept that a sensor network is a distributed network of endpoint functional nodes (F-nodes) which analyze the time synchronized signals and transmit the collected information (analysis results) to one or several collecting centers. In general form, information can be transmitted both via the wire and wireless networks. In case of application of wireless networks, owing to the self-organization property and application of special protocols, it becomes possible to transmit the information by retransmission from one node to another using the transit nodes (T-nodes) (Fig.1).

Fig. 2 shows a generalized scheme of SSAN complex F-node, including a mini-PC, measurement time synchronization unit, ADC, sensors, communication modules and optional field-programmable gate array (FPGA). In case of the absence of the necessity to use the FPGA, the ADC module is connected directly to a mini-PC. Depending on the conditions of application and accuracy requirements, the time synchronization unit may be realized on the basis of application of different types of satellite time synchronization GPS/GLONASS modules (NMEA 0183 signal comes to a mini-PC, PPS signal comes to ADC and FPGA input, the stabilized output of 10 MHz from GPS/GLONASS module comes to FPGA input) and/or the connectives of temperature-stabilized precise highly-reliable quartz oscillator with FPGA.

* Corresponding author: vmochalov@ikir.ru

DOI: 10.1051/e3sconf/20172002010

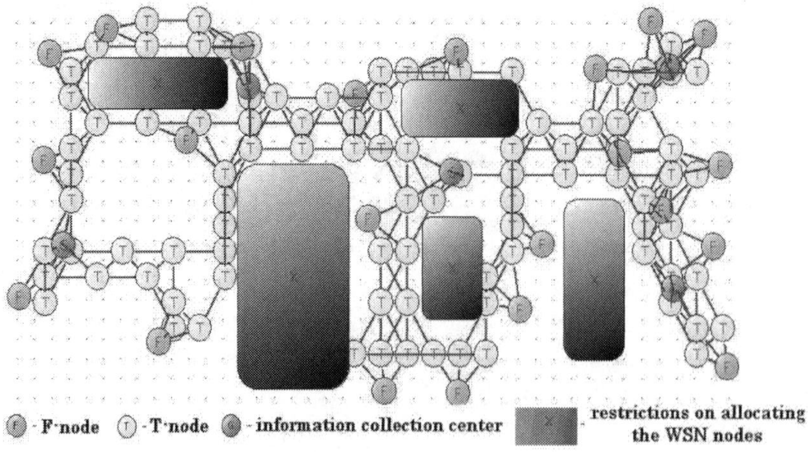

Fig. 1. Example of a distributed fail-tolerant system structure of a wireless sensor network [1].

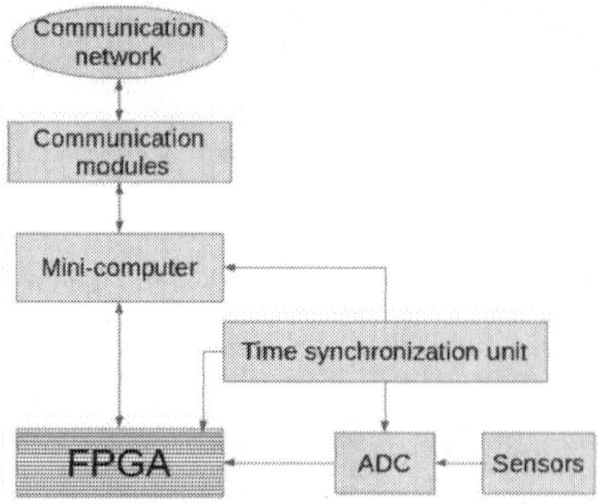

Fig. 2. Functional scheme of F-node of the software-hardware complex «Sensor signal analysis network».

Scope and field of application of SSAN complex:

• Projecting of sensor network structure taking into account different variants of network node location (ground, ocean, drones and spacecrafts),

• Application of universal software-hardware mini-PC nodes with time synchronization for fast implementation of sensor networks for the purpose of distributed analysis of time synchronized signals,

• Distributed complex analysis of electromagnetic radiation to investigate the magnetosphere dynamics, synoptic weather systems, seismic, volcanic activity etc.

Potential users of SSAN complex:

• The software-hardware complex can be applied for scientific investigations as well as for monitoring and automation of distributed object control, life-supporting system automation, detection and prevention of emergency situations, road monitoring (including railway roads, oil and gas pipe-lines, electrical and heat supply networks etc.) and so on.

SSAN complex features:

• Unique software has been developed [2] which allows us to build a distributed fault-tolerant sensor network structure in cooperation with a designer and to optimize the network node arrangement. Based on the developed algorithms [1, 3], the software arranges monitoring network nodes on a plane taking into account spatial limitations for node location and suitability evaluation functions. In the result of software operation, a list with the coordinates of distributed monitoring network nodes is formed and the image of network node arrangement is presented,

• The possibility to connect user modules for signal analysis, low cost of hardware for the mini-PCs of network computational nodes and low power consumption (~5 Watt) allow us to apply the hardware-software complex to solve a wide range of problems of distributed analysis of time synchronized signals,

• Application of a ready-made configured software-hardware complex will reduce organization facilities for the solution of some problems in registration and analysis of time synchronized signals.

2 Functional scheme of signal stream analysis by SSAN complex F-node on the example of monitoring of natural electromagnetic radiation sources based on the data of VLF range registration

The paper [4] describes one of the ways of realization of SSAN F-node (Fig. 3) which is used for monitoring of natural electromagnetic radiation sources based on the data of VLF range registration.

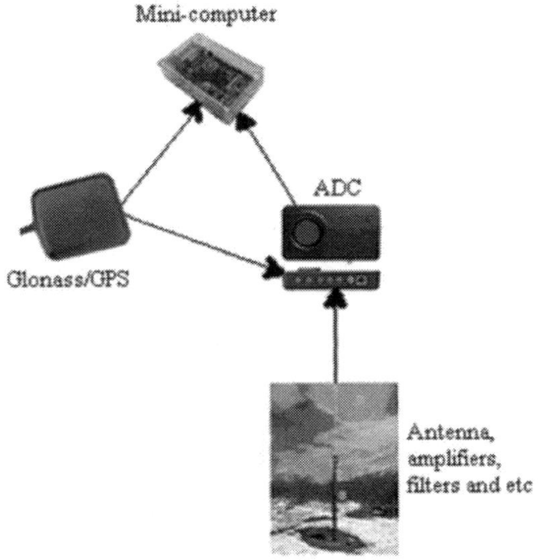

Fig. 3. Example of F-node for monitoring of natural electromagnetic radiation sources based on the data of VLF range registration [4].

Signal from the antenna comes to a pre-amplifier and then to a sound card linear input (applied as one of ADC variants). PPS signal from Glonass/GPS module also comes to the sound card linear input that allows us to make time synchronization of the measurements at different sensor network F-nodes. NMEA 0183 signal comes to a mini-PC to set the system time and to determine the F-node location. After digitization, the general signal from the sound card linear input comes to the mini-PC for stream processing. Several programs are

DOI: 10.1051/e3sconf/20172002010

launched on the mini-PC working under Linux operating system, they are: 1) program for data reading from ADC; 2) program of stream analysis (SA) of natural electromagnetic radiation sources based on the data registered in VLF range. SA program is capable of operating in two modes: 1) read mode of digitized data on different channels into a circular buffer inner structure and custom actions during every read record (almost real-time processing mode); 2) record mode of files with a given duration with the following processing of files by custom actions.

Fig. 4 shows a generalized functional scheme of signal stream analysis by SSAN complex F-node. Depending on the applied algorithms for event recognition, the following actions can be performed at the stage of primary processing: data normalization, calculation of signal spectrogram with different FFT window sizes (Fig. 5 and 7) and in different scales (linear and logarithmic) (Fig. 6 and 8), data filtration from noise (Fig. 9), estimation of different wavelet transforms (Fig. 10) and so on.

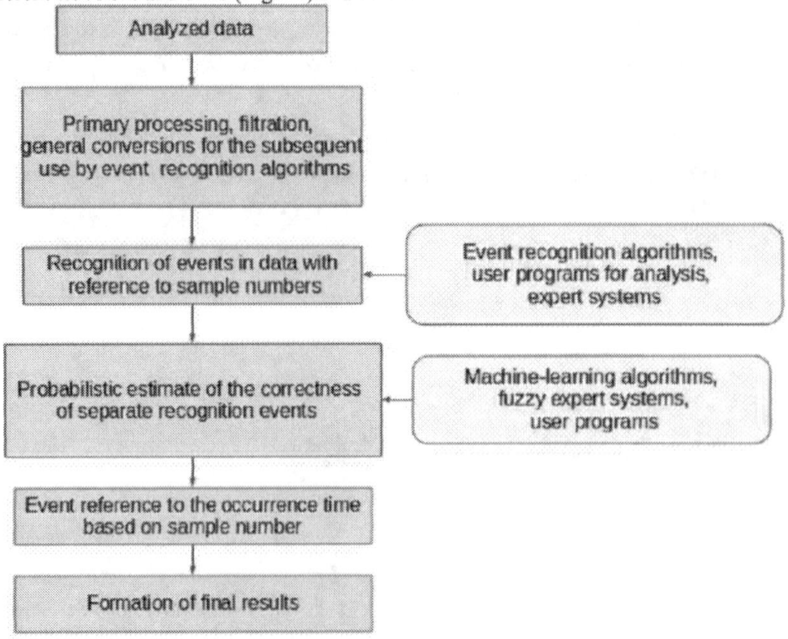

Fig. 4. Generalized functional scheme of signal stream analysis by SSAN complex F-node.

Fig. 5. Example of a spectrogram in linear scale on frequency with FFT window size of 256 and sampling rate of 44100 Hz. Whistlers, tweaks and atmospherics are seen on the spectrogram.

Fig. 6. Example of a spectrogram in logarithmic scale on frequency with FFT window size of 256 and sampling rate of 44100 Gz. Whistlers, tweaks and atmospherics are seen on the spectrogram.

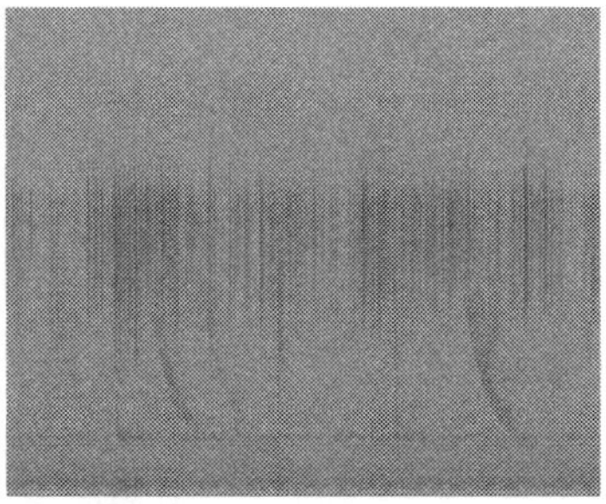

Fig.7. Example of a spectrogram in linear scale on frequency with FFT window size of 1024 and sampling rate of 44100 Hz.

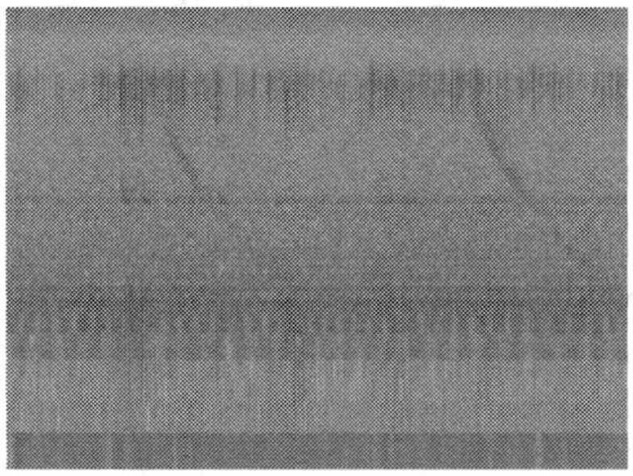

Fig.8. Example of a spectrogram in logarithmic scale on frequency with FFT window size of 1024 and sampling rate of 44100 Hz.

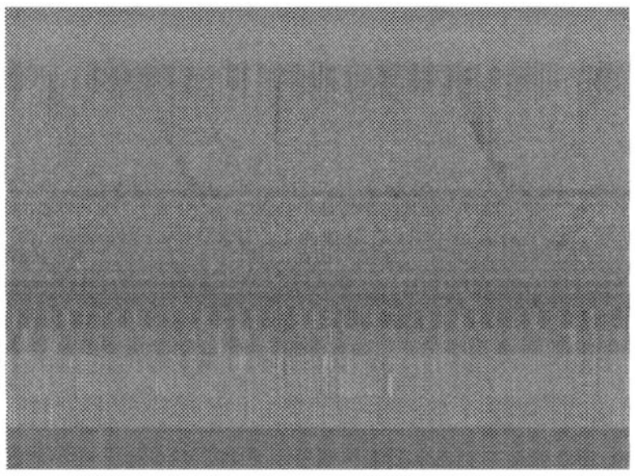

Fig.9. Example of a spectrogram filtered from strong atmospherics.

After the preliminary processing, filtration and other conversions, algorithms for event recognition (for example, atmospherics, tweaks and whistlers) are implemented. Different algorithms are applied to recognize different types of events. For example, a whistler can be recognized by a template (in the simplest case it is a line segment with different inclination angles) on a spectrogram filtered from strong atmospherics and plotted in a logarithmic scale on frequency. To recognize an atmospheric, we can apply analysis of its wave form (shape, amplitude, number of zero level intersections etc.), spectrogram analysis at different scales of FFT windows, hybrid approach of joint analysis of a wave form and a spectrogram and so on.

Fig.10. Example of continuous wavelet-transform with Morlet wavelet. Atmospherics, a long whistler and a tweak are clear in the figure.

Each algorithm of event recognition can be characterized by different criteria: operation quality (the probability of correct or false detection of an event, estimate of the number of correctly recognized events from the total number, stability against noises etc.), applied classification model and the necessity to train it, flexibility and restrictions of application, operation rate, main memory footprint and so on. It is reasonable to built the system for stream analysis of different signals in analogy with hybrid intrusion detection systems where expert system are used together with different algorithms of intrusion and attack detection. For example, having different expert rules and prepared program modules for event recognition, it becomes possible to generalize the results of operation of different algorithms, to make specialized data filtration (for example, removal of atmospherics and tweaks for better whistler recognition), to make probabilistic estimate of separate event detection correctness (applying, for example, fuzzy expert systems, machine learning algorithms) and so on. After generalization of the results of different algorithm operation, the detected events are bound to sample numbers, final results of the analysis are formed and the data are sent to the collecting centers.

Acknowledgments

The work was supported by the VarSITI Grant "Creation of a database for atmospheric and whistler events detected in the Russian Far East".

References

1. V.A. Mochalov, *Lecture Notes in Computer Science,* **9247**, 11-22 (2015).
2. V.A. Mochalov, Certificate of registration of computer software, **2016612039** (2016).
3. V. Mochalov, *Proceedings, The IEEE 17th International Conference on Advanced Communication Technology*, 34-42 (2015).
4. V. Mochalov, D. Sannikov, R.Karimov, B.Shevtsov, G. Drugin, N. Cherneva, A. Mochalova, J. Lichtenberger, V. Argunov, *Procedia Engineering*, **168**, 1721-1724 (2016).

E3S Web of Conferences **20**, 02011 (2017) DOI: 10.1051/e3sconf/20172002011
Solar-Terrestrial Relations and Physics of Earthquake Precursors

Prediction of geomagnetic indexes with the help of artificial neural networks

Irina Myagkova*, *Vladimir* Shiroky, and *Sergey* Dolenko
D.V.Skobeltsyn Institute of Nuclear Physics, M.V.Lomonosov Moscow State University, 119991 Moscow, Russia

Abstract. The results of prediction of geomagnetic indexes characterizing the state of the Earth's magnetosphere obtained with the help of artificial neural networks (ANN) for various prediction horizons are presented. The forecasts are based on multivariate time series including the values of the geomagnetic indices themselves, as well as data about the parameters of solar wind and interplanetary magnetic field, during several latest hours.

1 Introduction

The Earth's magnetosphere is one of the key space environment domains affected by solar wind (SW). Magnetic storms are one of the main manifestations of space weather. The term "space weather" is now accepted to understand the processes and phenomena occurring at the Sun, in the solar wind, magnetosphere and ionosphere of the Earth, which can influence the operation of airborne and ground-based technological systems, as well as the well-being and health of people. Strong magnetic storms can cause disturbances in the operation of telegraph lines and radio communications, pipelines, power lines and power networks [1].

Space weather effects upon the near-Earth environment are due to dynamic changes in the "Sun – solar wind – Earth" chain. Experimental studies show that geomagnetic storms have a significant impact on the near-Earth radiation environment, because after them the flux of the relativistic electrons of Outer Radiation Belt of the Earth usually increases for an order of magnitude or more (e.g. [2,3] and references therein). Extremely intense flux of electrons can disrupt operation of the electronic chips of equipment located on board spacecraft (e.g., [4]).

The main sources of disturbances of the Earth's magnetosphere are the coronal mass ejections, reaching the Earth's orbit, and high-speed streams of SW. A necessary (and perhaps a sufficient) condition for the emergence of magnetic storms is the presence of the south (negative) z-component of the interplanetary magnetic field (IMF Bz), which enables the transfer of energy from the SW into the Earth's magnetosphere [5]. So, for short-term prediction it is especially important to have operative information about the values of SW and IMF parameters.

The level of geomagnetic disturbances is usually estimated using geomagnetic indices - Dst, Kp, Ap, AE-AL (e.g. [1]).The Dst variation (Disturbance Storm-time variation) is an

* Corresponding author: irina@srd.sinp.msu.ru

index of magnetic activity derived from the network of near-equatorial geomagnetic observatories that measures the intensity of the globally symmetrical equatorial electrojet (the "ring current"). It has been calculated at the World Data Center WDC-C2 at Kyoto, Japan (Geomagnetic Equatorial Dst index Home Page, http://wdc.kugi.kyoto-u.ac.jp/dstdir/index.html) since the International Geophysical Year, 1957, using data from four observatories at low to mid-latitudes; its hourly values are available online. The planetary 3-hour-range index Kp is the mean standardized K-index from 13 geomagnetic observatories between 44 degrees and 60 degrees northern and southern geomagnetic latitude. The scale is 0 to 9 expressed in thirds of a unit, e.g. 5- is 4 2/3, 5 is 5 and 5+ is 5 1/3. The 3-hourly Ap (equivalent range) index is derived from the Kp index. The global index Kp and derivative of the equivalent amplitude Ap are used to estimate the geomagnetic activity in the auroral zone as a whole.

There are several different methods used to predict Dst index. For example, Space Research Institute from Russia provides an advance warning about the geomagnetic storm magnitude: real-time predictions of the geomagnetic storm magnitude are updated every hour and are published at http://spaceweather.ru [6]. The project of University of California, Berkeley (http://sprg.ssl.berkeley.edu/dst_index/welcome.html) produces a prediction of Dst index one hour ahead using data from ACE spacecraft, based on the modification of the empirical formula of Burton [7], using the same method as the above-mentioned studies in the Space Research Institute. The Swedish Space Weather Centre (http://src.irf.se/en/forecasts/) predicts the next hourly value of the Dst index – one hour forward in relation to the last entered data – using the recurrent Elman neural network. The project WINDMI Real-Time Dst and AL indices provides the Dst prediction, which is performed using a physical model based on the calculation of ring currents in the magnetosphere-ionosphere system, and which also uses as input the data from the ACE spacecraft [8]. A modern model used to predict the Dst index one hour ahead proposed in [9] is based on artificial neural networks (ANN) combined with an analytical model of the SW– Earth's magnetosphere interaction.

The authors of this paper have their own experience of the Dst index prediction using scientific models based on ANN technology [10-12], as well as experience of using adaptive (including ANN) methods for prediction of electrons flux in the outer radiation belt of the Earth [13]. It was shown that the best quality of the Dst-index forecast is achieved when constructing a neural network model that uses both the history of the Dst index and the parameters of the solar wind (its speed) and IMF (Bz component) as input data [10]. The website of the Space Weather Analysis Center at SINP MSU provides online prediction of Dst index 0.5-1.5 hours ahead by the parameters of SW and IMF measured by the ACE spacecraft using ANN (http://swx.sinp.msu.ru/models/dst.php).

However, one of the most urgent tasks of forecasting is forecast not for hours, but for a longer period, at least for the day ahead. One of the possible options is the transition from the forecast of the hourly values of Kp and Dst to the forecast of daily average and maximum values for the day.

The present paper is devoted to the comparison of the results of predicting the geomagnetic indices Dst, Kp and Ap with different horizons of the forecast among each other and with trivial models.

2 Data Sources and Preparation

The input data usually used for Dst index prediction are the parameters of SW plasma and of IMF, measured at the Earth's orbit by available spacecraft, and the values of Dst index itself. The data used in this study were obtained onboard ACE (Advanced Composition Explorer) spacecraft (http://www.srl.caltech.edu/ACE).

Time series (TS) of hourly values of the following physical quantities were used as input for the neural networks:

a) SW parameters in Lagrange point L1 between the Earth and the Sun: SW speed v (measured in km/s) and SW protons density nP (measured in cm^{-3}).

b) IMF vector parameters in the same Lagrange point L1 (measured in nT): Bx, By, Bz (IMF components in GSM system), and B magnitude (IMF modulus).

c) Equatorial geomagnetic index Dst (measured in nT).

The specific sources of data are listed elsewhere [12].

To account for possible daily and yearly variations of the Dst index, TS of sine and cosine values with daily and yearly period were also used as input data. Delay embedding of all TS for 24 hour depth was used to account for the previous history of input features.

In the forecasting, the SW and IMF data from the ACE (Advanced Composition Explorer), located at the Lagrange L1 point obtained by the SWEPAM and MAG instruments, were used http://www.srl.caltech.edu/ACE/ASC/. We used not pre-processed and cleaned-up data of the second level, intended for scientific research, but operational data. This is due to the fact that the forecasting system being developed is designed to work in an online mode, in which the quality of the received data corresponds to the operational data.

Due to the nature of the data transmitted from the spacecraft, there are a significant number of gaps in them. Delay embedding significantly increases the negative influence of data gaps on data volume. Therefore, in this study we used filling of gaps 12 or less hours long by linear interpolation (for time moments when the gap is already over) or by extrapolation of the latest known value. Such method of gap filling is based on the fact that rapid changes are not typical for virtually all kinds of data used. Also, there is no reason for use of more sophisticated estimates.

As the working array, the data since October 22, 1997 (the beginning of data receipt from ACE) until the end of 2016 was used. After delay embedding, the working array was divided into training, validation and examination datasets. The training set was used to adjust the weights during ANN training, the validation set for periodic validation in the process of training in order to prevent overtraining, the examination set for independent evaluation of the results. The data from November 1997 till the end of 2009 was used for the training and validation sets, to which it was randomly split at the ratio of 75% to 25%. The data for 2010-2016 was used as the examination set.

3 Architecture and parameters of ANN

The ANN architecture used in the present study was the multi-layer perceptron (MLP). For all neural network models the results of which are presented in this paper, a multilayer perceptron with a single hidden layer of 16 neurons was used. Previous studies have shown that using more layers and/or neurons does not lead to better predictive results. Logistic transfer function was used for all neurons. ANN was trained with the error backpropagation algorithm with learning rate of 0.01 and moment of 0.5. Training was terminated when during 500 epochs the result of network operation on the validation set did not improve.

For each model, 5 networks with various sets of initial weights were trained, their answers were averaged, and the assessments of the quality of the model were considered for the averaged result. This approach means using the simplest committee of homogeneous predictive models.

To construct neural network models for daily values, several architectures of neural networks were used in this study, including the multilayer perceptron (MLP) and the general regression neural networks (GRNN). Statistical indexes of the quality of the constructed models for different architectures and learning algorithms turned out to be approximately the same. For predictions of the daily average values, the results obtained using NeuroShell Predictor software from the American company Ward Systems Group, Inc. [14] will be presented below. This package uses the original amorphous neural network architecture TurboProp2 based on the well-known cascade correlation algorithm [15], which adaptively selects the optimal complexity and optimal structure of connections among neurons in the process of learning. This architecture is characterized by extremely fast learning compared to other architectures, with a comparable quality of results.

All the parameters of the neural networks during the computational experiments were fixed, as well as the separation of data patterns between the training sample (from November 1997 to the end of 2009) and the examination set (independent data, from start of 2010 to the end of 2016), which was used to compare the quality of the obtained neural network models. Only the sets of variables the values of which were used as input data were changed, which permits us to state that the observed effects are related exclusively to which physical variables and with what time delays affect the formation of the daily average value of the geomagnetic indexes.

4 Results and discussion

To assess and compare the quality of models, the following statistical indexes characterizing the deviation of the prediction from the real values of the predicted quantities were used in this study: the multiple determination coefficient R^2 (R squared), the root mean squared error (RMSE), the correlation coefficient(r).

Fig. 1 displays the values of multiple determination coefficient R^2 and correlation coefficient r as functions of prediction horizon (from 1 to 12 hours) for the average of predictions of 5 ANN with various sets of initial weights, and for the trivial models (prediction=latest value), for Dst and Kp indexes. For Kp index, its integer 10-fold multiplied value Kp·10 was used as is often done for the convenience of writing fractional values of the index.

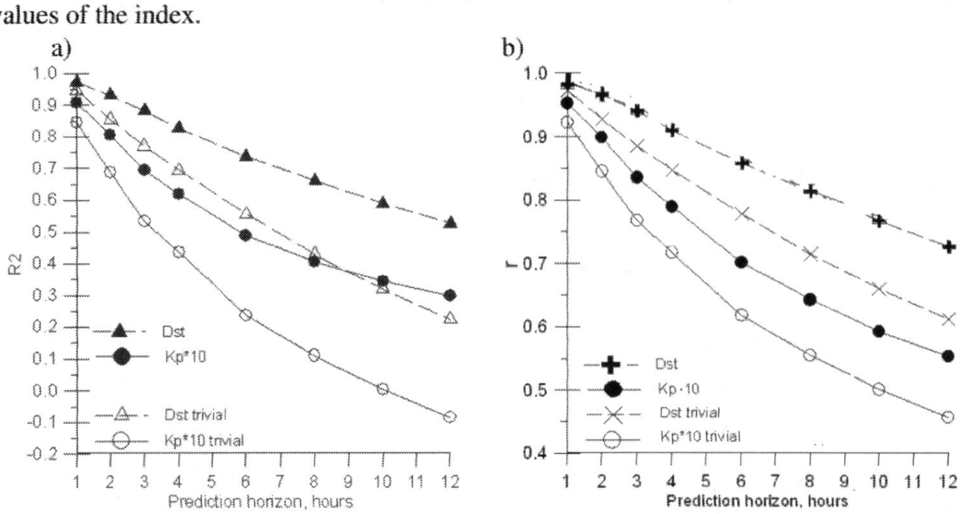

Fig. 1. Multiple determination coefficient R^2 and correlation coefficient (r) for Dst and Kp indexes vs prediction horizon in hours, for ANN and trivial models

Fig. 2 displays the values of root mean squared error (RMSE) for Dst and Kp indexes as a function of prediction horizon for the average of predictions of 5 ANN with various sets of initial weights, and for the trivial models.

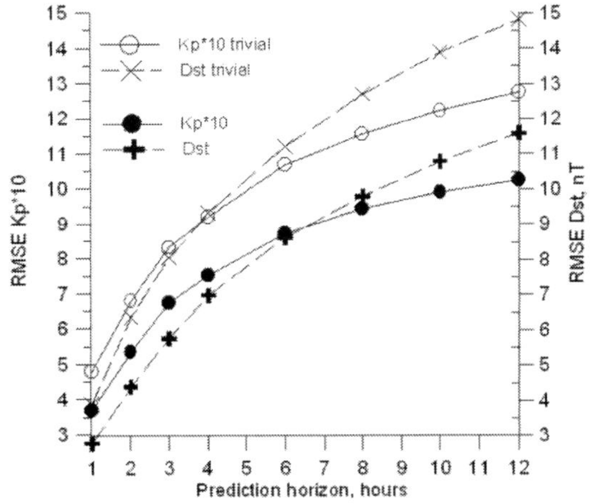

Fig. 2. Root mean squared error (RMSE) for Dst and Kp vs prediction horizon in hours, for ANN and trivial models.

Figures 1 and 2 show that the quality of ANN prediction, as expected, decreases with increasing prediction horizon for both geomagnetic indexes. It can be seen that sharper decrease is observed for Kp. If we compare the results obtained for geomagnetic coefficients for different horizons with the same dependence for the prediction of the electron fluxes of the outer Earth's Radiation belt (ERB) [13], we see that the quality of the electron flux prediction decreases with prediction horizon increasing from 1 to 12 hours slower than that for the predictions of Dst and Kp. This fact allows suggesting that the characteristic time for the development of processes in the Earth's magnetosphere that affect the electron flux of ERB is much larger than the characteristic time of the processes of development of a geomagnetic disturbance.

To confirm this hypothesis, one more goal of this paper was to obtain predictions for one day ahead for the daily average values of Kp and Ap indices, as well as for the daily maximum of the Dst amplitude (since the average daily value of the Dst-index has no physical meaning).

Table 1 displays the values of the same statistical indexes as were shown in Fig. 1,2 - the coefficient of multiple determination R^2, the root-mean-square error RMSE and the correlation coefficient r, obtained for the predictions of daily average geomagnetic indices Kp·10 and Ap, and of the Dst index daily maximum amplitude one day ahead, obtained with the help of the ANN model and the trivial model, which repeats the value measured for the previous day ("tomorrow as today"). From Table 1 it can be seen that best predicted is daily average Kp index, and worst of all, the daily average Ap index. The ANN models give better results than the trivial models. This is especially clear for R^2.

Table 1. *Statistical indexes for prediction of daily average values of Kp·10 and Ap and of daily maximum Dst amplitude one day ahead obtained with ANN and trivial models*

	R^2	RMSE	r	R^2	RMSE	r
	ANN model			Trivial model		
Kp	0.52	7.9	0.72	0.15	9.4	0.57
Dst	0.47	4.5	0.68	0.16	18.9	0.58
Ap	0.28	7.5	0.54	-0.09	9.2	0.46

 DOI: 10.1051/e3sconf/20172002011

Since the daily average Kp-index is a generally accepted index of the daily state of the Earth's magnetosphere, further attempts to improve the prediction have been made for it. Figures 3 and 4 show examples of the time series of the daily average Kp index (multiplied by 10), measured and predicted one day ahead by two different ANNs, from April 30 to September 2, 2015, and from May 9 to July 31, 2016, respectively.

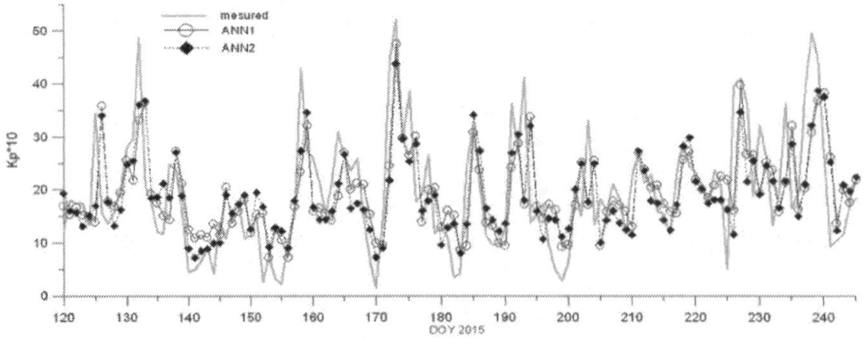

Fig. 3. An example of the predictions of the time series of daily average Kp-index one day ahead from April 30 to September 2, 2015 and their comparison with the measured values.

To construct the neural network model ANN1, the following 82 input features were used: sine and cosine values with one year period, current and prior values for the preceding 7 days ("rectangular" embedding of the multivariate time series) for the following quantities: daily minimum Dst index value, daily average and daily maximum Kp index values, daily average Ap index value, daily maximum and daily average values of the modulus of the IMF vector |B|, daily maximum absolute value of the IMF Bz component in the GSM system, daily average and daily maximum values of SW velocity v, and daily maximum SW density n_p. For the ANN2 model, the data was extended with the values of the same quantities, obtained at the previous turn of the Sun, 25, 26, and 27 days ago (to take into account the recurrence of solar wind fluxes from coronal holes).

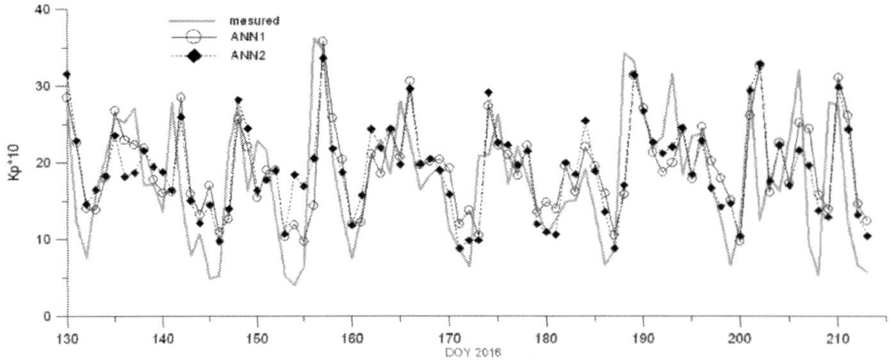

Fig. 4. An example of the predictions of the time series of daily average Kp-index one day ahead from May 9 to July 31, 2016 and their comparison with the measured values.

Figures 3 and 4 show that taking into account the data from the preceding rotation of the Sun brought some improvement, although it was not very significant. Further attempts to improve the prediction by adding data on the sunspot number (with the help of this it was planned to indirectly take into account the influence of the coronal mass ejections) and the ULF index of the IMF, did not lead to any noticeable extra improvement. The comparison of Figures 3 and 4 also shows that there is no clear dependence of the prediction quality on the phase of the solar activity cycle for the periods under study.

5 Conclusion

The results of prediction of geomagnetic indexes Dst and Kp obtained with the help of artificial neural networks (ANN) for various prediction horizons are presented. Forecasting of the hourly values of Dst and Kp as well as of the average daily values of these indices is based on the history of the values of the indices themselves and on the data on the parameters of the solar wind and the interplanetary magnetic field and their pre-history. It is shown that with the increase of the forecasting horizon, the quality of prediction decreases quite rapidly, especially for Kp index. This fact can be explained by the characteristic time of development of processes in the Earth's magnetosphere that affect the geomagnetic situation, as well as by the physical differences between the Dst and Kp indices. The direction of further research aimed at increasing the prediction horizon and improving the quality of forecasting should be taking into account the intraday hourly dynamics of the input parameters when predicting the daily average values of the geomagnetic indices.

This study was performed at the expense of the Russian Science Foundation, grant no. 16-17-00098.

References

1. L.L. Lasutin, *World and polar magnetic storms* (MSU, 2012)
2. R.H. Friedel, W.G.P. Reeves, T. Obara, J. Atmos. Solar. Terr. Phys. **64**, 265 (2002)
3. R. Kataoka, Y. Miyoshi, Ann. Geophys. **25**, 1335 (2008)
4. N. Iucci, A.E. Levitin, A.V. Belov et al., Space Weather, **3**, S01001 (2005)
5. B.T. Tsurutani, W.D. Gonzalez, F. Tang et.al., J. Geophys. Res. **A8**, 8519 (1988)
6. T.V. Podladchikova, A.A. Petrukovich, Space Weather: The Int. J. of Res.and Appl. **10**, CiteID S07001. (2012)
7. R.K. Burton, R.L. McPherron, C.T. Russell. J. Geophys. Res. **80**, 4204 (1975)
8. S. Patra, E. Spencer, W. Horton, J. Sojka. J. Geophys. Res. **116**, A02212, DOI:10.1029/2010JA015824. (2011)
9. M. Revallo, F. Valach, P. Hejda, J. Bochníček, J. of Atm. and Sol. Terr. Phys. **110**, 9 (2014)
10. S.A. Dolenko, Yu.V. Orlov, I.G. Persiantsev, Ju.S. Shugai, Lect. Notes Comp. Sci. **3697**, 532 (2005)
11. S.A. Dolenko, I.N. Myagkova, V.R. Shiroky, I.G. Persiantsev. Proc.10th Intl. Conf. "Problems of Geocosmos", 270 (2014)
12. I. Myagkova, S. Dolenko, V. Shiroky, N. Sentemova, I. Persiantsev, ACM Proceedings, 2015, article No.9, DOI: 10.1145/2797143.2797169 (2015)
13. A.O. Efitorov, I.N. Myagkova, N.S. Sentemova, V.R. Shiroky, S.A. Dolenko. Adv. Intel. Syst. Comp. **449**, 281 (2016).
14. NeuroShell Predictor – Advanced Neural Network and Genetic Algorithm Software. http://www.wardsystems.com/predictor.asp
15. S.E. Fahlman, Ch. Lebierre. Res. Rep. CMU-CS-90-100. School of Comp. Sci., Carnegie Melon University, Pittsburgh, USA. August 29, 1991

E3S Web of Conferences **20**, 02012 (2017) DOI: 10.1051/e3sconf/20172002012
Solar-Terrestrial Relations and Physics of Earthquake Precursors

Informational approach to the analysis of acoustic signals

Yuriy Senkevich[1,*], *Vyacheslav* Dyuk[2] , *Mikhail* Mishchenko[1], and *Alexandra* Solodchuk[1]

[1] Institute of Cosmophysical Research and Radio Wave Propagation FEB RAS, Russia, 684034, Kamchatka region, Elizovskiy district, Paratunka, Mirnaya str., 7
[2] St. Petersburg Institute for Informatics and Automation of RAS, Russia, 199178, St. Petersburg, 14 Line, 39

Abstract. The example of linguistic processing of acoustic signals of a seismic event would be an information approach to the processing of non-stationary signals. The method for converting an acoustic signal into an information message is described by identifying repetitive self-similar patterns. The definitions of the event selection indicators in the symbolic recording of the acoustic signal are given. The results of processing an acoustic signal by a computer program realizing the processing of linguistic data are shown. Advantages and disadvantages of using software algorithms are indicated.

1 Introduction

A method of signal processing that uses procedures of decomposition in non-orthogonal basis group of information processes' self-similarity mechanisms in the context of acoustic signals in cosmophysics is proposed. Scientific interest is caused by similarity of models of synergetics nonequilibrium systems with catastrophic events that occur in lithosphere and atmosphere. We expect to see this similarity in properties of signals caused by those events.

There is a feature of acoustic signals due to unsteadiness properties, notable episodic occurrence of energy anomalies, and frequent nonlinearity manifestation, which is generated by the dynamics of the nature of the source of such signals. Methods of preliminary processing that are traditionally applied to the signals, such as factorization, statistical accumulation and approximation methods often appear to be incorrect conversion operations or require strict restrictions on the dynamics of the frequency and time behaviour of the signal. During measurements caused by the seismic activity of acoustic signals, decomposition techniques in known basis functions are frequently used. In this case, a significant variability of the optimal set of decomposition elements from one measurement episode to another is detected due to previously named features of processed signals. In such cases, the incorrect application of spectral analysis often leads to erroneous conclusions [1]. In addition, the classical methods of isolating a useful time-localized seismic signal operate at a signal-to-noise ratio exceeding background noise several times. However, during the most part of the observation time it is impossible to detect weak signals that are harbingers

* Corresponding author: senkevich@ikir.ru

of seismic events. As a rule, physical events in the geosphere, accompanying the accumulation of critical values of the parameters prior to the moment of a catastrophic event, have a low amplitude, frequently due to the significant diversity in the space of the source and receiver of the signal and the conditions of its propagation. Comparable in intensity with background, desired signal that has similar probabilistic characteristics with a masking background does not allow the effective use of classical spectral and correlation methods as well as popular non-orthogonal decompositions and neural networks. Nevertheless, a mixture of useful signal and background obviously carries in itself useful information. It should be expected that the nature of this useful signal and background should depend on the state of the environment, where seismic processes constantly change its isotropic and homogeneous properties. This affects the conditions for the propagation of acoustic waves, which is reflected in a variety of ways on the characteristics of the signal itself and is expressed in the form of additive-multiplicative phase-frequency distortions. Due to the local distribution of the inhomogeneous radiation boundaries of the signal source in space, there are also additional phase delays with respect to the point receiver. In such conditions, to obtain useful information, the task of detecting and recognizing seismic events is better to be considered not as a task of restoring (isolating) the original signal or isolating it from the background. But as a task of estimating the extent of the environment-induced change in the received message at various consecutive measuring moments including both a distorted "useful" signal, and a change in the background obtained at the point of reception.

2 Method of investigations

To decipher the received signal mixture plus background, we propose a methodology based on the principles of synergetics that allow considering received signal as a function of reflecting the reaction of the system in the process of its probabilistic transitions in the network of possible states of this system. In this case, to assess the current state of the system, it is advisable to apply an information approach to data analysis. In the considered practical case, the analysis of the received acoustic signals is depending on whether it will be changed in a certain time window. A search is carried out for some information units characterizing the signal, as well as the identification of characteristic consecutive episodes of amplitude-phase variability of the signal (patterns). The degree of changes in signal structure can be estimated by cyclic calculation of the increment of information entropy values based on the extracted patterns.

The stated representations are consistent with the well-known approach to describing the behaviour of a certain class of stochastic dynamical systems in the form of symbolic dynamics [2]. If the set of admissible states of the system is $\{Qi\}$ ($i = 1, 2, \ldots$), then the result of consecutive N steps of the process can be written in the form of a Markov chain Qi_1, $Qi_2, \ldots$ Qi_N. Such a record is equivalent to describing its evolution in the language of "letters" $\{i_N\}$ of the corresponding "alphabet". Symbolic dynamics allows considering selected sequences as a set of "messages" or "texts" containing certain information.

For the use of symbolic dynamics in the presented study, the previously developed algorithms for the linguistic analysis of physiological signals were adapted [3]. These algorithms were used to convert an acoustic signal into a symbolic form and compose the message text.

To introduce the conducted research into the plane of practical application, we describe below the developed method of converting acoustic signals into text messages. Subsequently, the results of this transformation are guided by the search for general patterns conditioned by the cause-effect relationships of the electromagnetic processes events in the atmosphere with deformation processes that generate acoustic emission in the lithosphere. Earlier such a connection was confirmed [4, 5].

The method is realized by performing two stages. During the first stage, the acoustic signals are converted into text messages, and then, based on the results of processing, event objects are compiled. The transformation is carried out by using a priori information about the objectively established facts of a seismic event manifestation fixed at a certain point in time. The initial stage of the selection of event objects is performed by processing self-similar signal structures through the developed computer program of preprocessing the accumulated records of the acoustic signals of the IKIR archive. Preprocessing includes standard converting of files to an accessible data format, searching for characteristic signal patterns, removing redundant patterns, converting a signal into a message, highlighting an event alphabet, forming a linguistic event object.

The algorithm for searching for signal patterns is constructed from the well-known position of synergetics that near stable states the system describes self-similar trajectories that are expected to correspond to similar fragments of the reflecting behaviour of the signal system. For processing from the whole signal, we choose only local extremes, each of which can be described by two parameters: the amplitude and the time interval that has elapsed since the previous local extremum. Here we should agree that this choice of the signal description saves the most significant information about its amplitude-phase characteristic. Let some fragment of the signal with *N* extremes be presented by a set, consisting of pairs of numbers $x_1,\tau_1 \quad x_2,\tau_2 \quad ... \quad x_N,\tau_N$, where x_i is the value of the extremum amplitude and τ_i is the time shift with respect to the preceding extremum (τ_1=0). We calculate the ratios for each of the extremes as follows:

$$r_{i,i+m} = \begin{cases} 1, & x_i > x_{i+m} \\ 0, & x_i \le x_{i+m} \end{cases} \qquad \omega_{i,i+m} = \begin{cases} 1, & \tau_i > \tau_{i+m} \\ 0, & \tau_i \le \tau_{i+m} \end{cases} \qquad m = 1...M(M \le N-i) \tag{1}$$

where $r_{i,i+m}$ is logical comparison result of *i*-th and *i*+*m*-th values of the extremum amplitudes; $\omega_{i,i+m}$ is logical comparison result of *i*-th and *i*+*m*-th values of intervals between extremes. We order the series of such ratios in the form of square matrices for *M* ratios of amplitudes of extremes and time shifts between extremes. The resulting matrices will have a diagonal symmetry because of the algebraic symmetry property of the inequalities ($a > b$ implies $b > a$) and in this sense are redundant. Therefore, we will use only half of each of the constructed matrices, which we combine into one. Bearing in mind that, in accordance with rule (1), values on the main diagonal of each of the original matrices are zero, we obtain

$$D_i = \begin{pmatrix} 0 & \omega_{i,i+1} & \cdots & \omega_{i,i+(M-3)} & \omega_{i,(M-2)} \\ r_{i+1,i} & 0 & & & \omega_{i+1,i+(M-2)} \\ \vdots & & \ddots & & \vdots \\ r_{i+(M-3),i} & & & 0 & \omega_{i+(M-3),i+(M-2)} \\ r_{i+(M-2),i} & r_{i+(M-2),i+1} & \cdots & r_{i+(M-2),i+(M-3)} & 0 \end{pmatrix} \tag{2}$$

Matrix (2) represents a specific code for the selected (*i*-th) extremum in the signal. This code characterizes extremum's amplitude and time position with respect to neighboring extremum to a depth M to the right. As a consequence of the application of the rules of the relation (1), the transformation of the signal fragment from the *M* extremes into the matrix (2) has the property of invariance to displacement operations, as well as to amplitude and temporal transpositions of the original signal. This important property of the matrix follows from the basic property of inequalities:

$a > b \Rightarrow a + c > b + c, \forall c$ for the time shift operation of the signal, and

$a > b \text{ and } c > 0 \Rightarrow ac > bc$ for compression and stretching operations.

As a result, each of the resultant matrix (2) can be associated graphically with some invariant waveform (pattern).

When searching for patterns in a signal, the following rule is applied: if for two selected non-coincident (intersection of possible) sequences of local extremes the corresponding matrices of the form (2) of the same order M coincide, then the hypothesis is assumed that such a matrix describes a stable form invariant in the signal. The matrix itself can be interpreted as a definite sign – *symbol*. A more stringent constraint can be applied, given the need for more coinciding matrices for three or more episodes of the signal. We call a given restriction K the statistical threshold for the existence of a symbol in a signal, below which all the occurring combinations should be considered as insignificant or null-symbols (Z).

Figure 1 shows an example of the conversion of a signal fragment. With the rules described in this fragment of the alarm message, two symbols are allocated that have been assigned a conditional names a and b with dimensions 5 and 7, respectively. Below are the invariants of the symbols (templates) ordered in the amplitude of the extremum – II and the magnitude of the interval between the extremes – III.

The sequential application of the above algorithm for allocating patterns to each consecutive local extremum allows us to select a certain set of symbols – *alphabet*. Let's call *the size of the alphabet* $D = |A|$ the total number of alphabet symbols found in the message. For example, the alphabet of the alarm message shown in the figure 1 is $A = \{a, b\}$, $D = 2$.

The resulting alphabet allows you to convert the signal into symbolic form. Such a symbolic record will be referred to hereinafter as a message. For a symbol, the value equal to the dimension of the corresponding matrix M, will be called the *dimension of the symbol*. In figure 1, the message in the selected symbols of the alphabet A looks like "abba".

The next stage of data processing is the allocation of collected alphabets to groups of events of a certain class and the formation of event objects on their basis. The task is solved by methods of linguistic parsing of sequences of objects of events [6]. For this, a number of evaluation parameters are introduced.

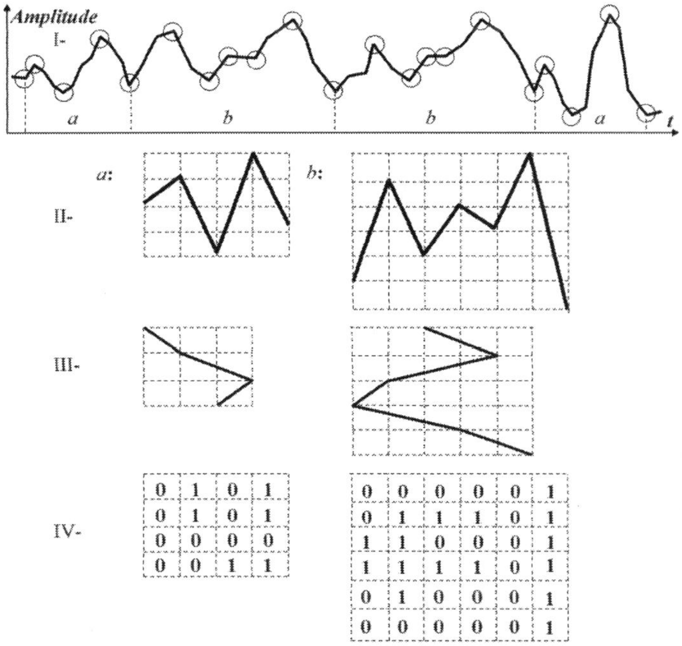

Fig. 1. Example of forming a matrix of relations for two patterns (a) and (b).

The frequency of the appearance of a symbol in the message is different and indicates the most preferable trajectory behaviour of the dynamic system. To estimate the frequency of the message, we will use the *probability of occurrence of the symbol a_i of the alphabet A*, which is equal to the ratio of the number of occurrences of the symbol in the message m to the total number of symbols in this message *N*.

$$P(a_i) = \frac{m}{N} \quad (3)$$

From the point of view of information theory, indicator (3) is also an estimate of the information load per symbol of the alphabet. In accordance with the recommendations of the authors of symbolic dynamics [1], one should try to find connections at the level of the Markov chain of high orders, calculating the conditional probabilities of occurrence of combinations of symbols: $P(b \mid a) = P(ab) / P(a)$; $P(c \mid ab) = P(abc) / P(ab)$ ... , where a, b, c are symbols of the alphabet $A = \{a, b, c\ldots\}$ isolated from the signal. If we accept that in a certain stable state the set of events, characterizing the state of the system does not change over a long period, then it is logical to assume that the composition of the alphabet *A* of the perceived message, reflecting this set of events, and its dimension *N*, should also not vary. Change is allowed in the order of appearance of each of the message symbols or the order of the sequences of symbols indirectly reflecting some fluctuations in the state of the system. In other words, a hypothesis is put forward that within each stable state the system is able to generate only a fixed set of alphabet symbols of a certain dimension. If there is a change in the state of the system, this should entail a change in the set of events, which in turn should cause qualitative and quantitative changes in the alphabet composition that can be detected using elements of the theory of pattern recognition. Following the recommendations of the theory of pattern recognition [7], when developing the analysis apparatus, it is necessary to provide for the possibility of learning based on the accumulation of standards, which are some fixed sets of parameters for subsequent comparisons with the current set of parameters. In the case under consideration, these will be sets of parameters, consisting of alphabets, information on the probabilities of the appearance of symbols of different dimensions, which will be stored in the memory block of the linguistic signal analysis circuit.

The next parameter for evaluating the state of the system is the *alphabet size N*, which is the total number of alphabet symbols found in the message. Set $A = \{a_i\}$ and its dimension *N* for the model of the dynamic system can be linked with its stable state, when the trajectory moves near one or several nearby attractors, without going beyond the boundaries of their action. This state can be represented, for example, by a certain period of relaxation after a catastrophic event. Violation of stability in the system under the influence of external forces or currents can cause the system to transition to another stable state, in which the system, undergoing a qualitative transformation, will generate symbols of the alphabet, characteristic of another state. The manifestation of the variability of the system state can change the statistics of the appearance of the symbols of the original alphabet, and can also supplement it with new symbols, and/or lose some of the previous composition of the symbols. This is a prerequisite for the conclusion about the transitional state of the system at the semantic level.

3 Results of investigation and discussion

To demonstrate the operation of the method, we selected a file of acoustic observations from the IKIR archive, obtained from a vector receiver installed on Lake Mikijah, from Nord direction with duration of 15 minutes, on which a seismic event is recorded in the form of a visually observable sequence of pulses on February 5. The digitization rate of the signal was

48000Hz. Three fragments in duration 120 sec 2017 (figure 2 at the top) were cut out of file, and for comparison, the signal spectra corresponding to the fragments are presented (figure 2 at the bottom). The first fragment covers the interval immediately before the appearance of a visually identifiable series of pulses associated with a seismic event (A). The second fragment covers the observed impulse bursts of the signal (B). The third fragment represents the interval after 5 minutes after the marked seismic activity (C).

The figure shows an increase in the spectral power in the frequency range from 3 to 8 kHz with a rise in the maximum value at a frequency of approximately 4.5 kHz from -78 to -68 dB. The spectra before and after the seismic event practically differ neither in energy nor in the contour of the spectrum pattern.

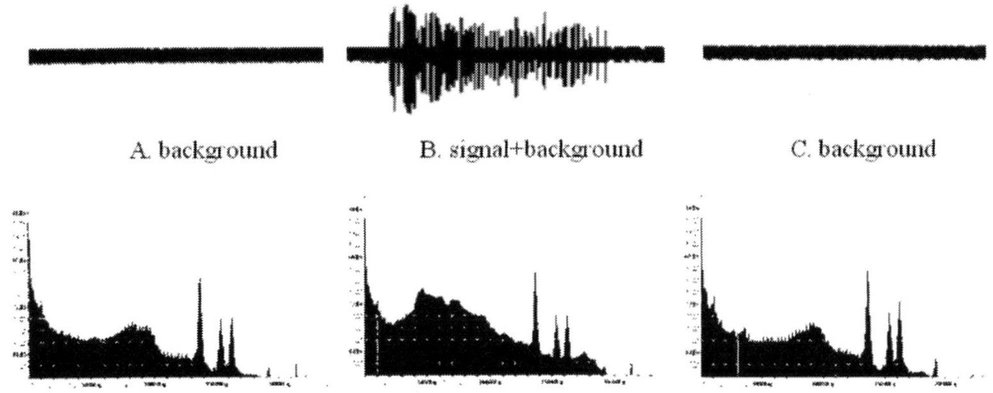

Fig. 2. Fragments of the recording of the channel N of the vector acoustic receiver, selected for linguistic processing at the top, and the corresponding Fourier spectra.

The linguistic processing of the presented signal fragments was carried out using the author's program of entropy-syntactic analysis of electrophysiological signals (ESAES ver. 3.0) [8]. The main window of the program with the result of processing the signal fragment is presented below (figure 3).

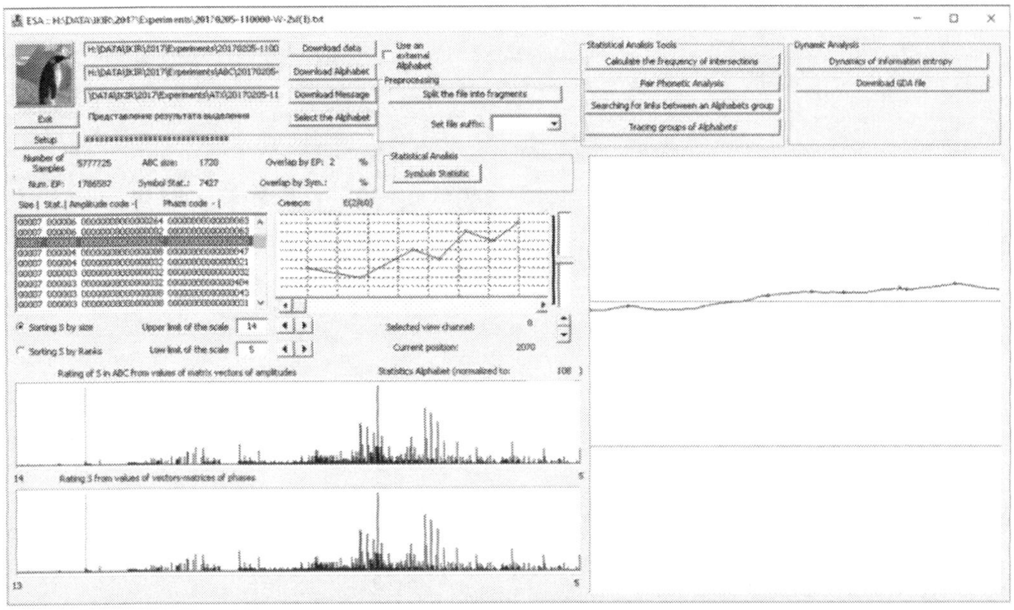

Fig. 3. Results of linguistic processing of a fragment of an acoustic background.

The program performs the decomposition of the signal into self-similar patterns, eliminates redundancy, calculates the statistics of the isolated symbols and compiles the alphabet. Additionally, it displays the ranking of the statistical values of the frequency of the alphabet symbols on the numerical axis of the symbol dimension values, which allows us to estimate indirectly the degree of randomness of the signal and to consider the most probable phase trajectories in the form of symbol templates of the selected alphabet.

The results of linguistic processing are tabulated.

Table 1. The results of the first stage of linguistic processing of the signal fragments presented in figure 2

	Fragment A (background 1)	Fragment B (signal+ background)	Fragment C (background 2)
Number of allocated local extremes	1083837	1073489	1085573
Dimension of Alphabets	1063	1092	1051
Sum of statistics	4358	4629	4537
Rank distribution of symbols of alphabets on values of symbols dimensions of	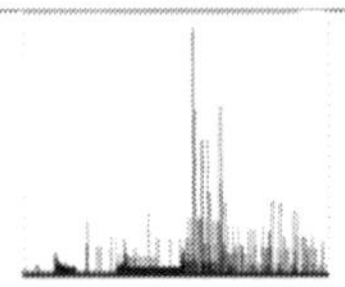	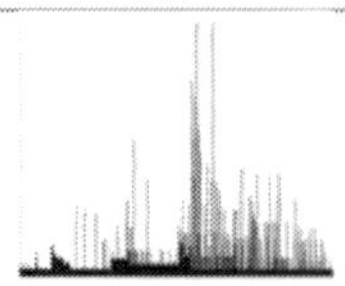	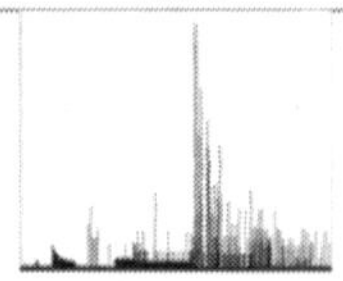

For a better understanding of the self-similar structure of the signal, in figure 4 are derived templates (patterns) of isolated alphabet characters of different dimensions.

Size	12	10	9
Pattern	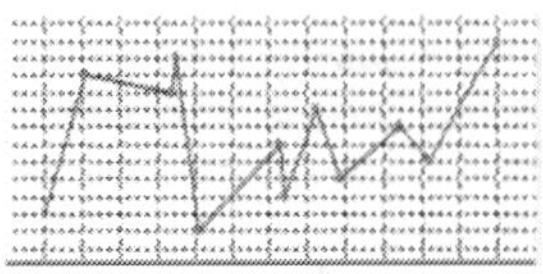	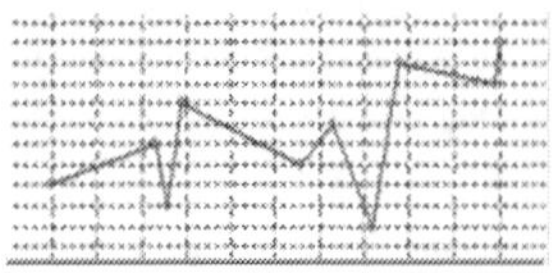	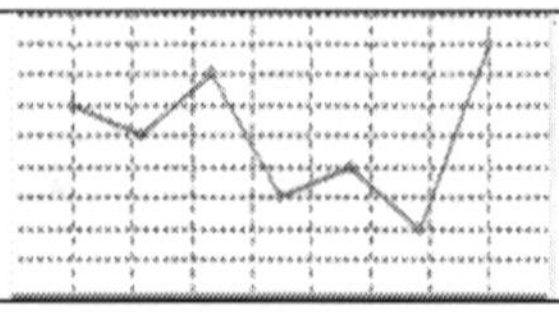
Code	0000000000000016 0000000000000036	0000000000000064 0000000000000256	0000000000000032 0000000000000191

Fig. 4. Representation of indicators of linguistic processing used for analysis.

For carrying out comparative linguistic analysis, a utility is built into the computer program. With its help, operations on sets of symbols of compared alphabets are carried out **A (background 1), B (signal+background) и C (background 2)**:

Table 2. Results of a comparative analysis of the signal messages alphabets presented in figure 2.

Operation	Formalization	Overlapping of sets in %	Illustration
Intersection A&C :	$\mathbf{C_1 = A \cap C}$	50% / 49%	Figure 5
Intersection B&C :	$\mathbf{C_2 = B \cap C}$	24% / 24%	Figure 6
Intersection B&A :	$\mathbf{C_3 = B \cap A}$	25% / 25%	Figure 7

It can be seen that the alphabets of "background 1" and "background 2" are much more repeated over the elements of the set (figure 5) than the comparative estimates for the sets of symbols of the alphabets "signal+background" – "background 2" (figure 6) and "signal+background" – "background 1"- (figure 7).

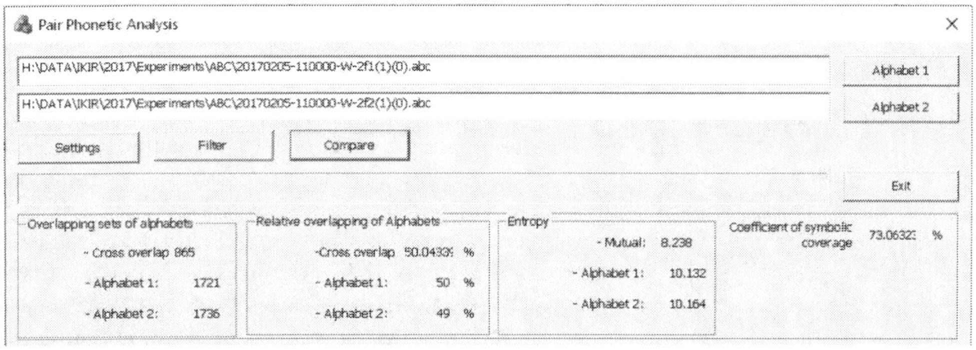

Fig. 5. Comparative analysis of the composition of the alphabets of background 1 and background 2.

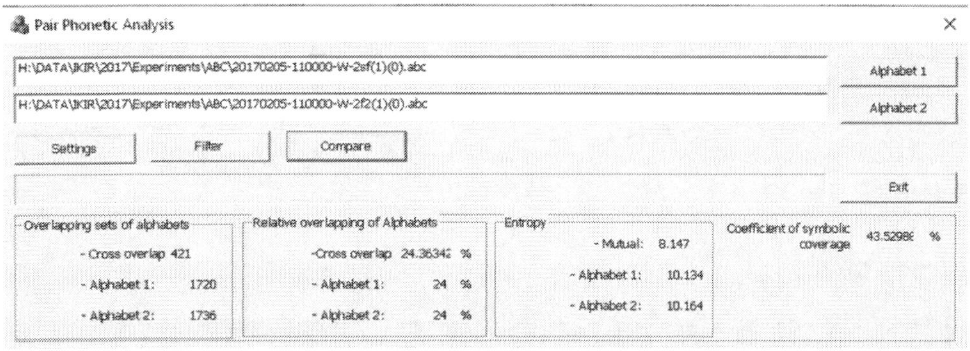

Fig. 6. Comparative analysis of the composition of the alphabets signal+background and background2.

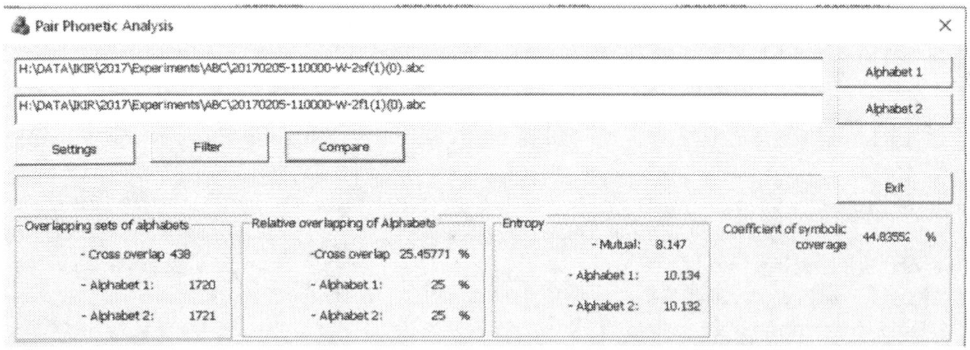

Fig. 7. Comparative analysis of the composition of the alphabets signal+background and background 1.

The presented statement of the foundations of the method, based on the information approach to the solution of the problem, corresponds to the idea of searching for the semantics of the detected structural changes in the signal, which is laid down by the phenomenon at the time of generation of this signal during a seismic event. The structure of the signal converted into a message is a reflection of the behavior of a nonlinear dynamic system, what a seismic event at least is, and the structure of the message is formed under the influence of the internal structure of a nonlinear dynamical system. Because of this, there is a way to recognize these influences and build models of system behavior. The last assumption allows us to consider cosmophysical processes from the standpoint of information theory, that is, to assess in detail the dynamics of the behavior of such systems, and, therefore, to expect a more successful solution of the problems of recognizing their current state.

4 Conclusions

The presented example of the initial stage of linguistic processing is intended to show the principal differences between the information approach and the classical methods of signal preprocessing. Self-similar patterns are not a consequence of the application of functional decomposition. These structural elements preserve the amplitude-phase relationships of repeatedly occurring episodes of the original signal without imposing any time or frequency limitations. Moreover, because of the design features of the patterning based on the ratios of the amplitudes of local extremes and the time intervals between these extremes associated with them, patterns have the invariance property, and therefore, when looking for similarity, do not require scaling, in contrast to, for example, wavelet transform. In the algorithm for searching and extracting such signal structures, the elements of self-organization of the formation of alphabets are included, the elements of which do not have a functional connection, which allows us to describe the signal without estimating its stationarity.

Of course, the search for such patterns requires enormous computational costs. Therefore, the implementation of the approach itself becomes possible only with the availability of powerful computing tools, which is an essential drawback of the algorithm of linguistic processing used here.

References

1. Fink A.M. *Signaly, pomekhi, oshibki...Zametki o nekotorykh neozhidannostyakh, paradoksakh i zabluzhdeniyakh v teorii svyazi [Signals, noises, errors ... Notes on some surprises, paradoxes and misconceptions in the theory of communication]* (Radio i svyaz, Moscow, 1984) 265 p.
2. Nikolis G., Prigozhin I. *Poznanie slozhnogo [Cognition of complex]* (Mir, Moscow, 1990) 344 p.
3. Senkevich Yu.I., Voprosy radioelektroniki, **4:1**, 88-95 (2008)
4. Druzhin G.I., Marapulets Y.V., Cherneva N.V. et al., Doklady Earth Sciences, **472:2**, 215-219 (2017)
5. Morgunov V. A., *Akusticheskaya, elektromagnitnaya emissii i deformatsionnyy protsess [Acoustic, electromagnetic emissions and deformation process]* (Nauka, Moscow, 1994) 167-185
6. Rusyn B.N., *Strukturno-lingvisticheskie metody [Structural-linguistic methods]* (Naukova dumka, Kiev, 1986) 127 p.
7. Fukunaga K., *Vvedenie v statisticheskuyu teoriyu raspoznavaniya obrazov [Introduction to the statistical theory of pattern recognition]* (Nauka, Moscow, 1979) 367 p.
8. Senkevich Y.I. *Program of entropy - syntactic analysis of electro-physiological signals (ESAES ver. 3.0)*(Computer program. Information card) 50200700165 certificate SFAP No. 7534 (RSTIC, Moscow)

Fractional models of seismoacoustic and electromagnetic activity

Boris Shevtsov[1,⋆] and *Olga* Sheremetyeva[1,2,⋆⋆]

[1] *Institute of Cosmophysical Research and Radio Wave Propagation FEB RAS, Paratunka, Russia*
[2] *Vitus Bering Kamchatka State University, Petropavlovsk-Kamchatskiy, Russia*

Abstract. Statistical models of the seismoacoustic and electromagnetic activity caused by deformation disturbances are considered on the basis of compound Poisson process and its fractional generalizations. Wave representations of these processes are used too. It is discussed five regimes of deformation activity and their role in understanding of the earthquakes precursors nature.

1 Introduction

Modeling of various regimes of seismoacoustic and electromagnetic emission caused by deformation disturbances, especially in the phase of their activity, is actual for understanding of earthquakes precursors nature. The changes in the regimes of the deformation process are of interest: unexpected decreases and increases in their activity, the duration of fading, and the discharging energy of accumulated stresses. All these features are detected in signals of seismoacoustic and electromagnetic emission, which are used for diagnostics of the deformation process.

The work of external forces supports a critical level of elastic stresses, as a result there are constantly growing plastic deformations. This process is characterized by the velosity of random dislocation changes, which are determined by the parameters L and u of the dislocation tensor - the spatial scale and the discontinuity of displacement vector (the Burger's vector). We suppose other parameters of the tensor are unchanged, but it is not difficult to take into account the full tensor description.

Note that with using the theory of dislocation changes, it is possible to consider discontinuities, moving along existing faults, repacking of grains or blocks, filtration of liquid in a porous medium, etc. in a wide range of scales L (up to 15 orders).

2 Fractional model

If the relative elastic deformations ε are constant, then we can take the following $u/L = \varepsilon$. In this case there is can considered random dislocation changes as a one parameter process $L_{V,T}(t) = \sum_{k=0}^{N} L(x_k, t_k)$, where the sum is taken over the all dislocations with coordinates x_k, t_k in a given volume V and time interval T.

⋆Corresponding author: bshev@ikir.ru
⋆⋆Corresponding author: olga.v.sheremetyeva@gmail.com

Random variables are x_k, t_k and $L(x_k, t_k)$. In the first approximation, they can be considered independent, then the process $L_{V,T}(t) = \sum_{k=0}^{N} L(x_k, t_k)$ is a compound Poisson process with the distribution of independent increments $p(L)$. For $N \to \infty$, we'll have a stable distribution $p(L_{V,T}, t)$ [8] with an increasing average that characterizes the process of plastic deformations in the volume V.

If the distribution of independent increments $p(L)$ is stable, then the distribution $p(L_{V,T}, t)$ will be stable at each step N with time growing parameters. This process is generalization of the Gaussian process and is the key to understanding of the universality and wide prevalence in nature of stable distributions having the power law properties. The Gutenberg-Richter law can also be approximated by these distributions on the assumption that dislocation changes are independent.

Independence of random variables x_k, t_k and $L(x_k, t_k)$ is violated in dislocation clusters, in which the dislocation density is so large that their radii of influence R_{inf} and T_{inf} depending on $L(x_k, t_k)$ are overlap [1, 2]. About modeling of zones of influence see [3, 4]. The difference between sparse and dense packing of dislocations in clusters may be distinguished.

Using the seismic process as an example we can considered the sequences of foreshocks and aftershocks as a sparse packing and the mainshock (fault) as a close packing of dislocations. The events cluster is a plastic deformation disturbance with a time growing packing of events. Obviously, that the dependence and connectedness of the events is increased with their density. The sequence of foreshocks can be considered as a phase of consolidation of certain scaled level events, which leads to the mainshock - event of more higher scaled level. A sequence of aftershocks can be regarded as the branching phase of the mainshock.

Such representation of the plastic deformation disturbance gives us the opportunity to see the interaction of scale levels. In the space time representation, the cluster of events is developed as a process of random walks that it is possible to compare with a fractional diffusion process [1, 2], in general, non-Markovian, nonlocal and with memory. It means that in the sequence of events all random variables $L(x_k, t_k)$ and x_k, t_k are dependent. Possible simplifications will be discussed below.

Influence radii R_{inf} and T_{inf} increase with $L(x_k, t_k)$ and become larger than the size of the area (V, T), this leads to that the large events always form a cluster in the considerational area [1, 2], so there are constraints to use the Poisson process for modelling large events.

For small events, there is an accumulating effect. Over time, the density of small dislocations increase, they consolidate and cause an event of more larger scale (mainshock), after which, as a rule, a backward wave arises down along a cascade of scales in the form of a branching process. With such cascade effects, events of different scales can not be considered independent.

Thus, we have the constraints on the use of the Poisson process as a model of dislocational deformations, both for large and small events. Nevertheless, for any of the chosen scales, assuming their equality, we can use the Poisson process as a first approximation or a background regime. Such events are referred as migrants for area (V, T) [5]. And it is necessary to consider deviations from the background regime to take into account more subtle effects.

Deviation from the background regime can occur in the hardening or relieving (weakening) medium (changes in rheology). When the medium is hardening there is arose the area in which the process of plastic deformation slows down and the flow of events decreases. Thus, for example, there is an area of seismic quiescence or gap (Fedotov), in which deformation heterogeneity is formed. The elastic energy accumulated in it is released when the stresses are discharged.

It is clear that such accumulation of energy can not be described by the Poisson process with a flow of independent events, because there is a high degree of localization of events that did not release in a seismic quiescence, but they will be realized in deformation disturbance. They can be regarded as adjourned events. Deficiency of events leads to an increase in elastic stresses, as a result the strengthening of the medium is overcome and the accumulated additional elastic energy is released. When the medium is hardening the sequential changes in the regimes of the plastic deformation process take place: normal (background) pulsations and their deceleration (fading), then activation (discharge) and again the background regim. This sequence of changes is called intermittency. When the medium is weaken, the terms "deceleration" and "activation" should be replaced one to another.

The arising deformation disturbance (in activation) can be considered as an event of a higher scale in relation to those events that did not take place during a quiescence caused by the local hardening of the medium. The energy of the deformation disturbance must correspond to the sum of the energies of the adjourned events.

Local hardening of the medium and the preparation of the deformation disturbance on one of the spatial scales L can be described with the fractional Poisson process, for which the average number of events in the time interval t is given by the expression

$$\overline{n}_\nu(t) = \frac{(\mu t)^\nu}{\Gamma(1+\nu)}, \tag{1}$$

where μ is the flow of events with the scale L without hardening of the medium, $\Gamma(1+\nu)$ – gamma-function, $0 < \nu \leq 1$ [6, 7]. For $\nu = 1$ we have the Poisson process with independent events.

The inverse $1/\nu$ of the fractal parameter ν can be considered as a hardening parameter of the medium. Indeed, with its increasing (decreasing of parameter ν), the event flow of $\overline{n}_\nu(t)$ slows down when $\mu t > 1$, and the events deficit

$$\overline{n}_\nu(t) - \overline{n}_1(t) \tag{2}$$

causes an increase in stresses. And the time of accumulation of adjourned events until the moment when hardening of the medium is overcome will determine the energy of the deformation disturbance.

It is noteworthy that the parameter ν is responsible for the fractal dimension of the distribution of events on the time interval [8]. Thus, hardening of the medium (its rheological property) changes the fractality of the process (event statistics) that is realized as a delayed relaxation caused by memory effects that have resulted from the hardening.

The slowing down of the fractional process is related to the behavior of probability to keep the initial state of process [6, 7]

$$P_0(t) = E_\nu(-(\mu t)^\nu), \tag{3}$$

this is the survival probability of the initial state, where $E_\nu(x)$ is the Mittag-Leffler function, the fractal generalization of the exponential, $0 < \nu \leq 1$. However, not a function $E_\nu(x)$, but a complex function $E_\nu(-(\mu t)^\nu)$ is called a fractal exponential, it is the relaxation characteristic of the process. The properties of its complexity, behind which there are two effects, will be discussed below.

For small values of the parameter ν, the relaxation of the initial state slows down, and the waiting times increase. The fading of pulsations is caused by this. The redistribution of probability is given by

$$\Delta P_0(t) = E_\nu(-(\mu t)^\nu) - E_1(-\mu t), \tag{4}$$

where $E_1(-\mu t) = e^{-\mu t}$.

Deviation of $P_0(t)$ from the usual exponential indicates the presence of aftereffects and memory, which are caused by the adjourned events. We'll considered the difference between memory and aftereffects in more detail.

When the hardening of the medium is overcome, the plastic deformation disturbance comes to replace of seismic quiescence. Disturbance has three phases: foreshocks, mainshock and aftershocks. In essence, these are the events that did not take place in seismic quiescence. The first and last phases of the disturbance are represented by a sparse packing of events and the mainshock is represented by dense one. Each event of the cluster affects on subsequent events, it is a aftereffect and at the same time, every event of the cluster is affected by previous events, it is a memory effect. The past impacts on the future through the present [8]. As a result, there is formed a chain of events so joint that the impacts of aftereffect and memory are difficult to separate, but below we will try to do it using different approximations.

Any event from the sequences of foreshocks, mainshock and aftershocks can be decomposed into the sum of background events with the scale of L. The total number of events with a scale L in the plastic deformation disturbance should roughly respective to the number of adjourned events during the quiescence of the process. Such representation of the deformation disturbance allows us to estimate the disturbance energy and to establish a relation between the scales: foreshocks – the wave of consolidation of events of the selected scale level L, the mainshock is the realization of the disturbance at the highest scale level, and aftershocks – the wave of branching of the mainshock. There are the waves up and down on the scales, and in the space time presentation, we have converging and diverging waves to and from the center which is the mainshock.

Note that in the quiescence phase there is only a divergent wave from the adjourned events. This is a slow diffusion process, while in the deformation disturbance, the process is faster by 3-4 orders. The fast process is the wave in the mainshock and the fractioal diffusion in the foreshocks and aftershocks.

Thus, the evolution of the packing of events of the scale L and the interaction of scales occurs in the deformation disturbance simultaneously as a result of consolidation and branching. Here we can note the analogy with turbulence in which the integration and fragmentation of vortices take place, an intermittency of regimes is observed. There is an analogy between the dislocation and vortex. We'll stay on their difference in spectra below.

The time form of the plastic deformation disturbance can be considered as a fractal analogue of the Berlage pulse with a variable packing density of events of scale L and variable fractal dimension. For the disturbance model, we'll use various modifications of the fractal exponential $P_0(t) = E_\nu(-(\mu t)^\nu)$, which was apllied to describe the phase of the quiescence.

For the probability of the mainshock emergence we'll use the expression

$$P_{t<t^*}(\tau) = E_\alpha(-(-\tilde{\mu}\tau)^\alpha). \tag{5}$$

This is an increase (time reversal) fractal exponential, where τ is the time with the starting point from the moment of the start of the mainshock t^*, $\tau = t - t^*$, $\tau < 0$, $\tilde{\mu}$ – is the average flow of events in the deformation disturbance, $0 < \alpha \leq 1$. The probability of absence of the mainshock will be

$$\tilde{P}_{t<t^*}(\tau) = 1 - P_{t<t^*}(\tau). \tag{6}$$

Differentiating this expression, we'll obtain the distribution of the increasing flow of foreshocks, which can be considered as the inverse Omory law. This process will be the initial stage of the fractal pulse of Berlage.

For the probability of the mainshock preserving, we'll use the expression

$$P_{t'<t}(\tau') = E_\beta(-(\tilde{\mu}\tau')^\beta). \tag{7}$$

This is decreasing fractal exponential, where τ' is the time with the starting point from the moment of the end of the mainshock t', $\tau' = t - t'$, $\tau' > 0$, the parameter $\tilde{\mu}$ – is the average flow of events in the deformation disturbance, $0 < \beta \leq 1$. The probability of the mainshock not preserving will be

$$\tilde{P}_{t'<t}(\tau') = 1 - P_{t'<t}(\tau'). \tag{8}$$

Differentiating this expression, we'll obtain the distribution of the decreasing flow of aftershocks, i.e. the Omory law, that can be considered as the attenuation stage of the fractal Berlage pulse.

To sew together the two obtained fractal expressions, we'll take into account that the flow of events in the phase of mainshock is tightly packed and constant, and the intencity of fractal flows can not exceed the tight packing. In the foreshock phase, the event package increase to the maximum possible, in the mainshock it persists, and in the aftershocks it decreases. Fractal parameters α and β are determined by the properties of the medium, its capacity to consolidate and to branch events. The average flow of events $\tilde{\mu}$ in the plastic deformation disturbance is determined by the phase of the quiescence, which is characterized by the parameters μ, ν and the time of quiescence. The time scales μ and $\tilde{\mu}$ can differ by 3-4 orders, since the perturbation is short-lived.

The phase of the foreshocks in the disturbance is usually very fast. This is a short front of the fractal Berlage pulse. So short in the compearison with the aftershocks phase that it can be neglected in the first approximation and only aftershocks as the branching of the mainshock can be considered (ETAS model) [5]. However, for the earthquakes forecasting, it's just the foreshock phase and the preceding quiescence stage are of interest. Using the length of quiescence time, it is possible to estimate the energy of the stress discharge and the phase of the foreshocks predicts the rate of the plastic deformation disturbance, that gives the time of the short-term forecast. In view of this, a new theory is needed.

3 Comparison with the ETAS model

It remains to show that our result corresponds to the ETAS model, if we neglect the effects of memory. For the probability of the initial state survivaling, the ETAS model gives [5]

$$P(\tau) = e^{-\nu\tau - b(\nu\tau)^{1-\theta}}. \tag{9}$$

Proceeding from the form of the exponential of this expression, we have two flows of events. The first of them corresponds to a Poisson process with a constant flow, and the second – to a process with a flow of events decreasing in time on power law. The second flow takes into account aftershocks as branching of the events of the first one. Neglecting the weaker first flow and differentiating the expression $1 - e^{-b(\nu\tau)^{1-\theta}}$ for the transition probability from the initial state to any other, we obtain the Weibull distribution for the aftershocks waiting times.

For the aftershock phase, our result gives expression

$$P_{t'<t}(\tau') = E_\beta(-(\tilde{\mu}\tau')^\beta). \tag{10}$$

Replacing in it the fractal exponential on expanded one

$$P_{t'<t}(\tau') \simeq e^{-(\tilde{\mu}\tau')^\beta}, \tag{11}$$

we get the result of the ETAS model, which agrees well with seismic data [5], however, more subtle effects are possible, which will be discussed below.

The differentiation of the expression

$$1 - P_{t'<t}(\tau') \simeq 1 - e^{-(\tilde{\mu}\tau')^\beta}, \tag{12}$$

which is the probability of transition from the initial state to any other, gives, as in the ETAS model, the Weibull distribution for the aftershocks waiting times. The values $0 < \beta < 1$ correspond to a decreasing flow of events (aftershocks), Omori's law. The value $\beta = 1$ corresponds to a constant flow of events – the Poisson process. In solid state physics parameter β is called the Weibull index.

From the comparison of the fractal approach with the ETAS model, we can conclude that the power law argument $(\mu t)^\nu$ of the fractal exponential $E_\nu(-(\mu t)^\nu)$ is responsible for the branching processes (aftereffects that determine the fractal slowing of the flow), and the function $E_\nu(x)$ itself describes the effects of memory. Both $(\mu t)^\nu$ and $E_\nu(x)$ give the delayed relaxation.

There are no memory effects in the ETAS model, and replacing the expression $E_\nu(x) \to e^{-x}$ changes the relaxation essentially. Thus, the fractal approach takes into account the history of the process and opens up more wider possibilities for describing the properties of the medium.

For example, to simulate the relaxation we can use a dependence $\tilde{P}_0(t) = E_{\tilde{\nu}}(-(\mu t)^\nu)$ with different exponents of branching ν and memory $\tilde{\nu}$. This approach corresponds to a fractional Poisson process with a time varying on power law flow of events. Wherein the number of fractional characteristics of the medium is doubled.

4 The physical significance of the exponent $b = 3/2$

We note that for foreshocks the power law argument of the fractal exponential is responsible for consolidation processes, but not the branching. The process of reverse diffusion (consolidation) has the opposite direction in a nonequilibrium medium. In the fault, two processes with opposite directions in the space of the scales are presented simultaneously. They interact in the system of opposing waves, forming a fractal wave of destruction (a pair of kink and antikink), which interacts in a complex way with the nonequilibrium medium, obtaining the energy of elastic deformations to produce fractures and generate elastic and electromagnetic oscillations. The model of the complex of fractal wave processes can be the subject of the following paper, but here we make only a few remarks.

In the coordinate system associated with the wave of fracture, the dislocation process can be represented as an oscillator. This is a well-known description of dislocation changes using the sine-Gordon equation. Fractal properties of the process are introduced to describe the interaction of the fracture wave with the medium. The period of the oscillations is determined by the time τ_L of formation of one dislocation of scale L in the fault. And the number of oscillations n_R is the ratio $n_R = \tau_R/\tau_L$, where τ_R is the time of formation of the discontinuity, the parameter n_R plays the role of a topological quantum number. Similar numbers can be introduced for foreshocks and aftershocks, $\check{n}_R$ and $\hat{n}_R$. Naturally, the sum of all the quantum numbers (the number of events of scale L in the cluster) must correspond to the deficit of events in the period of quiescence:

$$S_D = \check{n}_R + n_R + \hat{n}_R, \tag{13}$$

i.e. the quantum number of the deformation disturbance. This value, which is determined by the properties of the medium, characterizes the energy of the plastic deformation disturbance and the ratio of the scale levels.

The fractal structure of the deformation disturbance is opened by a wavelet transformation that is an expansion in the terms of the basis impulse functions, the form of which is consistent with the form of the disturbance under consideration. Such expansion was called the sparse approximation [9, 10], for which Berlage pulses are used as the basis functions. It is obvious that the number of wavelet summands can be compared to the quantum number of the deformation disturbance S_D (13). Each of the wavelet summands gives a detailed information how the fault is formed.

An analogy with quantum mechanics arises here, using which the structure of elementary particles can be represented as a deformation disturbance.

The formation of the fault can be regarded as a Poisson process, simple or fractal, and hence for a faults distribution on scales $p(L)$, which corresponds to the Gutenberg-Richter law, we will have a stable or fractional stable distribution. What are the differences between these distributions, see [8]. For us, the most interesting are fractional stable distributions, in which, as was shown above, the effects of branching and memory are taken into account.

 DOI: 10.1051/e3sconf/20172002013

In the formation of a fault, phased (coherent) elastic and electromagnetic oscillations arise [11]. This is due to the pair of forces and an element of current appear simultaneously in the dislocation, which serve as sources of two types of oscillations. And the fault can be represented as a flow of dislocations. It is interesting to find the relation between the energy and the spatial scale of the fault (and the perturbation as a whole), which until now is known only from observations. Let us turn again to an analogy with quantum mechanics.

Using the sine-Gordon equation, the formation of a discontinuity (fault) can be represented as the motion of a wave on a periodic potential with the addition of fracture energy and elastic deformations. Here we do not consider the loss of radiation. The Hamiltonian of these oscillations will have the form

$$E^2 = m^2(L) + p^2 - \sigma^2 L^2 + \varepsilon k L^3, \tag{14}$$

where E is the oscillation energy operator, $m^2(L)$ is the periodic potential (for the Gordon equation), p is the impulse operator, $\sigma^2 L^2$ is the energy of fractures, σ^2 is the specific energy of destruction, $\varepsilon k L^3$ is the energy of elastic deformations, ε is relative deformation (their critical level is supported by the work of external forces), k is the shear modulus.

A wave of unlimited destruction is possible if there is satisfied the Griffith condition $L > \sigma^2/(\varepsilon k)$. For $L \to \infty$ $(p \sim 1/L)$ we obtain the asymptotic expression $E^2 \simeq \varepsilon k L^3$, which gives the relation between the energy of elastic oscillations and the scale of the distubance $E \simeq \sqrt{\varepsilon k}\, L^{3/2}$. This expression is used to conversion the magnitude to the logarithm of the event energy, the exponent $b = 3/2$ (*b-value* in the expression $lgE = a + b\, lgL$, where $lgL = M$ is the magnitude). Until now, the value $b = 3/2$ was known only from experimental data, and the physical significance of the parameter $a = \sqrt{\varepsilon k}$ was not discussed till now. It should be specially noted that the value ε is a characteristic of the state of the medium and has the physical significance of critical elastic relative deformations.

5 Discussion and conclusions

Turbulent velocity pulsations satisfy the following relation $v \sim l^{1/3}$, and the energy $v^2 \sim l^{2/3}$, where l is the scale of pulsations. Turbulence has much in common with deformation disturbance (destruction), but their power law pulsation spectra differ in the exponents, which is due to differences in the initial dynamical equations.

Deformation disturbance have much in common with electrical breakdown and lightning discharges, with acoustic and electromagnetic fluctuations in the plasma. The developed statistical model will help to look from another point of view at the suddenness of precipitations from the radiation belts, the features of the aurora borealis and the nature of globe lightning. And the analogy with quantum mechanics opens new possibilities in studies of fractal dynamics of elementary particles and cosmic rays.

Plastic deformations have much in common with epidemics, see ETAS model [5], the prototype of which was the Markov-Poya process. Its fractal generalization will allow to take into account the effects of memory in the diseases.

The multiplicity of analogies between different physical processes indicates that they belong to the class of critical phenomena.

Plastic deformation disturbances develop on orders faster than the period of their preparation. Therefore, the disturbances are sources of abnormal increased seismoacoustic and electromagnetic emission, which acquires statistical characteristics of the related dislocational changes in clusters, that allows using anomalies in signals of emission to diagnose the activation of deformation processes. The intensity of emission in the background and during the seismic quiescence period is significantly lower than in activation phase. And the differences between the background and the quiescence period can be insignificant, therefore long observations and their comparative analysis will be required to identify a shortage (deficit) of events that will lead to a relaxation of stresses.

Emission of plastic deformation disturbance plays a special role in the processes of lithosphere-ionospheric interaction precisely because its intensity far exceeds the emission of the background regime. Ionospheric effects are observed using ionosonde, GPS technologies and lidars in the variations of the ionosphere parameters, the precipitation of charged particles from the radiation layers and the formation of light scattering layers in the ionosphere by excited atoms and ions. The comparable effects in the ionosphere can also be created by lightning discharges.

A fractal model of plastic deformation disturbance and associated seismoacoustic and electromagnetic emission is proposed.

The fractal model includes branching and memory effects. ETAS model – only branching.

The fluctuation spectra of the deformation perturbation and turbulence are differ on a reason of the differences in their dynamic equations.

A statistical justification of the Gutenberg-Richter law is provided using stable distributions for the background regime and fractional stable one for the phases of quiescence and disturbance. For a more detailed description, it is proposed to use the fractinalal Poisson process with flow a time varying on power law.

The topological quantum number arising in the representation of a deformation disturbance in the form of a fractal pulse Berlage is determined by the deficit of events at the stage of preparation of the deformation disturbance and when processing signal it is obtained with using a sparse approximation of the disturbance.

The topological quantum number defines the relation between scale levels and can be used as the logarithm base in scaling.

A dynamic relation was determined between the energy and the spatial scale of the deformation disturbance (dislocation, seismic event). Thus, it is given the justification for the transformation from the magnitudes to the logarithm of the event energy.

References

[1] B.M. Shevtsov, R.N. Sagitova, Doklady Earth Sciences, **426(2)**, 642-644 (2009)

[2] B.M. Shevtsov, R.N. Sagitova, Volcanology and Seismology, **2**, 56-66 (2012)

[3] A.S. Perezhogin, B.M. Shevtsov, R.N. Sagitova, G.M. Vodinchar, Matem. Mod., **19(11)**, 59-64 (2007)

[4] A.S. Perezhogin, B.M. Shevtsov, Computational technologies, **14(3)**, 48-57 (2009)

[5] A. Saichev, D. Sornette, J. Geophys. Res. **112** (2007)

[6] D.O. Cahoy, V.V. Uhaikin, W.A. Woyczyski, Journal of Statistical Planning and Inference, **140**, 3106-3120 (2010)

[7] N. Laskin, Communications in Nonlinear Science and Numerical Simulation, **8**, 201-213 (2003)

[8] V.V. Uchaikin, Uspekhi Fizicheskikh Nauk (UFN), **173**, 847-876 (2003)

[9] Y.V. Marapulets, A.B. Tristanov, B.M. Shevtsov, Akusticheskij zhurnal, **60(4)**, 427-435 (2014)

[10] Y.V. Marapulets, A.B. Tristanov, B.M. Shevtsov, Doklady Earth Sciences, **456(2)**, 705-708 (2014)

[11] V.N. Uvarov, Geophysical Journal, **38(6)**, 180-185 (2016)

Fractional properties of geophysical field variability on the example of hydrochemical parameters

Boris Shevtsov[1,*] and *Olga* Shevtsova[2,**]

[1] *Institute of Cosmophysical Research and Radio Wave Propagation FEB RAS (IKIR FEB RAS), 684034, Mirnaya str., 7, Paratunka, Elizovskiy district, Kamchatka, Russia*

[2] *V.I. Il'ichev Pacific Oceanological Institute FEB RAS, Vladivostok, Russia*

Abstract. Using the properties of compound Poisson process and its fractional generalizations, statistical models of geophysical fields variability are considered on an example of hydrochemical parameters system. These models are universal to describe objects of different nature and allow us to explain various pulsing regime. Manifestations of non-conservatism in hydrochemical parameters system and the advantages of the system approach in the description of geophysical fields variability are discussed.

Universal character and wide use of power laws in nature are associated with the fundamental summation of identically distributed random variables and the law of large numbers. We can list many examples of power law characterizing complicated systems of heterogeneous elements in different areas of natural sciences (astrophysics, seismology, turbulence theory etc.) including the same relations describing sea water chemical parameters system [1, 2]. The elements of this system are interacting between each other (directly and indirectly) and with the elements of other systems represented by hydrophysical fields and hydrobiology objects. All these interactions create a great diversity of variability factors affecting the chemical system behavior.

Fig. 1 represents the dependencies of standards σ on time average concentrations $\bar{\rho}$ of 16 chemical parameters measured at 13 points of diurnal observations in Amur Bay of the Sea of Japan (for initial data see [3]). Two regions are notable in the concentration range of $10^{-4} - 3.3 \times 10^{4}$ mg/kg, they are: with $\bar{\rho} < 10^{2}$ mg/kg for non-conservative parameters and with $\bar{\rho} > 10^{2}$mg/kg for conservative ones. All the data of non-conservative parameters (dissolved oxygen, nutrients, carbonic parameters) are approximated by a unified power law regression with the exponent $\gamma = 0.68 \pm 0.02$. The conservative parameter data (salinity and components of major salt composition) are approximated by a series of separate parallel regressions with $\gamma = 1$, each of which refers to a separate observation point.

It is obvious that for each concentration region there is a own set of peculiar basic factors of variability. In the parameters system under consideration, the fluctuation regime change corresponds to the boundary concentration of $\bar{\rho}=10^{2}$ mg/kg. In this paper, we consider how and what factors determine the behavior of a hydrochemical system.

Before we turn to the discussion of a statistical model of hydrochemical parameters system behavior, we shall adduce some analogies. For example, a flow of seismic events may be considered

*e-mail: bshev@ikir.ru
**e-mail: OV_shevtsova@mail.ru

as a system of heterogeneous elements which differ by energy and spatial scales and the well-known Gutenberg-Richter law is analogous to the power law illustrated in Fig. 1. Cosmic ray intensity distribution depending on their energies and turbulence spectrum are also presented by power laws. In all these cases dependence slopes characterize the system behavior on the whole.

Variability of hydrochemical parameters concentrations $\rho_i(x,t)$ is determined by their standards $\sigma_i(x,t)$, where $i = 1,..,N$ is a parameter number. We consider the dependence $\sigma_i(x,t)$ on average values of parameter concentrations $\bar{\rho}_i(x,t)$. In our statistical model of hydrochemical parameter system behavior, we take into account the main factors of variability responsible for parameters fluctuations. In this paper we consider a case when a sum $\rho_i(x,t) = \sum_{k=0}^{n(t)} \Delta\rho_i(x,t_k)$ of independent random increments $\Delta\rho_i(x,t_k)$ with distribution $p(\Delta\rho_i)$ is a compound (generalized) Poisson process. To take into account the memory effects of time increment dependence, we consider fractional compound Poisson process [4, 5].

When we impose the condition $\Delta\rho_i \geq 0$, we obtain an additive, positively directed process which describes substance concentration increase in an examination volume. To estimate the loss of substance concentration in a volume, it is necessary to introduce an alternating process, i.e. instead of the condition $\Delta\rho_i \geq 0$we assume that $\rho_i(x,t) \geq 0$.

The impact from turbulent fluctuations is formed by volume split and their random mixing. That results in a power law spectrum of fluctuations which is applied in the discussion of additive statistic model based on Poisson process.

To describe sea water tidal oscillations of we consider the changes of $\Delta\rho_i(x,t)$, proportional to the gradient of $\rho_i(x,t)$value and current speed. Correlation in tidal oscillations may be considered as memory effect which may be taken into account by fractional compound Poisson process.

For a standard compound Poisson process $\rho_i(x,t) = \sum_{k=0}^{n(t)} \Delta\rho_i(x,t_k)$ with flux $\nu$of independent random changes $\Delta\rho_i(x,t_k)$ at independent random time moments t_k (omitting the variables i and x) we have:

$$\bar{\rho}(t) = \overline{\Delta\rho}\,\bar{n}(t), \quad \bar{n}(t) = \nu t, \quad \overline{\rho^2}(t) = \overline{\Delta\rho^2}\,\bar{n}(t) + \overline{\Delta\rho}^2\left(\overline{n^2}(t) - \bar{n}(t)\right), \tag{1}$$

$$\overline{n^2}(t) = \bar{n}(t) + \bar{n}^2(t), \quad \sigma^2(t) = \overline{\Delta\rho^2}\,\bar{n}(t), \quad \sigma(t) = \sqrt{\overline{\Delta\rho^2}}\,\sqrt{\bar{n}(t)}. \tag{2}$$

The relation (2) results in a power law of chemical parameters variability with their average values with the exponent of 0.5. For relative variability $\sigma(t)/\bar{n}(t) = \sqrt{\overline{\Delta\rho^2}}\,/\,\sqrt{\bar{n}(t)}$ the exponent is -0.5.

Turbulence forms power law fluctuations which may have not the value $\overline{\Delta\rho^2}$. For this case, instead of the dispersion $\overline{\Delta\rho^2}$ we consider its fractional analogue:

$$\sigma^\alpha(t) = \overline{\Delta\rho^\alpha}\,\bar{n}(t), \quad \sigma(t) = \left(\overline{\Delta\rho^\alpha}\,\bar{n}(t)\right)^\gamma, \quad 1 \leq \alpha \leq 2, \quad 0.5 \leq \gamma = 1/\alpha \leq 1. \tag{3}$$

The obtained power law (3) agrees well with the experimental regression in Fig. 1 in region $\bar{\rho}<10^{\,2}$ mg/kg which has the exponent $\gamma = 0.68$. This value can be considered as a characteristic of non-conservative hydrochemical parameters system behavior on the whole. The physical meanings of the values α and $\gamma = 1/\alpha$ are turbulence fluctuation parameters in a system. When $\alpha = 1.5$, we have a reverse value of $\gamma = 1/1.5 = 2/3$. A value, close to this one, appear in many natural power laws.

To explain the break (cut off) of a power law in the region of boundary concentration $\bar{\rho}=10^{\,2}$ mg/kg, we should take into account the disturbance of random event independence during concentration change. The tidal gradient mechanism of fluctuations with their long correlations on the semi-diurnal period may be considered as a cause of this disturbance.

DOI: 10.1051/e3sconf/20172002014

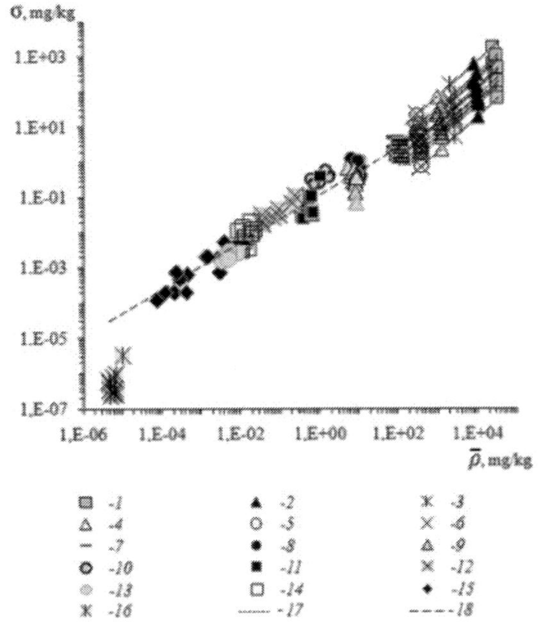

Figure 1. The relationship between the average concentration $\bar{\rho}$ and the standard σ for a complex of hydrochemical parameters at 13 horizons of four daily stations in various parts of Amur Bay. 1 – 16 – salinity, Na^+, SO_4^{2-}, Mg^{2+}, K^+, Ca^{2+}, HCO_3^-, CO_3^{2-}, O_2, SiO_3, CO_2, NH_4^+, NO_3^-, PO_4^{3-}, NO_2^-, a_{H^+}, respectively. 17 and 18 – power law regression for the

In Fig. 1 the fluctuation tidal mechanism manifests in the form of almost vertical grouping of data of salinity and components of major salt composition in the region of $\bar{\rho}>10^{2}$ mg/kg. Here the tidal mechanism prevails over the turbulent fluctuation one. As concentration decreases, the vertical grouping of data vanishes. The dependence of increments $\Delta\rho_i(x,t)$ is violated that results in the break of the power law.

In order to obtain a statistical model of fluctuations with memory, as an example, we consider the fractal Poisson process [4], for which we have:

$$\sigma_\mu^2(t) = \left[\bar{n}_\mu(t) + \bar{n}_\mu^2(t)\left(\frac{\mu\Gamma(\mu)\Gamma(1/2)}{\Gamma(\mu+1/2)} - 1\right)\right], \quad \bar{n}_\mu(t) = \frac{\nu t^\mu}{\Gamma(\mu+1)}, \tag{4}$$

where $\Gamma(\mu)$ is a gamma-function, μ is a fractional parameter of memory of fluctuations, $0 < \mu \leq 1$. When $\mu = 1$, the fractional process transforms into a usual one. The second turn in square brackets is responsible for the effects of memory. When $\mu = 1$, this turn is zero.

The fractional Poisson process has distribution for waiting time different from an exponential curve. Thus, fluctuations cannot be considered independent.

The fractional Poisson process differs from a standard one by power law distribution of waiting time for which long fading is possible.

When the substances concentration is low, the fractional Poisson process behaves as a standard one. When the concentration is high, it turns to the asymptotic $\sigma(t) \simeq \bar{n}_\mu(t)$ with the exponent 1. This is just the reason of the break (cut off) which is associated with memory effects.

For the fractional compound Poisson process [5] we have:

$$\sigma_\mu^2(t) = \left[\overline{\Delta\rho^2}\,\bar{n}_\mu(t) + \overline{\Delta\rho}^2\bar{n}_\mu^2(t)g(\mu)\right], \quad g(\mu) = \frac{\mu\Gamma(\mu)\Gamma(1/2)}{\Gamma(\mu+1/2)} - 1. \tag{5}$$

It is clear that the break point also depends on the moments of fluctuations $\overline{\Delta\rho^2}$ and $\overline{\Delta\rho}^2$. For high concentrations, we obtain the asymptotic $\sigma(t) \simeq \overline{\Delta\rho}\,g(\mu)^{1/2}\,\bar{n}_\mu(t)$. In Fig. 1 for high concentrations of $\bar{\rho}>10^{\,2}$ mg/kg, a set of parallel lines for different observation points corresponds to this asymptotic.

For the fractional compound Poisson process, which has a power law for waiting time, we have:

$$\sigma_\mu^\alpha(t) = \left[\overline{\Delta\rho^\alpha}\bar{n}_\mu(t) + \overline{\Delta\rho}^\alpha\,\bar{n}_\mu^\alpha(t)g(\mu)\right]. \tag{6}$$

When the concentration is low, this relation is analogous to (3). When the concentration is high, we have the same asymptotic as for (5).

Instead of the fractional one, we can take other generalization of the Poisson process with memory effects. In this case, the results which were presented above change slightly.

Finally, we discuss the influence of the variability factors associated with non-conservative behavior of hydrochemical parameters. These are chemical, physical-chemical and biochemical reactions which cause smoothing or intensification of turbulent fluctuation manifestations. As a result, experimental data may deviate from the power law regression to any extent. The complex of these deviations may be considered as a characteristic of non-conservative behavior of a hydrochemical parameters system.

Conclusions

1. Based on the most general assumptions on the fluctuation nature in the system of hydrochemical parameters, a model of power law of their variability on average concentration was obtained.
2. Power law break (cut off) was explained as a manifestation of variability tidal mechanism memory effect.
3. Features of gradient variability in power law were detected in the form of data vertical grouping.
4. The role of the non-conservative behavior of the hydrochemical parameters in the deviations of their variability on power law was shown.

References

[1] O.V. Shevtsova, *Increase in the relative diurnal variability of hydrochemical indices in Amur Bay, Sea of Japan*, in *Doklady Earth Sciences* (Springer Science & Business Media, 2015), Vol. 460, p. 73
[2] O.V. Shevtsova, Water Resources **40**, 181 (2013)
[3] O.V. Shevtsova, Water Resources **28**, 666 (2001)
[4] D.O. Cahoy, V.V. Uchaikin, W.A. Woyczynski, Journal of Statistical Planning and Inference **140**, 3106 (2010)
[5] N. Laskin, Communications in Nonlinear Science and Numerical Simulation **8**, 201 (2003)

Interactive system for geomagnetic data analysis

Igor Solovev[*]
Institute of Cosmophysical Research and Radio Wave Propagation FEB RAS, 684034 Kamchatka region, Elizovskiy district, Paratunka, Mirnaya str., 7, Russia

Abstract. The paper suggests the methods for analyzing geomagnetic field variations, which are implemented in "Aurora" software system for complex analysis of geophysical parameters. The software system allows one to perform a detailed magnetic data analysis. The methods allow one to estimate the intensity of geomagnetic perturbations and to allocate increased geomagnetic activity periods. The software system is publicly available (http://aurorasa.ikir.ru:8580, http://www.ikir.ru:8280/lsaserver/MagneticPage.jsp).
This research was supported by the Russian Science Foundation (Project No. 14-11-00194).

1 Introduction

In this paper, we described software tools for geomagnetic data analysis, based on the application of modern mathematical methods and information technologies. Implemented software methods are aimed at studying of complex dynamic processes in the magnetosphere. The state of the magnetosphere is determined by the parameters of the interplanetary space, which has a negative effect on satellite operation, radio communications and other important aspects of human life. Therefore, this field of research is of great scientific and practical interest today [1, 2].

Variations of the Earth's magnetic field contain important information on the processes in the magnetosphere that occur during the period of high solar activity. Currently, methods [3-6] and application software for the processing and analysis of geophysical data are being developed. They provide the users with convenient tools for conducting experimental and theoretical studies (eg. http://www.cosmos.ru/magbase; http://matlab.izmiran.ru/magdata/, https://www.ngdc.noaa.gov/; http://smdc.sinp.msu.ru/). Over the past 20 years, the network of terrestrial magnetic observatories has increased significantly, and a large amount of information has been accumulated. But at the same time, there are still unsolved problems, such as digitization of previously accumulated magnetograms [7-8], development of automatic systems for data collection and for initial data processing [9-11]. Due to the continuous accumulation of data and the transition to a second registration format, the relevance of these tasks is growing.

Anthropogenic interference arising in a signal does not allow us to monitor and to assess the Earth's magnetic field automatically. The authors of the works [2, 9-11], on the

[*] Corresponding author: kamigsol@yandex.ru

basis of DataMining applications, developed methods for automating the work of experts (creating so-called "electronic experts") to solve the problems of primary data processing, to detect industrial and physical anomalies and to form world databases. Moreover, an algorithm based on the application of continuous wavelet transform [12] and threshold functions, first proposed in [3], showed high efficiency in detection of industrial noise.

To use the geomagnetic data effectively and to obtain qualitatively new results, it is necessary to create adequate automated data processing and analysis systems. Difficulties in solving these problems are related to the nonstationary structure of the field variations, the presence of local features of various shapes and durations that carry important information about the state of the earth's magnetic field. Taking into account the complex structure of field variations, modern mathematical methods and technologies are currently used. Thus, for example, to solve the problems of automatic detection of the periods of magnetic storm initial phase [13], of removing noise, and excluding the periodic component of geomagnetic field variations caused by Earth rotation [14, 15], a wavelet transform was used. On the basis of this tool, the problem of automatic calculation of geomagnetic activity index [16] was also solved, which made it possible to reduce the error of its calculation in comparison with the traditional methods [17-19] by 20%. When solving the problems of the modeling of field variations [20] and prediction of the travel time of the interplanetary shock wave [21], neural networks are applied. The developed approach allowed us to improve the quality of the magnetic storm forecasting technique (by 9% of cases) in comparison with the method based on the logistic regression model [21]. In the papers [3, 22], based on the combination of neural networks and wavelet transform, a method for automatic estimation of the state of the geomagnetic field was proposed, and it was implemented as a software application.

The algorithms presented in this paper are based on the proposed geomagnetic signal model [23, 24], including methods for identification of a typical field component and various scale disturbances during the periods of high solar activity. The obtained numerical solutions make it possible to distinguish periods of weak and strong geomagnetic activity and to estimate the intensity of geomagnetic disturbances. Based on the processing of 1-second data, geomagnetic disturbances can be analyzed in the ranges of Pc1-Pc5 and Pi1-Pi3 pulsations. We also constructed wavelet portrait of the magnetic storm which shows the distribution of the disturbance intensity along the analyzed meridian of observations.

Numerical solutions formed the basis of the developed software component of "Aurora" system, which is in open access at http://aurorasa.ikir.ru:8580 (mirror link-http://aurorasa2.ikir.ru:8580) and http: //www.ikir.ru:8280/lsaserver/MagneticPage.jsp. The development of the program system was supported by the grant of the RNF No. 14-11-00194.

2. Geomagnetic data analysis methods implemented in the software system

2.1 Estimate of geomagnetic disturbances intensity is based on the following operations:

1. Wavelet transform of minute data is performed:

$$W_{\Psi} f_{b,a} := |a|^{-1/2} \int_{-\infty}^{\infty} f(t)\Psi\left(\frac{t-b}{a}\right)dt,\ f \in L^2(R), a,b \in R, a \neq 0, \qquad (1.1)$$

where a is the scale, Ψ is the basic wavelet.

In the decomposition (1.1) we use an orthonormal third-order Daubechies wavelet basis, which was determined by minimizing the error in data approximation [23]. Wavelet Ψ is a window function and has a zero mean and, when the scale a tends to be zero, coefficients $W_\Psi f_{b,a}$ describe the properties of the function f in the neighborhood of $t = b$.

2. Then the intensity of field perturbations at time $t = b$ is estimated by the formula [24]:

$$E_b = \sum_a e_{b,a} \tag{1.2}$$

where $e_{b,a} = |W_\Psi f_{b,a}|$.

Fig. 1. illustrates the result of calculating the intensity of geomagnetic disturbances based on operation (1.2) for the period December 19-21, 2015.

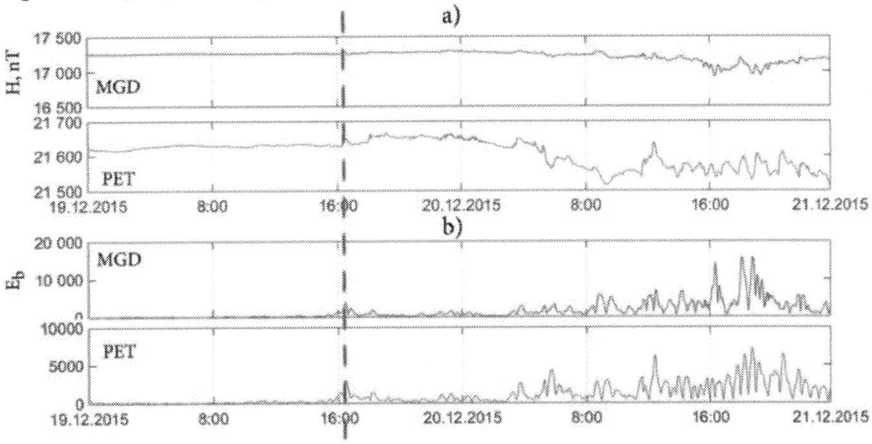

Fig. 1. The result of the operations (1.1), (1.2) for data stations "Magadan" and "Paratunka" during the period December 19-21, 2015; a) H-components of the Earth's magnetic field; b) calculation of the geomagnetic disturbances intensity, based on operation (1.2). The dashed line indicates the beginning of the magnetic storm

2.2 Isolation of the periods of weak and strong geomagnetic activity based on the application of threshold functions:

$$K_{T_{a,1}}(e_{b,a}) = \begin{cases} 0, если\ e_{b,a} < T_{a,1} \\ e_{b,a}, если\ e_{b,a} \geq T_{a,1} \end{cases}, \quad K_{T_{a,2}}(e_{b,a}) = \begin{cases} 0, если\ e_{b,a} < T_{a,2} \\ e_{b,a}, если\ e_{b,a} \geq T_{a,2} \end{cases}, \tag{1.3}$$

The threshold values of $T_{a,1}$ allow us to identify weak and strong perturbations, and the threshold values of $T_{a,2}$ allows us to identify strong perturbations. Identification of the thresholds $T_{a,1}$ and $T_{a,2}$ was performed by assessing a posteriori risk [25], the estimates used the data of the observatory "Paratunka" (Kamchatka region) for the period 2002-2010. The results of calculating the values of $T_{a,1}$ and $T_{a,2}$ for different scale levels a are shown in Fig. 2.

DOI: 10.1051/e3sconf/20172002015

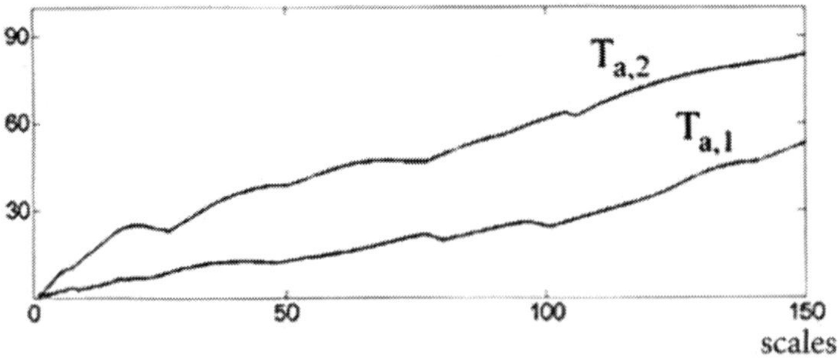

Fig. 2. The results of calculating the values of $T_{a,1}$ and $T_{a,2}$. On the axis oy, the values of $T_{a,1}$ and $T_{a,2}$ are shown

Fig. 3 shows the result of the application of threshold functions (1.3) to detect the periods of weak and strong geomagnetic activity for the period December 19-21, 2015.

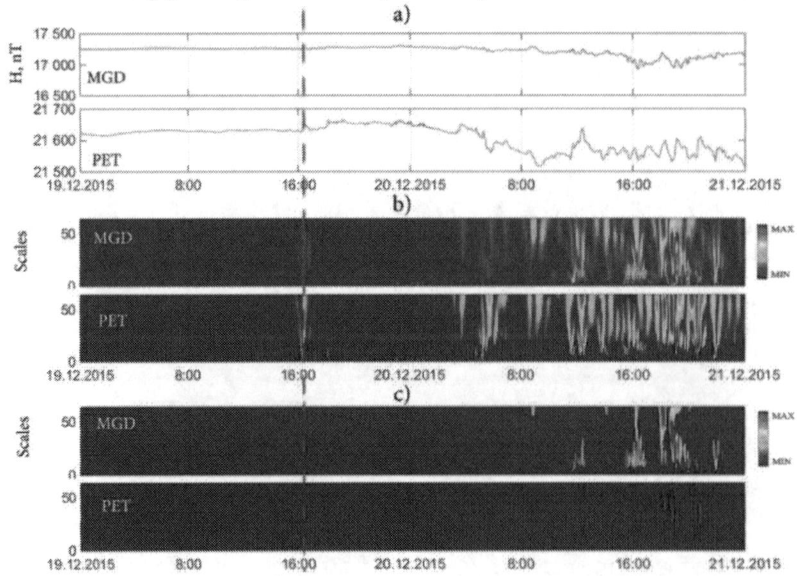

Fig. 3 Result of detection of periods of weak and strong geomagnetic activity according to the data of Magadan and Paratunka stations for the period December 19-21, 2015. a) H-components of the Earth's magnetic field; b) detected periods of weak and strong geomagnetic activity (the threshold $T_{a,1}$ was applied); c) detected periods of strong geomagnetic activity (the threshold $T_{a,2}$ was applied). The dashed line indicates the beginning of the magnetic storm

The software system supports the generally accepted IAGA-2002 data format and uses minute data for the processing. Fig. 4 presents the results of the minute data analysis of Paratunka and Magadan stations during the magnetic storm on August 15, 2015. According to space weather data (http://ipg.geospace.ru/) on August 15, 2015, a sharp increase in the solar wind speed was registered, caused by the coronal ejection of solar matter (CME from August 12) and high-speed solar wind from the coronal hole (reached the central meridian on August 11). At "Paratunka" and "Khabarovsk" stations there were sharp increases in the horizontal component of the field at 08:30 UT (Fig. 4a). Analysis of the geomagnetic disturbances intensity (Fig. 4b) shows that, approximately 12 hours before the beginning of the magnetic storm, short-term increases were observed. During the main phase of the storm (the period of a significant decrease in the Dst-index), the maximum values of the geomagnetic disturbances intensity were recorded (Fig. 4b). This example shows the effectiveness of the proposed software for geomagnetic data analysis tasks, which, in contrast to the analogues, allows obtaining accurate quantitative estimates of the degree of geomagnetic field disturbance.

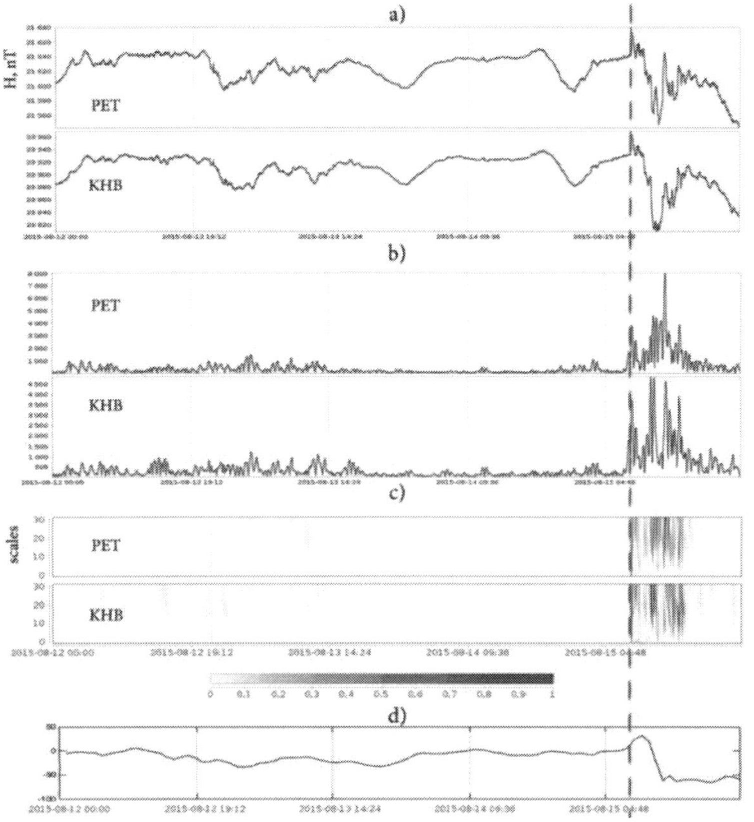

Fig. 4 Software processing of minute geomagnetic data during the period August 12-15, 2015 a) H-components of the Earth's magnetic field; b) calculation of the geomagnetic disturbances intensity, based on the operation (1.2); c) detected periods of weak and strong geomagnetic activity (the threshold $T_{a,1}$ was applied); d) Dst– index. The dashed line indicates the beginning of the magnetic storm

2.3 Estimation of the geomagnetic disturbances intensity in the range of geomagnetic pulsations

According to (1.2), the intensity of positive (increase in field variation relative to the characteristic level) and negative (decrease in the variation of the geomagnetic field with respect to the characteristic level) geomagnetic disturbances on the scale at a time point $t = b$

$$e^{\pm}_{b,a} = (W_{\Psi} f)(b,a). \tag{1.4}$$

Then the intensity of the field disturbance at a time point $t = b$ can be estimated:

 DOI: 10.1051/e3sconf/20172002015

$$E_b^{\pm} = \sum_{a=a_1}^{a_2} e_{b,a}^{\pm} \quad (1.5)$$

Fig. 5 shows the result of an estimate of the geomagnetic disturbances intensity in the range of Pi2 pulsations (periods from 45 to 150 seconds) based on operation (1.5) within the period from 7:00 UT to 10:00 UT on August 15, 2015.

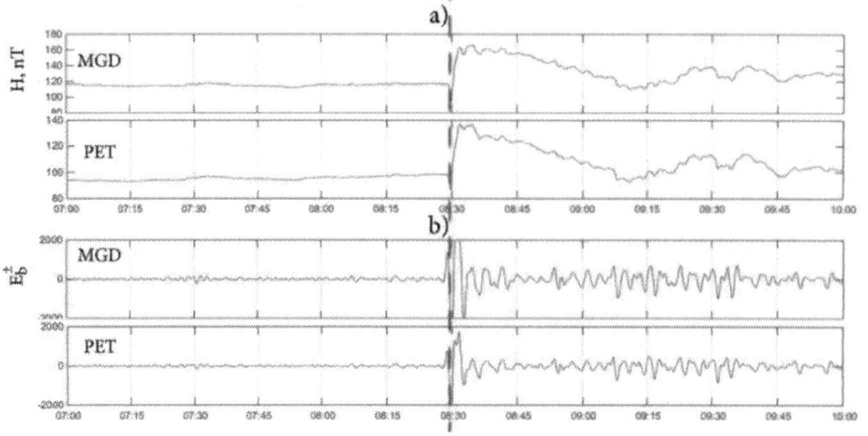

Fig. 5 The result of estimation of geomagnetic disturbances intensity in the range of Pi2 pulsations at Magadan and Paratunka stations on August 15, 2015 from 7:00 UT to 10: 00 UT. a) H-components of the Earth's magnetic field; b) calculation of geomagnetic disturbances intensity, based on the operation (1.5). The dashed line indicates the beginning of the magnetic storm

Geomagnetic disturbances in the range of pulsations are extracted on the basis of decomposition of variations of the horizontal component of the geomagnetic field into wavelet packets:

$$f_{(0,0)}(t) = \sum_n c_{j,k,n}\varphi_{j,k,n}(t) + \sum_{(j,k)\in P_i}\sum_n d_{j,k,n}\Psi_{j,k,n}(t) + \sum_{(j,k)\notin P_i}\sum_n d_{j,k,n}\Psi_{j,k,n}(t), \quad (1.6)$$

where $\varphi_{j,k} = \{\varphi_{j,k,n}\}_{n\in Z}$ is basis in the node (j,k) of a wavelet packet tree obtained by a scaling function; $\Psi_{j,k} = \{\Psi_{j,k,n}\}_{n\in Z}$ is the wavelet-basis in the node (j,k) of the wavelet packets tree; coefficients $c_{j,k,n} = \langle f, \varphi_{j,k,n}\rangle$; $d_{j,k,n} = \langle f, \Psi_{j,k,n}\rangle$; P_i is a set of index pairs, (j,k) is a node of the wavelet packet tree, where j is the scale, k is the component number, the subscript (0,0) indicates the origin of the discrete data.

Approximating component $\sum_n c_{j,k,n}\varphi_{j,k,n}(t)$ describes the trend of the geomagnetic field variation, detailing components $\sum_{(j,k)\in P_i}\sum_n d_{j,k,n}\Psi_{j,k,n}(t)$ describe pulsations from a range determined by a set of indexes P_i. Component $\sum_{j\notin D_i}\sum_n d_{j,k,n}\Psi_{j,k,n}(t)$ is taken as noise. Fig. 6 shows the scheme of decomposition into wavelet packets up to the 2nd level.

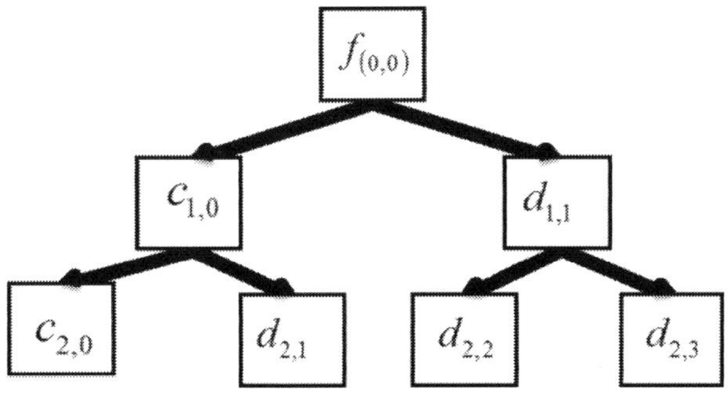

Fig. 6. Scheme of decomposition $f_{(0,0)}(t)$ into wavelet packets

In order to extract the pulsations in the ranges Pc1, Pc2, Pc3, Pc4, Pc5 we use the following sets of index pairs: $P_1^{Pc1} = \{(1,1);(3,3)\}$, $P_2^{Pc2} = \{(3,2)\}$, $P_3^{Pc3} = \{(3,1);(4,1),(6,3)\}$, $P_4^{Pc4} = \{(6,2);\}$, $P_5^{Pc5} = \{(7,1);(8,1),(10,3)\}$and to extract the pulsations Pi1, Pi2, Pi3 we use $P_6^{Pi1} = \{(1,1);(2,1);(2,2);(3,1);(4,1);(5,1)\}$, $P_7^{Pi2} = \{(6,1);(7,1)\}$, $P_3^{Pi3} = \{(8,1);(9,1)\}$, respectively.

2.4 Construction of a wavelet-portrait of a magnetic storm:

1. According to station network data on the basis of operation (1.2), geomagnetic disturbances intensity is calculated;
2. On the basis of operation (1.1) a wavelet spectrum of geomagnetic disturbance intensity values is constructed.

The second geomagnetic data of the ground station network are used to construct the wavelet-portrait. The obtained geomagnetic disturbances intensity wavelet-spectra show the disturbance distribution dynamics in the analyzed regions.

Fig. 7 shows the result of constructing a wavelet-portrait during the magnetic storm on August 15, 2015. The constructed wavelet-portrait shows a spatial-temporal pattern of the disturbance dynamics before and during the magnetic storm along the meridian of the observations. In the initial phase of the storm from 8:00 UT to 9:30 UT, disturbances were observed in a wide range of frequencies at all stations. During the main phase of the storm from 10:00 UT, the disturbances at MGD, PET, MMB stations were of a general nature, which characterizes large-scale processes in the magnetosphere. At the high-latitudinal KTN station, the maximum disturbances were observed within the period from 11:20 UT to 15:00 UT and coincided with the maximum values of the indices AE, AU, AL. Moreover, within the period from 12:30 UT to 13:10 UT, active regions were detected at the northern stations of KTN and MGD (marked by the oval in Fig. 7). They are likely to be associated with auroral processes.

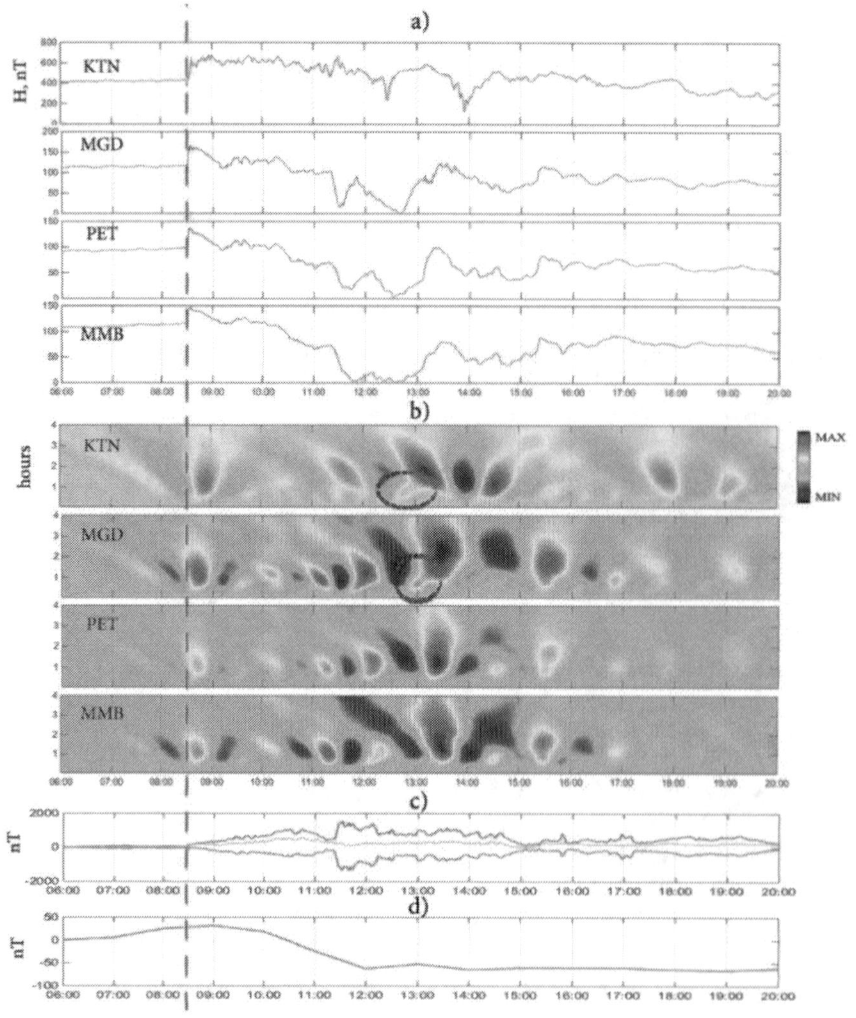

Fig. 7. Second Earth magnetic field data processing by the «Aurora» system during the magnetic storm on August 15, 2015 a) H-components of the Earth's magnetic field; b) wavelet-spectrum of the geomagnetic disturbances intensity; c) indices AE (blue line), AU (yellow line) and AL (red line); d) Dst– index. The dashed line indicates the beginning of the magnetic storm

Conclusion

The paper presents a software system for geomagnetic data processing and analysis that allows us to perform the detailed analysis of the occurrence and dynamics of geomagnetic field perturbations in the areas under consideration during the periods of increased solar activity. The implemented software modules are available in the Internet (http://aurorasa.ikir.ru:8580, http://www.ikir.ru:8280/lsaserver/MagneticPage.jsp). Comments and suggestions on the application of the software can be sent to aurora@ikir.ru.

The work was supported by the RNF Grant No. 14-11-00194. The author expresses gratitude to the organizations that are registering the geomagnetic data that were used in the work and to the Institute of Applied Physics for the continuous implementation and publication of a space weather survey in the Internet.

References

1. Yu.I. Yermolaev, M.Yu. Yermolaev, IZV ATMOS OCEAN PHY+, **46(7),** 799–819 (2010)
2. R.I. Krasnoperov, A.A. Soloviev, Gornyi Zhurnal **10,** 89-93, doi: 10.17580/gzh.2015.10.16 (2015)
3. O.V. Mandrikova, E.A. Zhizhikina, Computer Optics, **39**(3), 420-428 (2015, in russian)
4. O.V. Mandrikova, N.V. Fetisova (Glushkova), Riad Taha Al-Kasasbeh, D.M. Klionskiy, V.V. Geppener, M.Y. Ilyash, Annals of geophysics **58(5)**, (2015)
5. O.V. Mandrikova, N.V. Glushkova, Yu.A. Polozov, Pattern Recognition and Image Analysis, **25(3)**, 470–480, 2015.
6. O.V. Mandrikova, T.L. Zalyaev,11th International Conference "Problems Of Geocosmos", 77-78 (2016)
7. S.Yu. Khomutov, VarSITI Newletter (2014)
8. S.Yu. Khomutov, I.N. Khomutova, Vestnik KRAUNC. Fiz.-mat. nauki. **15(4)**, 112-117. (2016) (In Russian)
9. R.V. Sidorov, A.A Soloviev, Sh.R. Bogoutdinov, Izvestiya, Physics of the Solid Earth, **48(5)**, 410-414, (2016)
10. A.A Soloviev et al. Data Science Journal, **12** doi:10.2481/dsj.WDS-019 (2013)
11. Soloviev, A. Chulliat, S. Bogoutdinov, A. Gvishiani, S. Agayan, A. Peltier, B. Heumez, EPS, **64(9),** 743-752, (2012)
12. Daubechies *Ten Lectures on Wavelets* (CBMS-NSF Lecture Notes nr. 61, 1992)
13. A.G. Hafez, E. Ghamry, H. Yayama, K. Yumoto, ASR, **51,** 39-49 (2013)
14. Z. Xu, L. Zhu, J. Sojka, P. Kokoszka, A. Jach, J. Atmos. Solar–Terr. Phys., **70,** 1579-1588 (2008)
15. Jach, P. Kokoszka, J. Sojka, L. Zhu, J. Geophys. Res., **111(A9)** doi:10.1029/2006JA011635 (2006)
16. O.V. Mandrikova, S.E. Smirnov, I.S. Solov'ev, Geomagn. Aeron., **52(1)**, 111–120 (2012)
17. V.P. Golovkov, V.O. Papitashvili, N.E. Papitashvili, Geomagn. Aeron., **29**, 667–670 (1989)
18. K. Nowożyński, T. Ernst, J. Jankowski, Geophys. J. Int., **104,** 85-93 (1991)
19. M. Menvielle, N. Papitashvili, L. Hakkinen, C. Sucksdorff, Geophys. J. Int., **123**, 866-886. (1995)
20. K. Unnikrishnan, Journal of Atmospheric and Solar-Terrestrial Physics, **121**, 206-220 (2014)
21. K. Alielden, A. Mahrous, Solar Radio (CESRA 2016), 89 (2016)
22. O.V. Mandrikova, E.A. Zhizhikina, V.V. Geppener, XVIII International Conference On Soft Computing and Measurements,(2015)
23. O.V. Mandrikova, I. Solovjev, V.V. Geppenerc, A-KR. Taha, D. Klionskiy, DIGIT SIGNAL PROCESS, **23**, 329–339 (2013)
24. O.V. Mandrikova, I.S. Solovev, T.L. Zalyaev, EPS, **66.** doi:10.1186/s40623-014-0148-0 (2014)
25. B. R. Levin *Theoretical Basis of Statistical Radio Techniques* (Fizmatgiz Moscow, 1963).

E3S Web of Conferences **20**, 02016 (2017) DOI: 10.1051/e3sconf/20172002016
Solar-Terrestrial Relations and Physics of Earthquake Precursors

Comparison of the earth's crust geothermal field in Northern Tien Shan region with seismic characteristics

Andrey Vilayev[1,*], *Zhumabek* Zhantaev[1]

[1] National Center of Space Research and Technology, Institute of Ionosphere, 050020 Kamenskoe Plateau, Almaty, Kazakhstan

Abstract. The deep crustal temperatures of the Northern Tien Shan region are being calculated on the basis of direct measurements of the heat flow and empirical relation of geothermal parameters and seismic velocities. For this purpose the applied method of numerical solution of conductive heat transfer equation takes into account additional sources of heat release due to radioactive decay. Thermoelastic deformations, reaching 20-40% of the total lithostatic pressure at depths of 35-60 km, are determined. The criteria controlling the distribution of hypocenters of earthquakes have been determined in accordance with the morphology of the regions of excess temperatures and thermal stresses.

1 Introduction

The temperature field largely determines the physical state of the Earth's lithosphere material. The study of the thermal state of the earth's interior is especially important for areas where intensive tectonic movements occur, accompanied by strong and catastrophic earthquakes. Northern Tien Shan is one of such areas. This is evidenced by the earthquakes occurred in the region: Vernenskoye (Ms = 7.1, 1887), Chilik (Ms = 8.1, 1889), Kemin (Ms = 8.0, 1911), Zhalanash (Ms = 6.8, 1978), Bisorun (Ms = 5.9, 1990), etc. The aim of this paper is to examine the distribution of the deep thermal field of the Earth's crust of Northern Tien Shan to assess the seismic activity of the region.

2 Methods and Materials

The method used for the research implied numerical modelling of the Earth's crust temperatures based on the results of heat transfer equations with the use of experimental thermophysical parameters - heat flow, thermal conductivity coefficient, radiogenic heat generation. A comparative analysis of the parameters of the calculated geothermal model with the distribution of earthquakes of the territory has been performed.

* Corresponding author: vilayev@gmail.com

Mathematical model. The basic equation of stationary conductive heat transfer for long-term geological processes in a heterogeneous anisotropic elastic medium can be expressed as [1]:

$$\nabla k\, \nabla T + k\nabla^2 T + A(x, y, z) = 0 \tag{1}$$

where k is the thermal conductivity of the rocks at the point with coordinates (x,y,z), T is the temperature, and A (x,y,z) is the radioactive generation of heat in a unit volume. The symbol ∇ denotes the differential Laplace operator in three-dimensional space

$$\nabla = \left(\frac{\partial^2}{\partial x^2} + \frac{\partial^2}{\partial y^2} + \frac{\partial^2}{\partial z^2}\right)T \tag{2}$$

The coefficient of thermal conductivity at the point (x, y, z), taking into account the temperature dependence, was calculated for the Northern Tien Shan region as follows:

$$k = 0.01157Vp^3 * (-0.9 + 0.00246T + 343T^{-1})^{-1} \tag{3}$$

where Vp - velocity of seismic waves (km/s), T - initial temperature in ^{o}K.
Radiogenic heat release in the Earth's crust due to the decay of radioactive elements ^{40}K, ^{238}U and ^{232}Th was calculated from the empirical relationship between the heat generation A (μW/m^3) and the velocity Vp (km/s) for the earth crust of the Northern Tien Shan:

$$A = 0.3125**(10^{-5})\cdot exp(1.61*Vp) \tag{4}$$

Equation (1), with allowance for (3,4), is solved using the numerical method of finite differences on the grid 41x31x20 in the variable direction scheme [2].
The heat flow determined from well measurements was taken as a boundary condition for heat transfer on the surface; according to meteorological data, the average annual temperature of the earth's surface is 10 °C at a depth of 1 m. The initial temperature distribution corresponded to a linear geothermal gradient. The lateral boundary conditions the heat flow to zero through the vertical boundaries of the model. At the lower boundary of the modelling region, the mantle component of the heat flow was determined as the difference between the flow on the surface and the radiogenic component from the decay of the U, Th, K-containing rocks, as well as constant temperature, assuming the presence of the rock at the depth of 120 km with Solidus temperature (1370 °C).
To calculate the effective thermal stresses in the earth's crust, the working formula is applied [3]

$$\tau_{zz} = \rho g z + 3\beta K\,(T\text{-}T_0) \tag{5}$$

where – 3β is the volume coefficient of thermal expansion of the rocks, K is the bulk modulus, ρ is the density, g is acceleration of gravity, and (T-T_0) is the temperature deviation from the initial linear distribution. The product 3βK (T-T_0) denotes thermal stresses caused by a combination of a nonlinear temperature change. The bulk modulus K is calculated from the ratio of the P-waves and S-waves velocities in the conditions of natural bedding of the rocks:

$$K = \rho(V^2_p - 4/3\ V^2_s) \tag{6}$$

where Vp/Vs = 2.04 stands for the rocks of the Northern Tien-Shan,
Density of rocks is defined in each node of the grid model according to the law of linear approximation:

$$\rho = (0.315V_p + 0.74)*10^3 \quad (7)$$

Taking into account the sphericity of the surface and the considerable spatial dimensions of the model, equation (1) was solved in spherical coordinates relative to the center of the Earth:

$$\nabla T = \frac{1}{r^2}\frac{\partial}{\partial r}\left(r^2\frac{\partial T}{\partial r}\right) + \frac{1}{r^2 \sin\theta}\frac{\partial}{\partial \theta}\left(\sin\theta\frac{\partial T}{\partial \theta}\right) + \frac{1}{r^2 \sin^2\theta}\frac{\partial^2 T}{\partial \varphi^2} \quad (8)$$

3 Results and Discussion

Thermophysical model. Geologically, the region is a part of the Alpine neotectonic belt and includes the longitudinally extending mountain ranges Kungei and Zailiyskiy Ala-Tau separated by intermontane depressions [4]. From the east, the region is bordered by spurs of the Dzhungar Ala-Tau mountain ranges, and from the west by the eastern end of the Chu-Ili Mountains. In the earth's crust, according to geophysical data, sedimentary, granite (upper and lower), and basalt layers are distinguished. The thickness of the earth's crust varies from 42 km to 55-60 km, increasing southeastward towards the mountain structures of the Tien Shan and Dzungaria.

Structure of seismic velocities field. The initial experimental base of the deep structure of the earth's crust was P-waves velocities data to the depth of 100 km within the region with

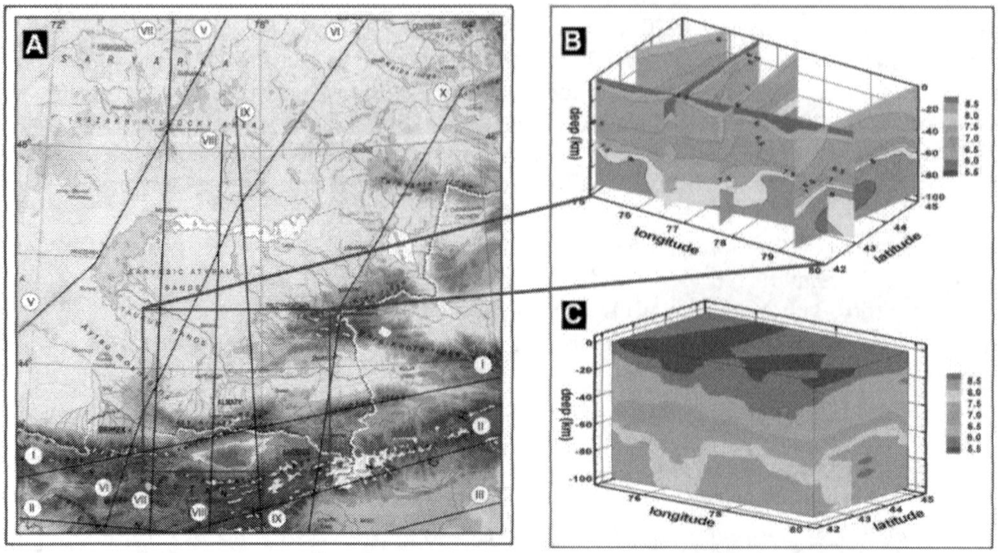

coordinates determined by the methods of the DSS, the DSS-MOVZ [4,5]. (Fig. 1).

Fig. 1. Geotraverse system of P-waves velocity lithospheric modelling: A - planned position of geotraverses, B - 2D models P-waves velocity by profiles, C - 3D velocity model in study area

Thermal conductivity. The average values of thermal conductivity coefficient of the sedimentary layer are determined in the range 1.5÷1.7 W/m·K, granite - about 2.6, dioritic - 2.2÷2.3 W/m·K, basaltic - 3÷3.1 W/m·K. For the upper mantle, the thermal conductivity coefficient corresponds to 3.2 W/m·K.

Radiogenic heat generation. A comparison analysis of the estimates of the heat generation values of different layers of the earth's crust according to the data of [6-8] and equation (4) showed satisfactory convergence. Magmatic formations developed in the North Tien Shan and Dzungar blocks are characterized by a higher level of heat generation. The highest concentration of radioactive elements is observed in the upper part of the granite layer at the depths 5 - 30 km.The heat release of the granite layer is from 1.0 to 2.0 μW/m^3, sedimentary layer - from 0.2 μW/m^3 to 1.0 μW/m^3. With depth, the radioactive heat release decreases unevenly. The lower part of the granite layer of the mountain areas is characterized by an intensity of 0.6 ÷ 1.1 μW/m^3, the platform part is 0.5 ÷ 1.0 μW/m^3. The heat generation of the basalt layer is 0.5 μW/m^3.

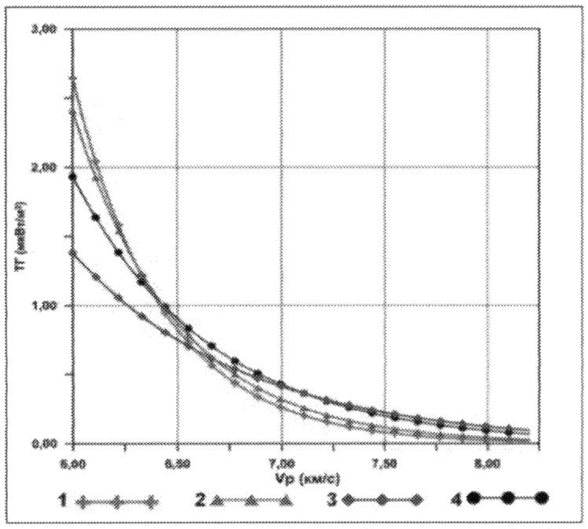

Fig. 2. Dependence of the heat release rate intensity in the crust rocks from elastic waves velocity: 1 – [6], 2 – [7], 3 – [8], 4 – equation (4).

Heat flow. The surface heat flow (HF) map is compiled from the data of [9,10]. In addition, 249 well measurements performed by the authors were taken into account. The average density of the points of heat flow determination is 1 measurement per 18 km^2

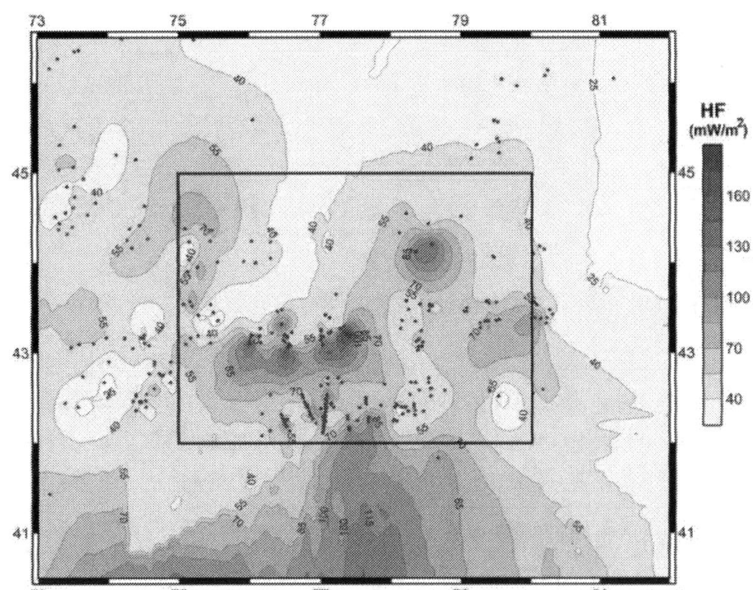

Fig. 3. The heat flow in the northern Tien Shan region (mW/m^2): o - points of measurements in the wells, black outline - area modeling.

Distribution of the HF in the region corresponds to the data for the areas of Cenozoic tectonic activity. Three areas with different heat flow regime can be distinguished: - high-altitude areas, platform part and modern intermontane depressions. High-altitude areas are characterized by an increased level of heat flow from 54 mW/m^2 to 266 mW/m^2 with an average value of 63 mW/m^2. The heat flow density in intermountain depressions varies from 31 mW/m^2 to 119 mW/m^2 with an average value of 54 mW/m^2. The heat flow of the platform part of the area is 40÷45 mW/m^2. The contribution of the radiogenic component of the heat flow to the total HF is from 60% to 90%.

Geothermal model. The distribution of heat flow, the coefficient of thermal conductivity of rocks and radiogenic heat generation were used to calculate the field of deep crustal temperatures of the Northern Tien Shan (Fig. 4). The structure of the thermal field was studied by dividing the calculated temperatures into a stationary (linear) and anomalous. The anomalous (excess) temperatures are determined by subtracting the stationary component corresponding to a linear geothermal gradient of 12 °C/km from the calculated field.

There is a linear decrease in temperatures from the south to the north from the mountain structures to the platform part of the area. The horizontal temperature gradients decrease from 1.5 °C/km for mountain ranges to 0.5 °C/km in this direction. The absolute temperature is 600-800 °C at the depth of 40 km, 850-1050 °C at the depth of 80 km and 1200 °C at the depth of 100 km. In the depth interval from 20 km to 45 km, a temperature anomaly with increased values of excess temperatures up to 250 °C relative to the linear distribution was determined. The anomaly of excess temperatures is elongated in the sub latitudinal direction just above the boundary of M.

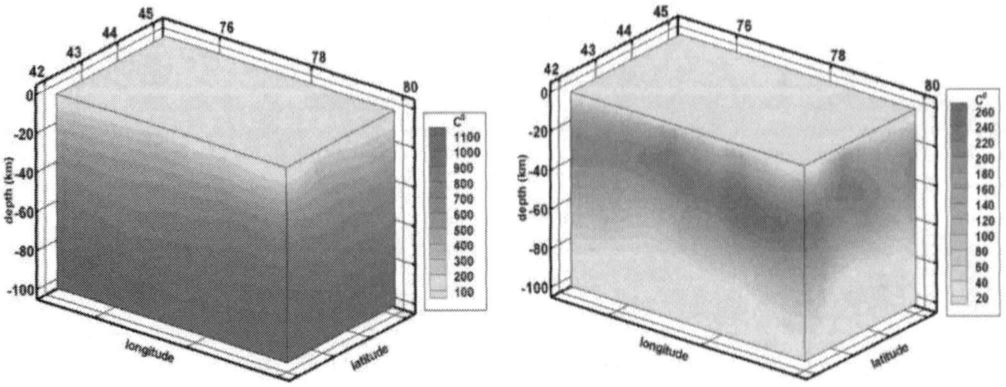

Fig. 4. Results of temperature distribution modelling in the Earth's crust of the Northern Tien Shan region. Left - integrated temperature field (°C), right - anomalous (excess) temperatures (°C).

Thermoelastic stresses. The operating pressure is calculated as the sum of the lithostatic pressure resulting from the mass of overlying rocks and thermoelastic stresses due to uneven heating and inhomogeneity of physical-mechanical properties (equation 5).

At depths of 30÷50 km, the contribution of temperature stresses is 12%÷27% of the lithostatic pressure and reaches 0.45 GPa (Fig. 5). The ratio of the total pressure (with allowance for thermobaric) and lithostatic pressure in the region of anomalous temperatures is 1.3÷1.4. The calculations are in the range of the experimental data of mineral equilibrium, showing that the tectonic pressure can reach up to 100% of the lithostatic and the values of the first gigopascals [11].

 DOI: 10.1051/e3sconf/20172002016

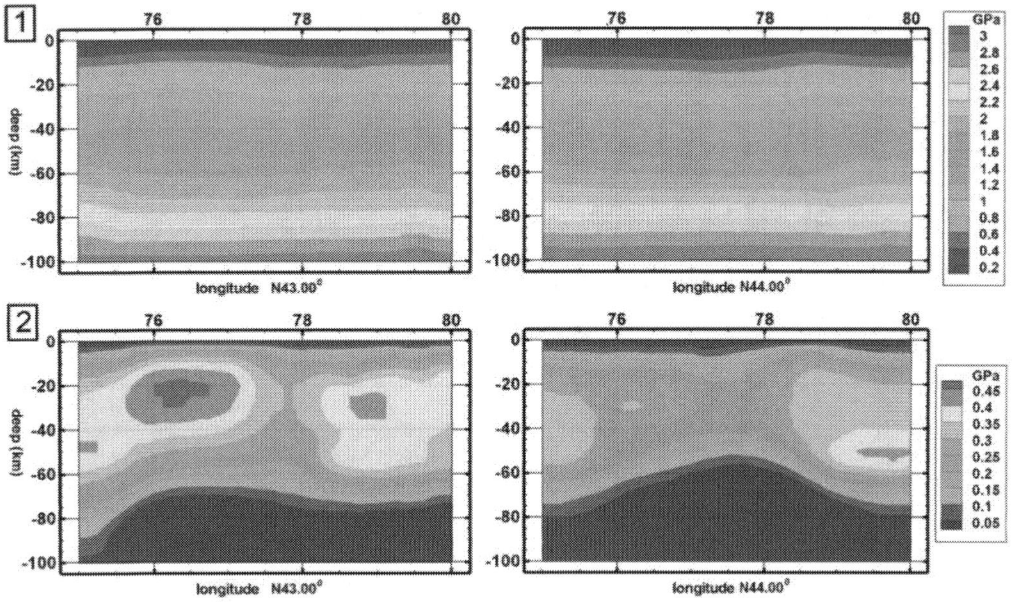

Fig. 5. Depth distribution of thermal stresses in the latitude of N43^0 и N44^0: 1 - total lithostatic pressure and thermal stress, 2 - thermal stresses.

Comparison of geothermal constructions with seismic regime. The seismogenic zone of the Northern Tien Shan is characterized by a seismoactive layer up to 35 km thick. The lower part of the crust practically does not contain foci of earthquakes. The isotherm morphology of excess temperatures of 230 °C in the area of anomalous heating and the isobaric surface of thermoelastic stresses of 0.35 GPa can be attributed to the criteria that controls the distribution of earthquake hypocenters in the earth's crust (Fig. 6).

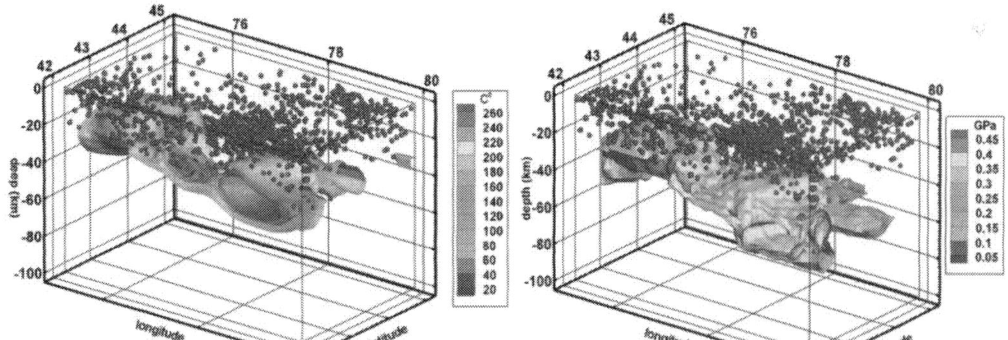

Fig. 6. Isothermal surface of anomalous temperatures of 230 °C (left) and isobaric surface of thermoelastic stresses of 0.35 GPa (right) that control the earthquakes' hypocenters distribution in the earth's crust of Northern Tien Shan.

A comparison of the heat flow distribution and earthquakes of the average energy class (9.0-10.5) distribution on the plan shows that the epicenters are concentrated mainly in the regions of low HF values (Fig. 7). The increased density of epicenters here can be explained by the lower viscosity properties of the earth's crust and increased fragility of the rocks. We can assume a fragile mechanism of the earthquakes formation by reducing the plasticity limit of the material of the crust. The question requires further study.

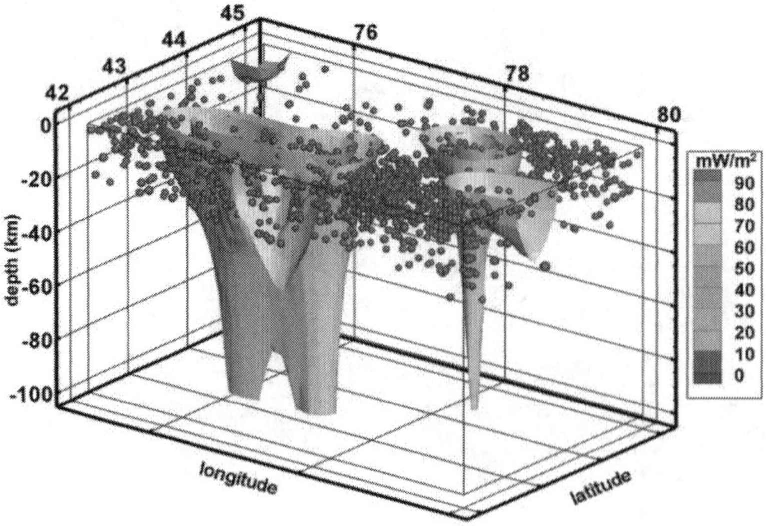

Fig. 7. Model of the heat flow of the Northern Tien Shan with hypocenters of earthquakes

However, the catastrophic earthquakes of the last century (Vernenskoye, Ms = 7.1, 1887; Chilik, Ms = 8.1, 1889; Keminskoye, Ms = 8.0, 1911) occurred in the areas of increased heat flow values. We assume that in these regions the phase transformations of eclogite-basalt under the influence of high-temperature fluids (more than 650 °C) could take place. These temperatures are reached at depths of 35-60 km. In the melting regions, the volume of rocks increases and their density decreases, and as a result, vertical movements within the crust are activated above the melting zone. The mountains are growing up due to the increase in the volume and the growth of the mountains "roots" down through the melting of basalts. Thus conditions for catastrophic earthquakes are being formed.

4 Conclusion

Geothermal modelling allows us to conclude that temperature regime has the prevailing influence on the development of the seismic processes in continental conditions. One of the crucial aspects is determining the high-temperature regions with excess temperatures of 100-250 °C above the solidus, which can lead to partial melting of rocks.

The accumulation of thermoelastic stresses in the earth's crust can be considered as one of the criteria of seismic activity observed in the Northern Tien Shan region. The distribution of the average energy class earthquakes hypocenters is controlled by the region of excess temperatures up to 230 °C and the isobar of thermal stresses less than 0.35 GPa.

The combination of numerical simulation of the earth's crust temperature regime in the Northern Tien Shan region and seismic data provides additional information for predictive results on the possible mechanisms and physical parameters of the seismoactive environment of the region.

References

[1] P. Vincenzo, V. Massimo, Ch.Paolo Geothermics. Heat Flow in the Lithosphere, (2014)
[2] A. Samarskii, A. Gulin Chislennye metody (1989) *(rus)*

[3] J. Jaeger, N. Cook, and R. Zimmerman Fundamentals of rock mechanics (2007)
[4] A. Timush Seismotektonika lithosphery Kazakhstana (2011) *(rus)*
[5] Seismicheskoe rayonirovanie Respubliki Kazakhstan (2000) *(rus)*
[6] L. Rybach, G. Bantebarth Relationships between the petrophysical properties, density, seismic velocity, heat generation and mineral logical constitution, Earth and Planet Sc. letters, **57**, 367-376 (1982)
[7] A. Smyslov, U. Moiseenko, T. Chadovich Teplovoy rezhym i radioaktivnost Zemli (1979) *(rus)*
[8] V. Gordienko, O. Zavgorodniaia, N. Iakobi Teplovoy potok kontinentov (1982) *(rus)*
[9] J. Davies Global map of solid Earth surface heat flow, Geochemistry, Geophysics, Geosystems, doi:10.1002/ggge.20271 (2013)
[10] W. Gosnold, B. Panda The Global Heat Flow Database of The International Heat Flow Commission, http://www.und.edu/org/ihfc/index2.html (2011)
[11] J. Pleuger and Y. Podladchikov A purely structural restoration of the NFP20-East cross section and potential tectonic overpressure in the Adula nappe (central Alps), Tectonics, **33**, 656–685 (2014)

 DOI: 10.1051/e3sconf/20172002017

Symbolic Computations in Simulations of Hydromagnetic Dynamo

Gleb Vodinchar[1,2,*] and *Lyubov* Feshchenko[1]

[1] *Institute of Cosmophysical Research and Radio Wave Propagation FEB RAS, Paratunka, Kamchatsky Kray, Russia*
[2] *Vitus Bering Kamchatka State University, Petropavlovsk-Kamchatsky, Russia*

Abstract. The compilation of spectral models of geophysical fluid dynamics and hydromagnetic dynamo involves the calculation of a large number of volume integrals from complex combinations of basis fields. In this paper we describe the automation of this computation with the help of systems of symbolic computations.

1 Introduction

Many problems of geophysical fluid dynamics reduce to the solution of the Navier-Stokes equations in a rotating coordinate system in an area having spherical symmetry [1]. This area is usually a sphere or a spherical shell. In the most general case we can speak of a spherical shell (inner radius is zero in sphere case). The research of the planetary or stellar dynamo suggests the study of an even more complex system describing magneto-convection. For these systems, exact solutions are unknown and various numerical methods are being intensively used.

The spectral approaches are very popular, when solutions of the non-stationary problem are represented as a linear combination of a finite number of stationary basis fields (modes) with time-dependent amplitudes. For these amplitudes, using one of the weighted residuals methods, a dynamic system is composed. This system together with a set of modes, is simplified model of the initial spatially distributed hydrodynamic or magnetohydrodynamics system [2]. The implementation of this technique involves first construction the dynamic system, and then its numerical study. This system will be quadratically nonlinear with constant coefficients, which are determined by the values of the control parameters of the model and the selected modes.

It should be noted that the construction of the system may require much greater computational effort than its subsequent solution. This is due to the fact that the calculation of the coefficients of the model reduces to the calculation of volume integrals over the spherical shell from various multiplicative combinations of the modes themselves and/or the operator ∇. The number of these coefficients increases in proportion to the third power of the number of modes used. Even for very rough models, including ~ 10 modes, the coefficient count goes to thousands.

However, most of the coefficients of system are zero. The reason is that the spherical symmetry of the region determines the representation of the modes in latitudinal-longitude direction with the

*e-mail: gvodinchar@ikir.ru

help of spherical harmonics. The orthogonality of the spherical harmonics and some related functions on the surface of the sphere determines the vanishing of a large number of coefficients. Sometimes this orthogonality can be shown in a general form, sometimes it is veiled and manifested only in the process of integration, when the integral over the surface of the sphere is first computed, and then along the radius.

The value of every coefficient is the measure of interaction of the corresponding modes in the process, described by term of equation. For this reason exact equality to zero some coefficients, established by analytic integration, can carry considerable physical information. To automate the process of compiling the described dynamic systems, the systems of analytical computations are convenient. In any such system, we can programmatically implement the following calculation scheme:

1. Analytical expressions are construct for each mode.
2. Analytical expressions are construct for all multiplicative combinations (integrands).
3. Analytical integration over the sphere surface is performed. This integration gives zero or some expression depending on the radial variable;
4. If the integral over the sphere surface is non-zero, analytic or numerical integration along the radius is performed.

It is clear that the third and fourth point can be performed by standard means of any system of analytical calculations, because any such system allows you to do analytical and numerical integration. To implement the first and second items, it is necessary to calculate the operations of vector analysis in analytical form in spherical coordinates. For this, it is convenient to use systems of analytical computation, containing vector analysis command packages. For example, the Maple system contains powerful package of such calculations VectorCalculus [3, 4]. However, the necessary minimum set of vector analysis operations is easy to programmed in any system.

Now, we consider the described range of problems and their solution in more detail

2 The original equations and basic systems

The motion of an incompressible fluid in a rotating spherical shell Ω, typical of geophysical fluid dynamics, is described by the Navier-Stokes equations in a dimensionless form [5]:

$$\begin{aligned} &\frac{\partial \mathbf{v}}{\partial t} + (\mathbf{v}\nabla)\,\mathbf{v} = \mathrm{Re}^{-1}\Delta\mathbf{v} - \nabla p - \mathrm{Ro}^{-1}\mathbf{e}_z \times \mathbf{v} + \mathbf{f}, \\ &\nabla\mathbf{v} = 0, \end{aligned} \tag{1}$$

where $\mathbf{v}$ is the velocity filed, p is the effective pressure (including the centripetal term), $\mathbf{f}$ is the mass density of the external forces (the specified field), $\mathbf{e}_z$ is the unit vector of the rotation axis, Re is the Reynolds number, and Ro is the Rossby number.

In the future, we always use the standard connected Cartesian (x, y, z) and spherical (r, θ, φ) coordinate systems, whose origin O is placed in the center of the shell Ω, and the z-axis is directed along the rotation axis.

The system (1) is supplemented by the boundary conditions for the fields $\mathbf{v}$ and p on the inner $r = r_i$ and outer $r = 1$ boundaries of the shell. In case $r_i = 0$, the internal conditions assume the requirements for the finiteness of the fields at the center. The concrete form of the boundary conditions does not yet play a role, we shall only consider them linear and homogeneous.

In the problems of planetary and stellar dynamo, the system (1) is supplemented by equations for the magnetic field [6]:

$$\frac{\partial \mathbf{B}}{\partial t} = \nabla \times (\mathbf{v} \times \mathbf{B}) + \mathrm{Rm}^{-1} \Delta \mathbf{B}, \qquad \nabla \mathbf{B} = 0, \tag{2}$$

where Rm is the magnetic Reynolds number, and on the right side of the first of equations (1) there is also a Lorentz term $(\nabla \times \mathbf{B}) \times \mathbf{B}$. For magnetic field homogeneous linear boundary conditions are also formulated.

Since the velocity and magnetic induction fields has zero divergence, they can be represented as a sum of the toroidal and poloidal components [1], and the inhomogeneity of the fields in the (θ, φ) – direction can be expanded on a spherical harmonics. Then the toroidal basis modes for velocity and magnetic induction will have the form:

$$\mathbf{v}^T_{knm} = \nabla \times \left(u^T_{knm}(r)\, Y^m_n(\theta, \varphi)\mathbf{r} \right), \qquad \mathbf{B}^T_{knm} = \nabla \times \left(b^T_{knm}(r)\, Y^m_n(\theta, \varphi)\mathbf{r} \right), \tag{3}$$

and the poloidal basis modes will have the form:

$$\mathbf{v}^P_{knm} = \nabla \times \left[\nabla \times \left(u^P_{knm}(r)\, Y^m_n(\theta, \varphi)\mathbf{r} \right) \right], \qquad \mathbf{B}^P_{knm} = \nabla \times \left[\nabla \times \left(b^P_{knm}(r)\, Y^m_n(\theta, \varphi)\mathbf{r} \right) \right]. \tag{4}$$

In these formulas $Y^m_n(\theta, \varphi)$ are the spherical harmonics, $\mathbf{r}$ is the radius-vector, the index k corresponds to the quantization of the spatial spectrum in the r – direction, and the specific form of the radial functions u^T_{knm}, b^T_{knm}, u^P_{knm}, b^P_{knm} can be different. For example, they can be chosen so that the modes (3, 4) are modes of free dissipation of the fluid and the magnetic field in the shell. Also, the form of the radial functions depends on specific boundary conditions. We shall further assume that these functions are known.

The solutions of problem (1, 2) are construct in the form of linear combinations of modes (3, 4) with time-dependent amplitudes. The solenoidal nature of the velocity and magnetic induction fields, as well as the boundary conditions, will be satisfied automatically. We shall further exclude the pressure field from the solution. Also we will assume that in the field of external forces the temporal and spatial variables are separated, i.e. $\mathbf{f}(\mathbf{r}, t) = \xi(t)\mathbf{f}_0(\mathbf{r})$.

So, we will use decompositions

$$\mathbf{v} = \sum_i x_i(t)\mathbf{v}_i(\mathbf{r}), \quad \mathbf{B} = \sum_i y_i(t)\mathbf{B}_i(\mathbf{r}), \tag{5}$$

where $\mathbf{v}_i(\mathbf{r})$ and $\mathbf{B}_i(\mathbf{r})$ are any of the fields (3, 4), and the multi-index $i = (\varepsilon, k, n, m)$, $\varepsilon \in \{T, P\}$.

3 The equations of the model

We supplement the magnetohydrodynamics system (1, 2) with the Lorentz term and used decompositions (5). To obtain the model equations in accordance with the methods of weighted residuals, it is necessary to multiply these equations by some weight functions and integrate over the Ω [2]. Let us consider this using the example of the Galerkin method, when the basis modes are used as weight functions.

First we consider the question of excluding the pressure field. In the most general case, it is sufficient to simply apply the rotor operation to both parts of the Navier-Stokes equation. Clearly, this will give not just a transition to the equation-consequence, but it will be an equivalent transformation. However, in the case using of zero boundary conditions for the velocity field, the pressure disappear as a result of the Galerkin procedure itself. Indeed, multiplying the Navier-Stokes equation by mode $\mathbf{v}_l(\mathbf{r})$ and integrating, we obtain for the term with the pressure (taking into account the solenoidal nature of velocity and the divergence theorem)

$$\iiint_{\Omega} (\nabla p)\,\mathbf{v}_l\,dV = \iiint_{\Omega} \nabla\,(p\mathbf{v}_l)\,dV - \iiint_{\Omega} p\nabla\mathbf{v}_l\,dV = \oiint_{\partial\Omega} p\mathbf{v}_l\,dS = 0. \tag{6}$$

Therefore, we assume that a term with pressure is excluded. These two cases do not differ in principle, the difference is only in the technical details of the subsequent calculations, so we confine ourselves to the second case. The implementation of the Galerkin procedure will lead to the system:

$$\begin{aligned}
&\sum_i A_{li}\frac{dx_i}{dt} = \sum_{i,j} B_{lij}x_i x_j + \mathrm{Re}^{-1}\sum_i C_{li}x_i + \mathrm{Ro}^{-1}\sum_i P_{li}x_i + \sum_{i,j} L_{lij}y_i y_j + F_l\xi(t),\\
&\sum_i H_{si}\frac{du_i}{dt} = \sum_{i,j} W_{sij}x_i y_j + \mathrm{Rm}^{-1}\sum_i D_{si}y_i,
\end{aligned} \tag{7}$$

In this system the coefficients are calculated by the following formulas:

$$\begin{aligned}
&A_{li} = \iiint_{\Omega} \mathbf{v}_i\mathbf{v}_l\,dV, \quad B_{lij} = -\iiint_{\Omega} \left(\mathbf{v}_i\nabla\mathbf{v}_j\right)\mathbf{v}_l\,dV,\\
&C_{li} = \iiint_{\Omega} (\Delta\mathbf{v}_i)\,\mathbf{v}_l\,dV, \quad P_{li} = -\iiint_{\Omega} (\mathbf{e}_z\times\mathbf{v}_i)\,\mathbf{v}_l\,dV,\\
&L_{lij} = \iiint_{\Omega} \left[(\nabla\times\mathbf{B}_i)\times\mathbf{B}_j\right]\mathbf{v}_l\,dV, \quad F_l = \iiint_{\Omega} \mathbf{f}_0\mathbf{v}_l\,dV,\\
&H_{si} = \iiint_{\Omega} \mathbf{B}_i\mathbf{B}_s\,dV, \quad W_{sij} = \iiint_{\Omega} \left[\nabla\times\left(\mathbf{v}_i\times\mathbf{B}_j\right)\right]\mathbf{B}_s\,dV,\\
&D_{si} = \iiint_{\Omega} (\Delta\mathbf{B}_i)\,\mathbf{B}_s\,dV.
\end{aligned} \tag{8}$$

The main problem is precisely in the computation of these integrals. It should be noted that the calculations are carried out in a spherical coordinate system, where the expressions are very cumbersome even for the basic differential operators like gradient, curl, and divergence. It is very cumbersome, even in view of the solenoidal nature of velocity modes, the invariant representation [7]:

$$(\mathbf{v}_i\nabla)\,\mathbf{v}_j = \frac{1}{2}\left[\nabla\left(\mathbf{v}_i\mathbf{v}_j\right) - \nabla\times\left(\mathbf{v}_i\times\mathbf{v}_j\right) - \mathbf{v}_i\times\left(\nabla\times\mathbf{v}_j\right) - \mathbf{v}_j\times(\nabla\times\mathbf{v}_i)\right].$$

The transition from such invariant representations through the gradient and curl operators to the tensor formalism in spherical coordinates does not simplify the situation. If we take into account that the modes of speed and induction themselves have a very complicated form (3, 4), it becomes clear that analytical calculations without the use of computer systems are unreal. Even taking into account the obvious symmetries $A_{li} = A_{il}$, $P_{li} = -P_{il}$, $H_{si} = H_{is}$, $C_{li} = C_{il}$, $D_{si} = D_{is}$ situation varies little. The last two symmetries are a consequence of the self-adjointness of the Laplace operator.

The situation will be somewhat simplified if the normalized eigenmodes of the free dissipation of velocity and magnetic induction are used as the basis modes. Then $A_{li} = H_{li} = \delta_{li}$ and matrices with elements C_{li} and D_{si} become diagonal. However, the most difficult to calculate are the coefficients of the nonlinear terms of the system (7).

Consider, for example, the integrand in the coefficient W_{sij} for the case when all three modes are poloidal:

$$\left[\nabla \times \left(\mathbf{v}_i \times \mathbf{B}_j\right)\right] \mathbf{B}_s =$$

$$= \left[\nabla \times \left[\left(\nabla \times \left(\nabla \times \left(u^P_{k_i n_s m_s} Y^{m_i}_{n_i} \mathbf{r}\right)\right)\right) \times \left(\nabla \times \left(\nabla \times \left(b^P_{k_j n_j m_j} Y^{m_j}_{n_j} \mathbf{r}\right)\right)\right)\right]\right] \cdot \left(\nabla \times \left(\nabla \times \left(b_{k_s n_s m_s} Y^{m_s}_{n_s} \mathbf{r}\right)\right)\right).$$

It is clear that after substituting expressions for spherical functions and radial functions, we obtain a construction for which even a simple calculation of all curles and vector products is not possible manually. But we still need to take the integral.

From all of the above, it follows that there is simply no alternative to the use of systems of analytical computation for compiling models of type (7). Here it is also necessary to take into account that after selecting the modes and calculating the model it may turn out that it does not describe all the necessary processes and its interactions inherent in the dynamo-system (1, 2). For example, if all the coefficients are zero, we find that the model (7) does not contain information about the rotation of the shell. In these case, we have to change the set of modes and, at least in part, recalculate the coefficients.

4 The use of symbolic computations for the compilation of model equations

In the Introduction, we mentioned that it is convenient to compile models (7), if the system of symbolic computations used contains a library of functions of vector analysis, and supports work in spherical coordinates. But even if there is no such library, it's easy to describe a minimal set of user-defined structures and functions, sufficient to solve the described problems. For convenience, we will use names of data types and functions from the VectorCalculus library of the Maple package [3, 4].

First of all, we need to define the data structure VectorField, representated of vector fields. This structure must obviously be described as a structural type with three components corresponding to three projections of the field onto the vectors of the local spherical basis. The type of each component is a character expression. We deliberately avoid using the term «field» to refer to components of structural type, so that there is no confusion with scalar and vector fields. If the system used supports an object-oriented paradigm, then VectorField is defined as a class. Each vector field will be a variable of the VectorField type or an object of the VectorField class.

Next, we describe functions or methods that implement a set of necessary operations:

1. addition and subtraction of vector fields (using the object approach it is better to overload the addition and subtraction operations for class objects);
2. product of a vector field and a scalar field – ScalarVectorProduct;
3. scalar product of two vector fields – DotProduct;
4. vector product of two vector fields – CrossProduct;
5. divergence of vector field – Divergence;
6. curl of vector field – Curl;

7. gradient of scalar field – Gradient.

Formulas for the symbol representation of these operations (formulas of vector analysis in spherical coordinates) can be found in many sources, for example, in the popular handbook [7].

Then we need to specify analytic expressions for spherical functions, unless, of course, they are standard in the system used. To determine them we use known recurrence relations for Legendre polynomials and associated Legendre functions [7]. We can simply define the spherical harmonics $Y_n^m(\theta, \varphi)$ in the system as a function of arguments (n, m, θ, φ) by recursion over integer variables n and m, but then each of its calls will take a long time. Therefore, it is better to simply define a two-dimensional array Y[n,m] of symbolic expressions with variables (θ, φ), having previously determined the largest expected value of index n.

Similarly, having determined the maximum assumed value of the radial index k, we form the expression arrays for the radial functions u^T_{knm}, b^T_{knm}, u^P_{knm}, b^P_{knm}. Then, we need to determine the specific type of basis modes. For example, if the modes of free dissipation are used for them, then all radial functions do not depend on index m, and are some linear combinations of power functions and spherical Bessel functions [8]. Expressions for such functions are also easily obtained from known recurrence relations [7].

So, let the expressions for all radial functions and spherical harmonics be constructed and written into the corresponding arrays.

We now define two arrays of multi-indexes for the velocity and magnetic field modes. Their elements are the main input data of the described program of symbolic computations. Based on the values of the elements of these two arrays, we compose arrays of VectorField type, calculating their elements by the values of the multi-index and formulas (3, 4), using certain or standard vector analysis operations. But, at the same time, it is still better not to use specific expressions for radial functions, simply defining them as arbitrary functions of the argument r. This will make subsequent calculations more efficient.

After filling the array of fields, we can proceed directly to the calculation of the arrays of coefficients (8) in cycles.

At each step of the cycle an integrand is constructed, which is multiplied by the product of the scale factors of the spherical coordinates. This expression is first analytically integrated over φ in the range from $-\pi$ to π. If the integral is zero, then the corresponding coefficient is assigned a value of zero. Otherwise, the obtained expression is integrated over θ in the range from 0 to π. If the integral is zero, then the corresponding coefficient is assigned a value of zero.

With this calculation technique, coefficients that are exactly zeroed out at the stage of integration over the sphere are determined. And specific expressions for radial functions are not needed. The system only needs to know, that these are some functions of the argument r. Due to this, the system does not waste computer time and RAM for processing expressions for radial functions. And only if the integral over the sphere turns out to be nonzero, it is necessary to substitute specific expressions for radial function in place of abstract functions and carry out integration over the r. And this last integration can already be performed numerically, because, as a rule, expressions for radial functions contain parameters that are determined numerically. In such a situation, analytical integration is inexpedient.

An exception is the work in the sphere, since at the zero lower limit part of the integrals along the radius become improper. In this case, it is necessary to first set the expressions for the radial functions without specifying the numerical coefficients, to try to perform analytical integration, and, if successful, substitute the values of the coefficients already in the final expression.

5 Conclusion

In this paper, the technology of using the systems of symbolic computation in some problems of geophysical hydrodynamics is described. It was implemented by the authors in the Maple12 package and was successfully used to develop and study some models of geodynamo, for example [8–11]. The technology allows them to automate the routine operations for calculating the coefficients of Galerkin approximations in hydrodynamic problems and significantly increases the reliability and speed of computations.

References

[1] S. Chandrasekhar, *Hydrodynamic and Hydromagnetic Stability* (Dover Publ. Inc., New York, 1981)

[2] C. A. J. Fletcher, *Computational Techniques for Fluid Dynamics* (Springer-Verlag, Berin Heidelberg New York, 1988)

[3] *Maple User Manual* (Maplesoft, Waterloo Maple Inc., 2016)

[4] https://www.maplesoft.com/support/help/maple/view.aspx?path=VectorCalculus

[5] H. Greenspan, *The Theory of a Rotating Fluids* (University Press, Cambridge, 1968)

[6] Y. B. Zeldovich, A. A. Rusmaikin, D. D. Sokoloff, *Magnetic Fields in Astrophysics. The Fluid Mechanics of Astrophysics and Geophysics* (Gordon and Breach, New York, 1983)

[7] G. A. Korn, T. M. Korn *Mathematical Handbook for Scientists and Engineers* (Dover, New York, 2000)

[8] G. M. Vodinchar, A. N. Godomskaya, O. V. Sheremetyeva, Bulletin KRASEC. Physical and Mathematical Sciences, **9(2)**, 27–33 (2014)

[9] G. M. Vodinchar, L. K. Feshchenko, Magnetohydrodynamics, **52(1)**, 287–299 (2016)

[10] G. M. Vodinchar, L. K. Feshchenko, E3S Web of Conferences, **11**, 00025 (2016).

[11] G. M. Vodinchar, A. N. Godomskaya, O. V. Sheremetyeva, E3S Web of Conferences, **11**, 00024 (2016).

DOI: 10.1051/e3sconf/20172003001

Estimation of the efficiency of the earthquake prediction on the basis of the analysis of ionospheric parameters

Vadim Bogdanov[1,*] and *Aleksey* Pavlov[1,**]

[1] *Institute of Cosmophysical Research and Radio Wave Propagation FEB RAS, Paratunka, Russia*

Abstract. A method of the operational forecast of strong earthquakes is considered. Exceedance of the current values of ionospheric critical frequency f_0F_2 over the median values during the disturbed state of the magnetosphere is considered to be a precursor. The reliability, validity and the efficiency of the precursor were evaluated by A.A. Gusev's and G.M. Molchan's methods for the period of 2009-2016. Earthquakes with the magnitude of $M \geq 5.0$ occurring within the zone of precursor appearance determined by Dobrovol'skiy's radius were considered as forecasting earthquakes. It was shown that the precursor has the best predictive efficiency for the seismic events with the magnitude of $M \geq 6.5$.

1 Introduction

Based on the results of vertical radio sounding of ionospheric parameters, it was noted that for Kamchatka one of the most informative parameters is the critical frequency f_0F_2 of the regular F_2 layer [1]. On the eve of the seismic event in the dynamics of the behavior of the ionospheric parameters, the following anomalies are possible:

1. the shift of the maximum of the critical frequency f_0F_2 of the regular layer F_2 to the morning hours from midnight (UT) 3-5 days before the seismic event, which coincides with the results of other authors;

2. the excess of the current values of f_0F_2 over the median values, which is formed against the background of the perturbed magnetosphere, indicating that there is an additional ionization source in the F_2 layer associated with the electric field penetrating the ionospheric heights and resulting currents.

The complexity of observing ionospheric earthquake precursors is the difficulty in isolating lithospheric effects against the background of magnetospheric disturbances, this can only be done for strong events.

*e-mail: vbogd@ikir.ru
**e-mail: pavlov@ikir.ru

DOI: 10.1051/e3sconf/20172003001

2 The algorithm of operational forecast of earthquakes

The algorithm of the operational forecast of earthquakes, presented in this paper is based on a joint analysis of three parameters: the relative values of the deviations $\Delta f_0F_2/f_{med}$ of the critical frequency f_0F_2 from its median values f_{med}, the total values of the K-index for the day and the phases of the moon (Fig. 1). As conditions for declaring the beginning of the alarm period t_{al}, the following criteria are selected: $\Delta f_0F_2/f_{med} \geq 0.2$ or the daily maximum of f_0F_2 falls within the interval 0-6 (UT), the sum of the K-index values for the day $\sum K \geq 20$, with the above two mentioned criteria in the intervals 7 days before and 7 days after the new moon phase. The duration of the alarm period is chosen equal $t_{al} = 5$ days.

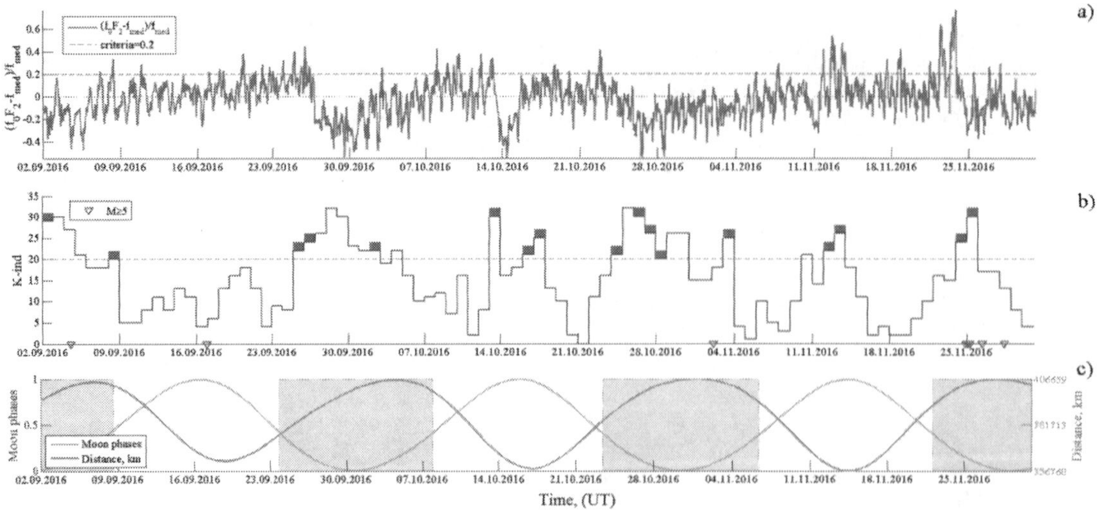

Figure 1. The temporal series of the the relative values of the deviations $\Delta f_0F_2/f_{med}$ (a). The temporal series of the daily values of the K-index (b). The temporal series of the values of the phases of the moon and its distances to the Earth (c). The red rectangles indicate the beginning of the alarm periods t_{al}.

As predicted seismic events, earthquakes from the magnitudes $M \geq 5.0$, $M \geq 5.5$, $M \geq 6.0$, $M \geq 6.5$, occurred at depths of up to 100 km were considered. The radius of the earthquake preparation zone was determined according to the Dobrovolsky formula [2]

$$\lg r = 0.43M$$

and amounted to 140 km, 230 km, 380 km and 620 km for the considered ranges, respectively (Fig. 2) from the point of registration of ionospheric observations of IKIR FEB RAS in Kamchatka.

To analyze the efficiency of the considered precursor, the earthquake catalog of the Kamchatka branch of the Federal Research Center "United Geophysical Service of the RAS" [3] for 2009-2016 was used.

Reliability of the precursor R was determined as the ratio of the number of earthquakes $n(E_A)$ for which the precursor was isolated to the number of all earthquakes $n(E)$:

$$R = n(E_A)/n(E)$$

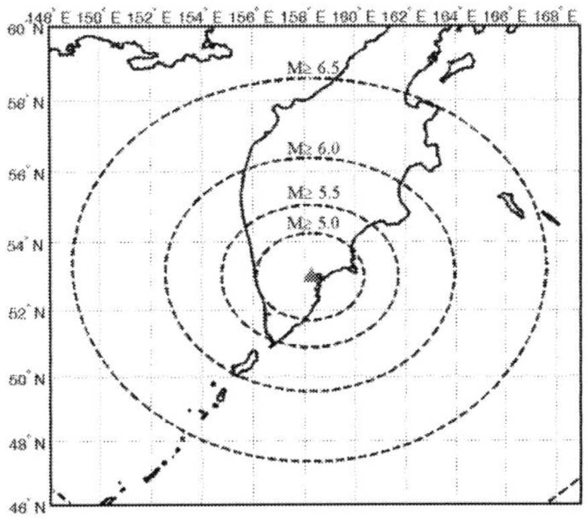

Figure 2. Areas of earthquake preparation with magnitudes $M \geq 5.0$, $M \geq 5.5$, $M \geq 6.0$, $M \geq 6.5$, determined by the Dobrovolsky radius.

The validity of the precursor V was defined as the ratio of the number of predictive anomalies $n(A_E)$ to the total number of identified anomalies $n(A)$:

$$V = n(A_E)/n(A)$$

Table 1. Dependence of the reliability and validity of the precursor from the lower threshold of the considered earthquake magnitudes.

The predicted magnitude	$M \geq 5.0$	$M \geq 5.5$	$M \geq 6.0$	$M \geq 6.5$
r, km	140	230	380	620
Reliability of the precursor R	0.28	0.11	0.38	0.625
Validity of the precursor V	0.057	0.049	0.065	0.04

According to the results presented in Table 1, the reliability of the considered precursor is higher for earthquakes of higher magnitudes than for smaller ones. For earthquakes with a magnitude $M \geq 6.5$, the reliability R was 0.625. The validity of the precursor V for all the magnitudes under consideration is low and its values do not exceed 0.065. Since it is not uncommon for a precursor to be assigned several earthquakes (a group) and they need to be taken into account when deciding whether to implement a prediction and estimating the alarm time, the following algorithm was applied: 1) the forecast is considered to be the first earthquake that was implemented, corresponding to the forecast of the magnitude interval, 2) the alarm is removed, 3) the following earthquakes, which could formally correspond to the same predictive anomaly, are declared to be missed ("skipping the target") [4].

To evaluate the efficiency of the considered precursor, approaches by A.A. Gusev [5] and G.M. Molchan [6] were used.

Efficiency of the forecast according to A.A. Gusev's technique is calculated for a particular spatial region and a certain energy range of earthquakes according to the formula:

$$J_G = \frac{N_+/T_{alarm}}{N/T}$$

where T is the total time of monitoring the seismic situation; N_+ – the number of earthquakes that correspond to a successful prediction in time T; N – the total number of earthquakes that occurred (having spatio-temporal characteristics, similar to the predicted ones) that occurred during the time T; T_{alarm} – total alarm time (the total duration of all time intervals in which the forecast for the estimated method operated during the total monitoring time). The statistical significance of α is determined by the probability of obtaining the efficiency values of J_G in the absence of an earthquake-precursor relationship. In the absence of such a connection, i.e. when random guessing, the efficiency of J_G is 1.

Table 2. Estimation of the efficiency of the precursor according to the A.A. Gusev method.

The predicted magnitude	$M \geq 5.0$	$M \geq 5.5$	$M \geq 6.0$	$M \geq 6.5$
r, km	140	230	380	620
N_+	7	6	8	5
N	25	51	21	8
T (day)	2922	2922	2922	2922
$T_{expected}$ (day)	426	419	421	429
Efficiency J_G	1.97 ± 0.72	0.82 ± 0.33	2.64 ± 0.93	4.25 ± 1.9
α	0.08	0.84	0.006	0.0025

As can be seen from the results presented in Table 2, the efficiency of J_G taking into account the standard deviation value assumes the largest values for the range of predicted magnitudes $M \geq 6.0$ and $M \geq 6.5$. At the same time, the statistical significance of α, which is 0.006 and 0.0025, respectively, demonstrates a low probability of randomly obtaining such values in the absence of earthquake connection with the forerunner in question.

The efficiency of the precursor J_M by the method of G.M. Molchan is defined as $J_M = 1 - \nu - \tau$, where $\tau = T_{alarm}/T$ – alarm measure; $\nu = 1 - N_+/N$ – the share of the misses of the target. For a random prediction $J_M = 0$, and for the ideal (without a miss target and with zero alarm time) – $J_M = 1$.

Table 3 shows the values of J_M for four values of the threshold magnitude of earthquakes $M \geq 5.0$, $M \geq 5.5$, $M \geq 6.0$, $M \geq 6.5$. For high magnitudes threshold ($M \geq 6.5$), the efficiency is substantially (almost 3.6 times) higher than for low ($M \geq 5.0$).

Table 3. Estimation of the efficiency of the precursor according to the G.M.Molchan method.

The predicted magnitude	$M \geq 5.0$	$M \geq 5.5$	$M \geq 6.0$	$M \geq 6.5$
r, km	140	230	380	620
τ	0.146	0.14	0.14	0.147
ν	0.72	0.88	0.62	0.375
Efficiency J_M	0.13	-0.025	0.237	0.478

Figure 3 shows the Molchan diagram for the magnitude thresholds under consideration. On these diagrams, the abscissa of the point is defined as the anxiety measure τ, and the ordinate is defined as the fraction of the passes of the target ν. The diagonal of the Molchan diagram $\tau + \nu = 1$, connecting the points (0; 1) (the "optimist" point) and (1; 0) (the "pessimist" point) corresponds to a random prediction. For this diagonal, it is possible to construct a confidence interval with a given significance α. In this case, only the lower branch of this confidence interval is of interest. As follows from Fig. 3a and Fig. 3b, the values of (τ, ν) obtained for magnitudes $M \geq 5.0$ and $M \geq 5.5$ lie above the lower limit of 99% of the confidence interval, which indicates the absence of a connection between the forerunner under consideration and earthquakes with given magnitudes that occurred on Distances of 140 km and 230 km from the observation point, respectively. The values (τ, ν) obtained for earthquakes with magnitudes $M \geq 6.0$ (Fig. 3c) and $M \geq 6.5$ (Fig. 3d) lie below the obtained confidence interval boundary, which can be interpreted as a high degree of reliability of the revealed connection of the forerunner in question with earthquakes of this magnitude range.

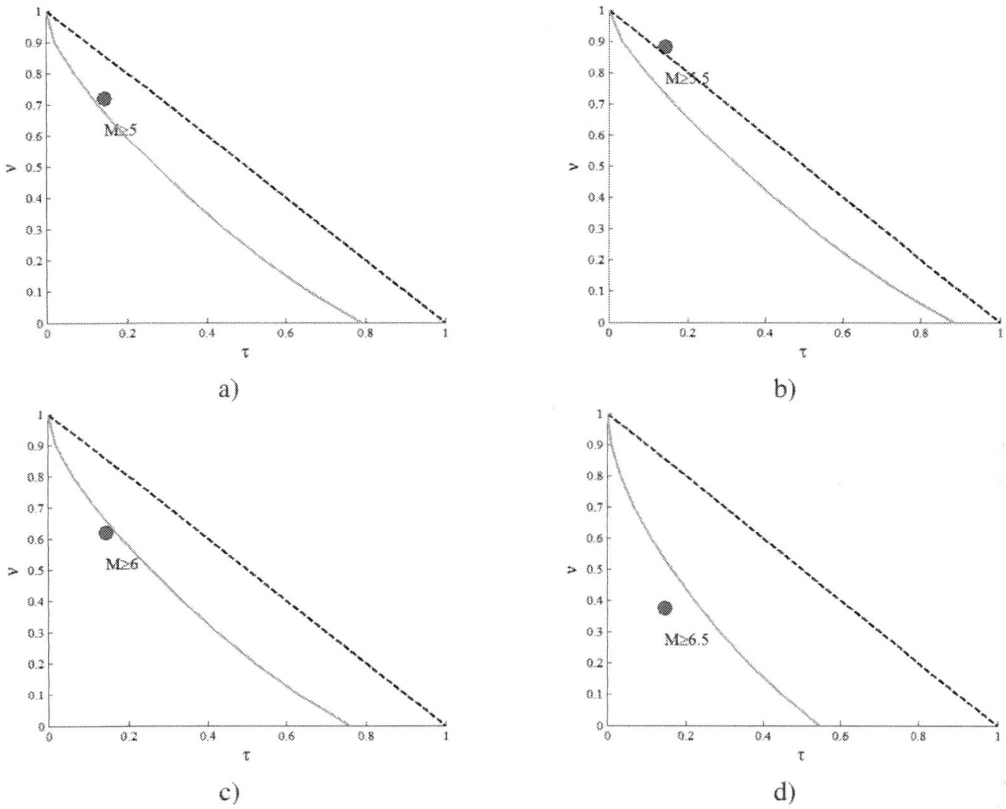

Figure 3. Molchan diagrams for different thresholds of magnitude M. The lower bounds of the confidence interval of the random prognosis with the significance level $\alpha = 0.01$ are noted.

3 Conclusion

An algorithm for the operational prediction of earthquakes is presented, in which the current values of the critical frequency f_0F_2 of the ionospheric layer F_2 exceed the median values during the periods of the disturbed state of the magnetosphere and the periods of the new moon as a precursor. For the predicted ranges of magnitudes $M \geq 5.0$, $M \geq 5.5$, $M \geq 6.0$, $M \geq 6.5$, the reliability R and the reliability of V have been calculated, and the efficiency of the precursor by the methods of A.A. Gusev and G.M. Molchan for the period 2009-2016. It is shown that for predicted earthquakes with a magnitude $M \geq 6.5$ the reliability of the precursor R reaches the highest value and is 0.625, and the reliability of V for all considered magnitudes has low values and its value does not exceed 0.065. The efficiency of J_G by the A.A. Gusev assumes the values 2.64 and 4.25 for the ranges of the predicted magnitudes $M \geq 6.0$ and $M \geq 6.5$, respectively. In this case, the statistical significance of α, which is 0.006 and 0.0025, respectively, demonstrates a low probability of randomly obtaining such values

DOI: 10.1051/e3sconf/20172003001

in the absence of earthquake connection with the forerunner in question. The analysis of efficiency, conducted by the method of G.M. Molchan also shows a high degree of reliability of the revealed connection of the precursor in question with earthquakes from the magnitudes $M \geq 6.0$ and $M \geq 6.5$.

References

[1] V.V. Bogdanov, A.V. Buzevich, A.V. Vinitsky, G.I. Druzhin, A.V. Kuptsov, I.N. Poddelsky, S.E. Smirnov, B.M. Shevtsov, *Complex seismological and geophysical research of Kamchatka. To the 25th anniversary of the Kamchatka Experimental-Methodical Seismological Department of the GS RAS* (KEMSD GS RAS, Petropavlovsk-Kamchatsky, 2004) 259–278

[2] I.P. Dobrovolsky, S.I. Zubkov, V.I. Myachkin, *Modeling earthquake precursors* (Nauka, Moscow, 1980) 7–24

[3] http://www.emsd.ru/sdis/earthquake/catalogue/catalogue.php

[4] V.A. Saltykov, Physics of the Earth **2**, 84–96 (2017)

[5] A.A. Gusev, *Seismicity and seismic forecast, properties of the upper mantle and their relation to volcanism on Kamchatka* (Nauka, Novosibirsk, 1974) 109–119

[6] G.M. Molchan, Physics of the Earth and Planetary Interiors **61**, 84–98 (1990)

Electromagnetic and acoustic radiations before Kamchatka earthquake

Yuriy Marapulets[1,*], *Gennadiy* Druzhin[1], *Nina* Cherneva, and *Alexsandra* Solodchuk[1]

[1] Institute of Cosmophysical Research and Radio Wave Propagation FEB RAS

Abstract. The most probable cause of simultaneous anomalous radiations in electromagnetic and acoustic fields, different in nature, is intensification of deformation processes during strong earthquake preparation. To verify this fact, simultaneous observations of electromagnetic signals and acoustic emission in the frequency range from 0.1 Hz to 11 kHz were carried out in Kamchatka. In the result of the experiment, radiation bursts appeared about 24 hours before a seismic event with MLH=5.4. Such bursts were not observed on other days within this time of a day in August-September, 2013.

1 Introduction

Investigation of acoustic and electromagnetic fields during increased seismic activity were carried out in different regions [1, 2], however, there are just a few simultaneous observations in natural conditions. The authors know only some cases of registration of anomalous radiations of these fields before seismic events with MLH≥5 [3]. In the laboratory research [4] it was registered that during rock deformation stress, acoustic and electromagnetic radiations are generated. To detect natural electromagnetic and acoustic emissions assuming common deformation nature of their occurrence, simultaneous observations of these fields were carried out in the seismically active region of Kamchatka peninsular in August-September, 2013.

Observations of electromagnetic field variations were carried out at «Karymshina» site (52.82N, 158.13E). Noise radiations were registered by a multichannel VLF-detector applying magnetic and electric antennas. The signals were recorded in a digital form. The sensitivity of VLF-detector magnetic component was not less than 2.10-7 nT/Hz½, and that of electric component was not less than 6.10-8 V/(mHz½).

We should note that the main noise sources of natural electromagnetic radiations in the North-East of Russia [5] are world lightning centers located in the near-equatorial region of the Earth.

Signals from the regional lightning were registered by VLF-detector at «Paratunka» site (52.97N, 158.25E), located at the distance of 18 km from «Karymshina» site when the threshold level of 1 V/m was exceeded [6].

* Corresponding author: marpl@ikir.ru

To register the acoustic emission, a complex based on a piezoceramic receiver (hydrophone) was used. It is installed at «Mikizha» site (52.99N, 158.23E) at the distance of 20 km from «Karymshina» site by the bottom of a lake [7]. The sensitivity of the hydrophone together with the pre-amplifier is the first hundreds of mV/Pa. The results of investigation of the acoustic emission in Kamchatka showed that acoustic radiation increase in the high-frequency range from hundreds of Hz to the first tens of kHz is determined by the intensification of rock deformations in the observation area including those associated with earthquake preparation [8]. Thus, high-frequency acoustic emission can be applied as a sensitive indicator of intensification of the deformation processes preceding seismic events [8, 9].

2 Detection of anomalies during electromagnetic and acoustic observation analysis

In the course of data analysis it was discovered that radiation bursts almost simultaneously occurred in electromagnetic and acoustic fields in the frequency range from 200 Hz to 11 kHz (Fig. 1, at the bottom) on September 1, 2013 between 04:00 and 09:00 UT at «Karymshina» and «Mikizha» sites. Such bursts were not observed on other days within the same period of a day. Azimuthal distribution of lightning strokes showed (Fig. 1, at the top), that their number increased significantly from the south-western direction (~ 2000 – 2400). There were also radiations from the eastern direction (~ 450 – 1350). It was assumed that the increased radiations were, probably, associated with deformation processes in the Earth crust. But there were doubts that the detected bursts are associated only with the intensification of lightning activity.

To determine the real cause of anomaly occurrence, noise radiations registered at «Karymshina» site were compared with the direction-finding data obtained at «Paratunka» site and with the data of the World Wide Lightning Location Network (WWLLN) [http://wwlln.com]. An on-line Earthquake Catalogue of Kamchatka Branch of Geophysical Service RAS [http://www.emsd.ru] was also considered.

Fig. 2 shows the direction-finding observations at «Paratunka» site. Vertical lines mark the «anomalous» period which coincides with that in Fig. 1. It is clear from the graphs that the increased level of radiations is observed from the direction of 1800 – 2700. Lightning locations from August 31 to September 2, 2013 were determined by WWLLN (Fig. 3).

In order to check the reliability of the determined directions to the radiation sources, the real azimuths of «Alfa» radio station working at the frequencies of 11.90, 12.65 and 14.88 kHz were compared with those obtained during signal reception at «Karymshina» and «Paratunka» sites. The error did not exceed 3 degrees. Moreover, we also used the lightning sources registered by WWLLN, including the lightning source near Kluchevskaya sopka and Siveluch volcanoes which was observed on August 31, 2013 (Fig. 3).

 DOI: 10.1051/e3sconf/20172003002

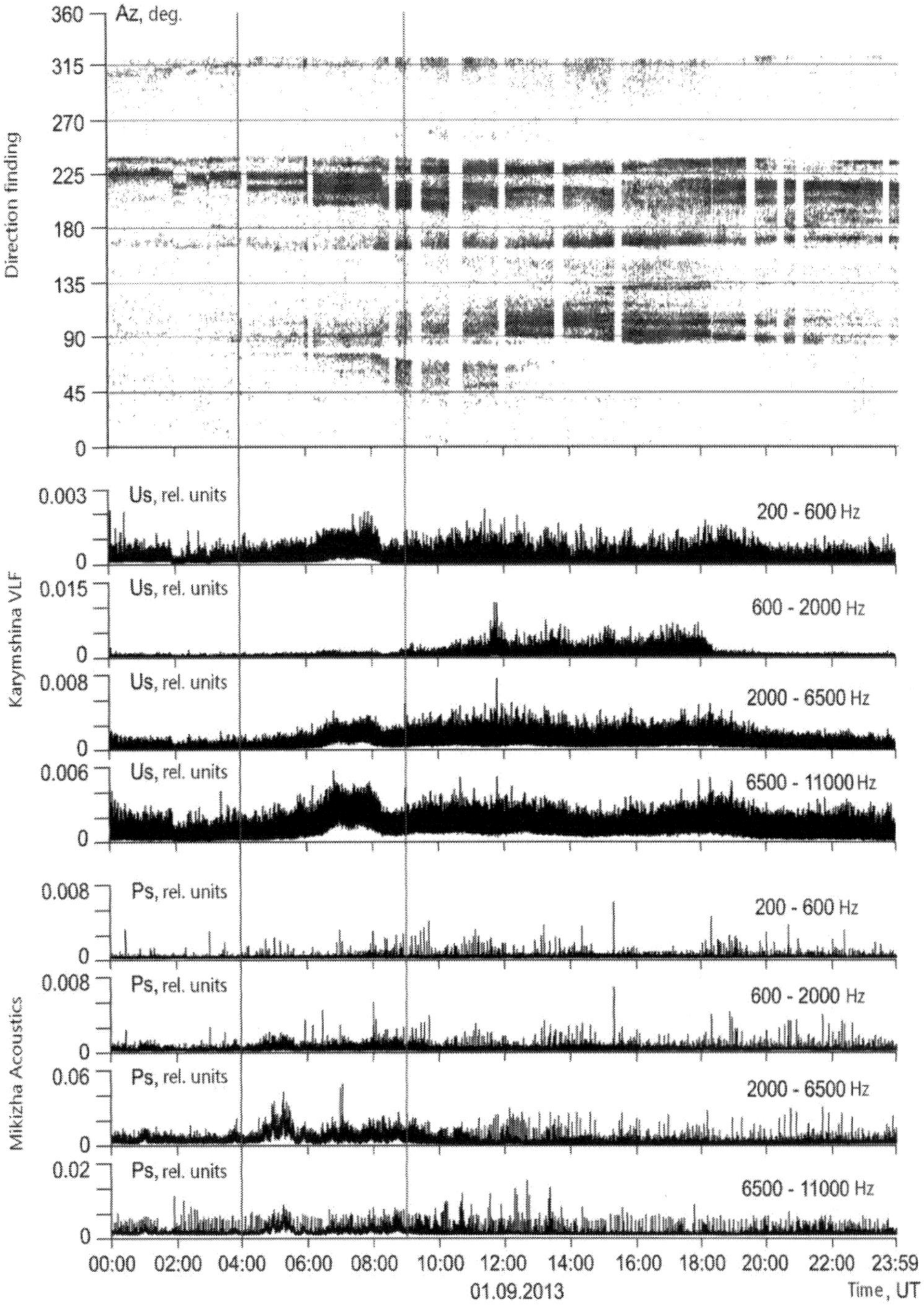

Fig. 1. Simultaneous observations of acoustic and electromagnetic radiations on September 1, 2013 at «Karymshina» (VLF) and «Mikizha» (acoustics) sites. In the upper part of the figure is the azimuthal distribution of lightning stroke number AZ (dots) within a day. The azimuth for «Karymshina» site is counted clockwise from the northern direction. In the bottom part of the figure are the amplitude variations of the envelope of VLF-radiation US acoustic emission PS, in four frequency sub-ranges. Vertical lines indicate the time interval from 4:00 till 9:00 UT, within which anomalous bursts of radiation were recorded.

DOI: 10.1051/e3sconf/20172003002

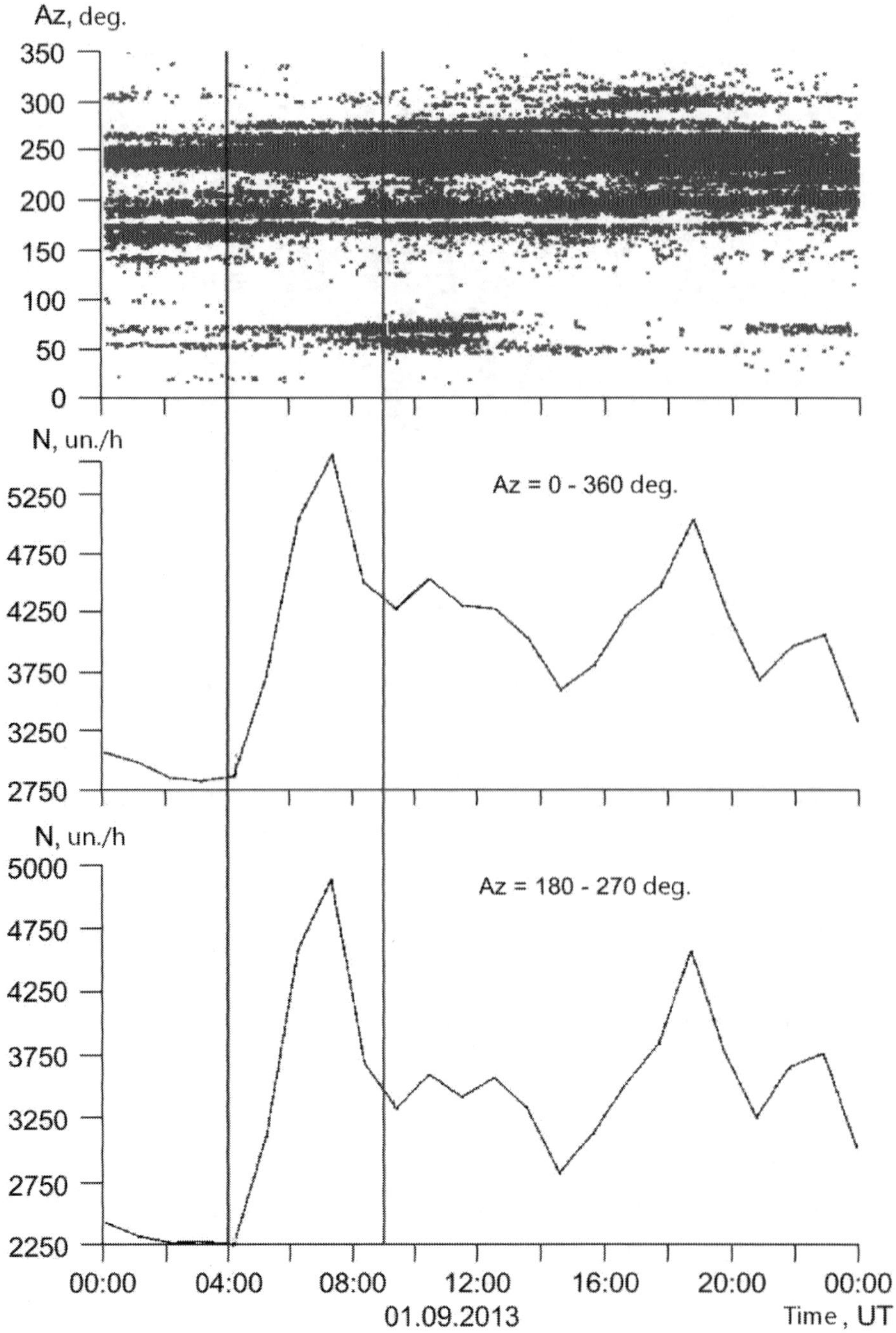

Fig. 2. Direction-finding observations of electromagnetic radiations on September 1, 2013 at «Paratunka» site. Azimuthal distribution of the sources registered at «Paratunka» site is shown in the upper part of the figure. Hourly numbers of lightning strokes the radiations from which were received from the directions of 0 – 360 degrees and 180 – 270 degrees are in the bottom. Vertical lines indicate the radiation area from 4:00 to 9:00 UT.

 DOI: 10.1051/e3sconf/20172003002

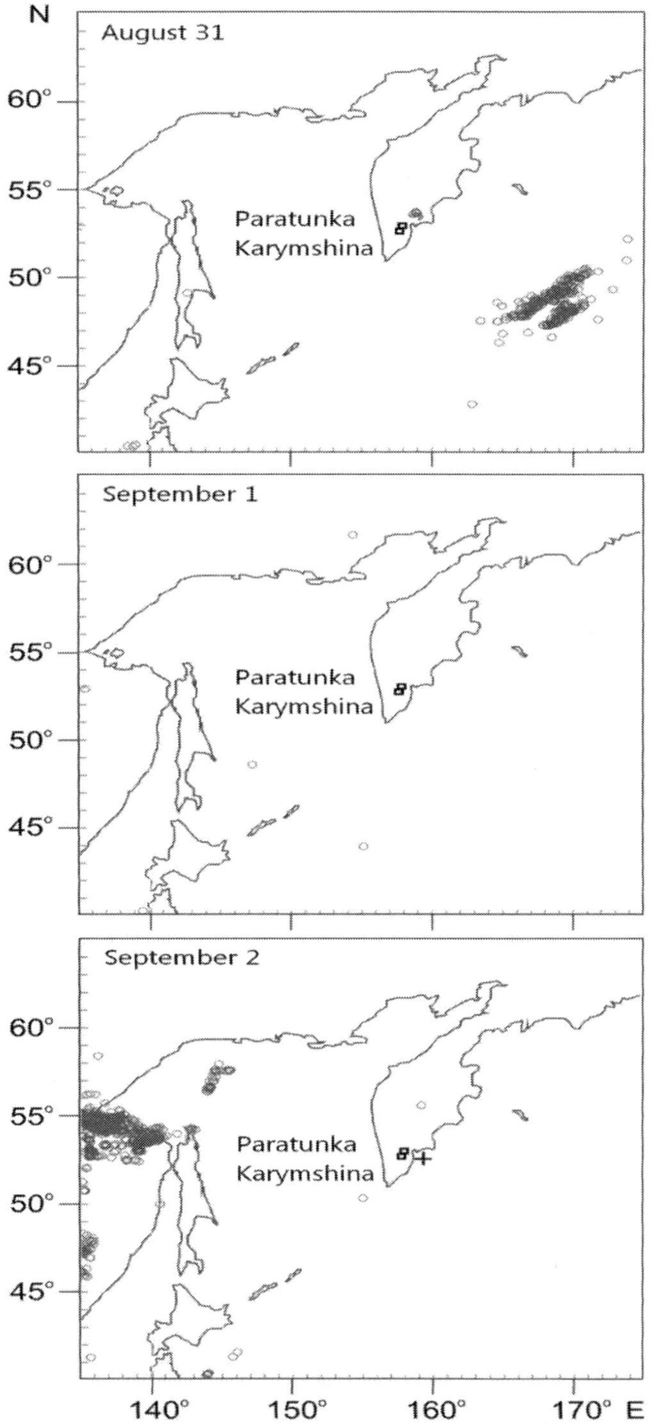

Fig. 3. A map with lightning strokes (circles) which occurred near Kamchatka from August 31 to September 2 and the earthquake (+) which occurred on September 2, 2013. Site locations are indicated by squares.

It is clear from Fig. 3 that lightning activity was quite high on August 31, on September 1 (during the anomaly) the lightning was not almost observed, and on September 2 it appeared again. The inconsistency of the number of received signals from lightning by WWLLN network and «Karymshina» and «Paratunka» sites is explained by the difference in instrumentation and radiation registration methods. It was established by the researches which had been carried out before [6] that in spite of the indicated differences, quite good agreement of the data was observed when lightning azimuths were determined. Thus, during the period of observation of anomalous bursts on September 1, there was no strong lightning.

It is quite possible that some local radiation sources appeared, the power of which is sufficient to be detected by closely located electromagnetic sensors located at «Karymshina» and «Paratunka» sites but insufficient to be registered by remote sensors of WWLLN network due to signal strong attenuation.

3 Discussion of the obtained results

Based on the data of the on-line earthquake catalogue, analysis of the seismic state revealed that the earthquake with the energy class of K=12.3 (MLH=5.1) occurred on September 2, 2013 at 8:19 UT, that is about one day after the anomalous radiations, at the epicentral distance of 140 km from «Karymshina» site. For the earthquake energy classification, we used classes K according to S.A. Fedotov's scale [10]. The relation of K to magnitude MLH is determined by formula $MLH=(K-4.6)/1.5$). We should note that this was the only earthquake with $MLH \geq 5$, which occurred in the first half of September in the south-eastern part of Kamchatka.

Taking into account all the abovesaid, the following scenarios of anomalous acoustic and electromagnetic radiations are the most probabilistic:

1. A powerful remote source (not associated with the seismic event) became active. The radiation from it came from the south-western direction, in the result, VLF signal level increased during the time period indicated by the vertical lines shown in Fig. 1 and Fig. 2. But in this case, the cause of acoustic emission anomalies is unclear (Fig. 1).

2. Remote source activity changed insignificantly but in the result of deformation processes, the propagation conditions changed (for example, the Earth's crust conductivity near the observation site changed) that caused intensification of the signals received from lightning. This fact may take place since considerable increase of electromagnetic field pulse flux was observed near the Earth crust fractures [11].

3. A closely located source of electromangetic radiation, associated with deformation processes in the earth crust, appeared. This scenario is quite possible as long as radiations were observed at the same time both in acoustic and electromagnetic fields.

Simultaneous anomalies of acoustic and electromagnetic radiations are often registered 1-2 days before quite strong earthquakes in Kamchatka. Based on the on-line earthquake catalogue, 11 seismic events with $MLH \geq 5$ occurred at the hypocentral distances up to 200 km from «Karymshina» site in 2013. From them, except for the case described above, radiation bursts were recorded before 6 seismic events (Table) that results in 63.6% from all the earthquakes with $MLH \geq 5$. Examples of anomalies are shown in Fig. 4.

Table. Simultaneous anomalies of acoustic and electromagnetic radiations before earthquakes with MLH≥5.

No.	Anomaly duration, hours	Earthquake advance time, hours	Earthquake data, date, time, M_{LH}
1	3.8	38.6	09.03.13, 14:56 UT, M_{LH}=6.0
2	2.9	22.8	03.05.13, 06:10 UT, M_{LH}=5.6
3	7.2	30.6	28.05.13, 16:25 UT, M_{LH}=5.4
4	5.0	33.5	15.07.13, 14:43 UT, M_{LH}=5.6
5	0.9	23.8	28.07.13, 07:21 UT, M_{LH}=5.3
6	3.0	29.8	21.09.13, 06:58 UT, M_{LH}=5.4

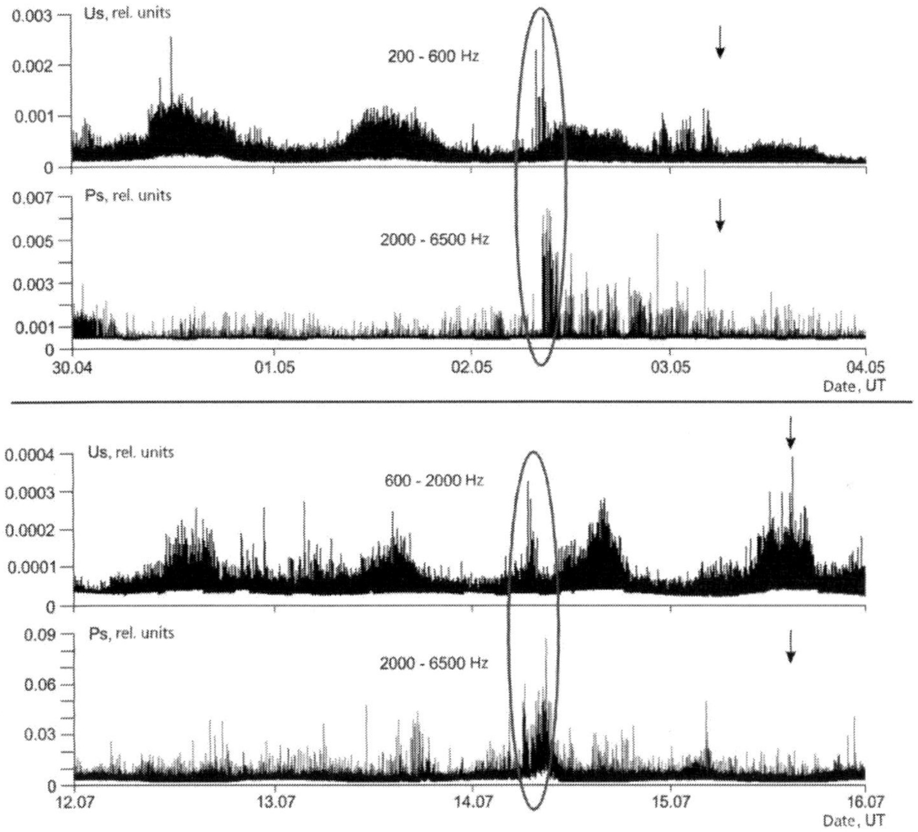

Fig. 4. Simultaneous observations of electromagnetic (US) and acoustic (PS) radiations before earthquakes on May 3 (at the top) and on July 14 (at the bottom). Anomalous bursts of radiations are marked by ellipses. Arrows indicate the earthquake times.

4 Conclusions

Thus, in the result of observations of acoustic and electromagnetic fields in 2013, it was discovered that more than a half of seismic events with MLH≥5 at the hypocentral distance up to 200 km are proceeded within 1-2 day interval by simultaneous bursts of acoustic and electromagnetic radiations in the frequency range from 200 Hz to 11 kHz. To the author's opinion, the most probable reason of occurrence of anomalous radiations in the fields different in nature is the intensification of deformation processes during earthquake preparation.

References

1. B.M. Gokhberg, V.A. Morgunov, O.A. Pokhotelov, [Seismoelectromagnetic phenomena] (Nauka, Moscow, 1988) 174 p.
2. M.A. Sadovskiy, [Electromagnetic precursors of earthquakes] (Nauka, Moscow, 1988) 89 p.
3. V.A. Gavrilov, Doklady Earth Sciences, **414**(1), 638-641 (2007)
4. Y. Mori, Y. Obata and J. Sikula, Acoustic Emission, **27**, 157-166 (2009)
5. G.I. Drugin, V.I. Shapaev , Geomagnetism and Aeronomy, **28**(1), 81-86 (1988)
6. G.I. Drugin, N.V. Cherneva, A.N. Melnikov, Russian Meteorology and Hydrology, **36**(7), 447-452 (2011)
7. A.V. Kuptsov, Izvestiya, Physics of the Solid Earth, **41**(10), 825-831 (2005)
8. G.I. Dolgih, A.V. Kuptsov, I.A. Larionov, et. al., Doklady Earth Sciences, **413**(1), 281-285 (2007)
9. Yu.V. Marapulets, B.M. Shevtsov, I.A. Larionov et. al., Russian Journal of Pacific Geology, **6**(6), 457-464 (2012)
10. S.A. Fedotov, [Energy classification of the Kuril-Kamchatka earthquakes and the problem of magnitudes] (Nauka, Moscow, 1972) 116 p.
11. E.L. Afraymovich, G.A. Zherebtsov, [Seismoionospheric and seismoelectromagnetic processes in the Baikal rift zone] (SB RAS, Novosibirsk, 2012) 304 p.

Gasgeochemical indicators seismic activity

Anatoly Obzhirov

V.I.Il'ichev Pacific Oceanological Institute (POI) FEB RAS, Russia, 690041, Vladivostok. Baltiyskaya str. 43

Abstract. Laboratory of Gasgeochemistry of POI FEB RAS is studying gas distribution in lithosphere, hydrosphere and atmosphere from 1977 years. Method consist is sampling from its in expedition, take gas from samples of sediment, water and atmosphere to use method degassing and analysis gas in chromatograph, to measure CH4, C2-C4, O2, N2, H2, He and some time Rn. Gas is using like indicators to search oil-gas deposits, gas hydrate, mapping zones faults, to determine seismic activity, to calculate green house gas (CH4, CO2). The next geological, geophysics and hydro-acoustics characteristics assist which help to explain to form methane bubbles fluxes and gas hydrate in the Okhotsk Sea. The methane fluxes are mostly located in the zones faults and it increase in period seismic activity.

1 Introduction

The first discovery of methane fluxes and gas hydrate field in the Okhotsk Sea was in the 1988 year. After it many international expedition were provided in Sakhalin shelf and slope of the Okhotsk Sea (fig.1) to search flux of methane and gas hydrate. It carried out in frame international projects – KOMEX (Russian-Germany, 1998-2004), HAOS (Russian-Japan-Korea, 2003, 2005-2006), SAKHALIN (Russian-Japan-Korea, 2007-2015), GAS-HYDRATE (Russian-China, 2016-2018). During the process of gas geochemistry investigations in period 1988-2016 in the Okhotsk Sea were found 500 gas fluxes, 17 areas with gas hydrate. Gas bubbles migrate via zones faults from sediment to water and some part to atmosphere. The next component of gas explain fellow. Methane and heavy hydrocarbon (C2-C4) show content oil-gas deposit and gas hydrate in the structure, CO2 is indicator to presence intrusion complex and H2, He show appear deeper flux of gas from bottom lithosphere or mantel via zone fault because of seismic activity. Concentrations and quantity gas components relationship with seismic-tectonic activity or period stability of it. In Far-Eastern regoin this process to start from 1988. and it to continue now. In zones faults with fluxes gas sediment in bottom Sea to disturb, to appear pits and knoll (valleys and mountains) 10-20 m size and to appear layers of gas hydrates, carbonate concretions, benthos and other modification. It is very important that quantity and power of methane fluxes increase in period seismic activity of the Far Eastern Seas region. It may use like indicator seismic activity and earthquake prognosis. Methane is going from deep layers to

Corresponding author: obzhirov@poi.dvo.ru

up in surface sediment and water column via zone fault. Sources of methane are oil-gas bearing layers and more deeper horisonts of litosphere

2 Method of investigations

It is very important to use geological complex to study methane flaxes and gas hydrate. 12 international Russian-Japan-Korea expeditions from 2003 to 2015 years were conducted on the East Sakhalin slop of the Okhotsk Sea using the Russian research vessel "Akademik M.A.Lavrentyev". All geological, geophysical, gas geochemical, hydro-acoustic, morpho-structure complex are very important indicator to search methane fluxes and gas hydrate and to understand regularity to form and to destroy of its. Multidisciplinary data were obtained during these expeditions: hydro-acoustic profiling using single-beam echosounders, seafloor imaging using side-scan sonar (SSS), seismic reflection profiling using a sub-bottom profiler and sparker-source seismic system, CTD profiling, water column sampling from bottom to surface using a Rosette sampler and near-bottom sediment and gas hydrate sampling using a gravity and hydro corer
Gas bubbles in the water column have high impedance contrast relative to water and appear on echograms as hydro-acoustic sound-scattering anomalies like vertical flux (gas "flares") in water column from Sea floor to up in surface (figures 1a and 1b). The hydroacoustic systems 'ELAC' and/or 'Sargan-EM' were used simultaneously with frequencies of 12, 20, and 135 kHz to detect gas flares [2, 10, 14].

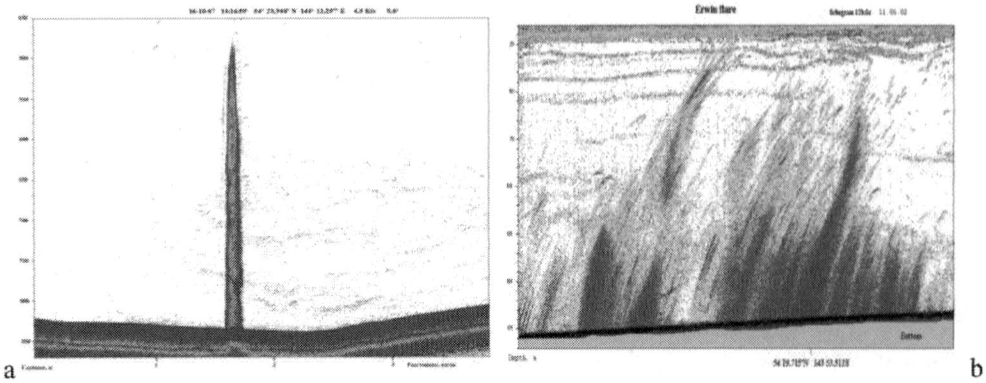

Fig. 1. a) Hydro-acoustics image of methane babbles flux from sediment to water in the Sakhalin slope of Okhotsk Sea (depth 820 m), b) Gas flux in the shelf, depth 175 m

Deep-tow side-scan sonar surveys were conducted using the 'SONIC-3' system with a frequency of 30 kHz in a medium-range mode and a swath range of 800–3200 m. Width range is 500–2,000 m, and the best resolution is 2.0 m. The 'SONIC-3M' system is also equipped with a sub-bottom profiler with a frequency of 8 kHz (maximum resolution = 0.3 m). The sonar "fish" was towed at 90–120 m above the seafloor and ship speed was 2–4 knots [2].

High-frequency seismic data were obtained using the sparker system made in Russia [14]. The system used a sparker with a frequency range of 200–1,200 Hz and an energy of 500–2000 J as a source, and a single-channel streamer as a receiver. The system provides a resolution of 2–5 m for sediment layer thickness and a bottom penetration of 50–300 m.
To measure concentrations `methane, heavy hydrocarbon (C2-C4), CO2, H2, O2, N2 in the water column and sediment was used method Head Space and degassing. Water for gas analyze was taken from Niskin bottles in glasses volume 128 ml with added chlorhexidine

 DOI: 10.1051/e3sconf/20172003003

(0.05%) to preserve natural gas from microbial change some concentration of it and immediately close with contact of air. Pure helium was used as the gas phase and it was added into glasses with water sample. Gas sample is injected into the chromatograph after intensive manual shaking and stirring on the mixer LS110 during four and more hours. Gas chromatograph "Cristall – Lux 4000M" (Russia) equipped with a flame ionization detector and two thermal conductivity detectors (FID/2 and TCD) used for gas chromatographic analysis the gas composition on board (figure 2a). Methane concentration was calculated as the sum of the dissolved fraction [14] and methane gas phase. The precision of the analysis was 5%.

The sediment samples to measure gas were taken from core with 12 ml plastic syringes via 20 cm interval (figure 2b). It transferred into 68 ml glasses with a saturated water solution of NaCl containing 0.5ml of chlorhexidine (0.05%). Then following the procedures described to measure gas in water samples o use method of Head space. Calculation of methane in sediments produced without the dissolved fractions of gas in water solution of NaCl.

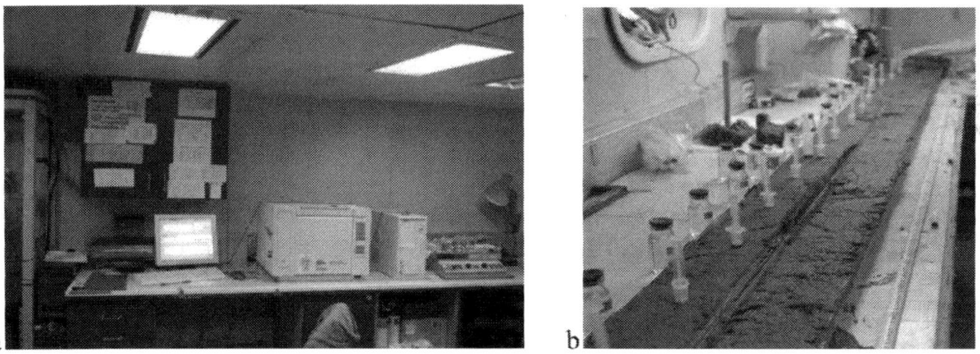

a b

Fig. 2. a) Gas chromatograph "Cristall – Lux 4000M" (Russia), b) Sampling sediment from core to measure gas.

3 Results of investigation and discussion

Sea of Okhotsk located between different lithosphere plates – Amur, Eurasian, North-American, Bering and Pacific (figure 3a). Relationship between lithosphere plates, regional blocks and active of asthenosphere in this area to support to form rifts, magma, faults and fluxes of gas-fluid (figure 3b). In the Okhotsk Sea took place riftogenesis that had 5 periods activity - early Paleocene, middle-later Eocene, later Oligocene, middle-later Miocene and Pliocene-Quaternary [4]. In these periods gorst- and graben- structures were formed. They creates good condition to form sediment basins. This process leaded to appear fold, flexure, zone fault in sediment basins especially in period global compression. To form of the folds and other structures in sediment basins was in different periods volcano and seismic activity. Thickness sediment in basin create about 7-8 km that was great condition to form oil-gas deposit in the Okhotsk Sea. It is very important that in East Sakhalin shelf and slop area of the Okhotsk Sea present zone destruction of the earth's crust and slide faults. These structures are zones for gas-fluid fluxes from, possible, asthenosphere and oil-gas-bearing layers to surface sediment.

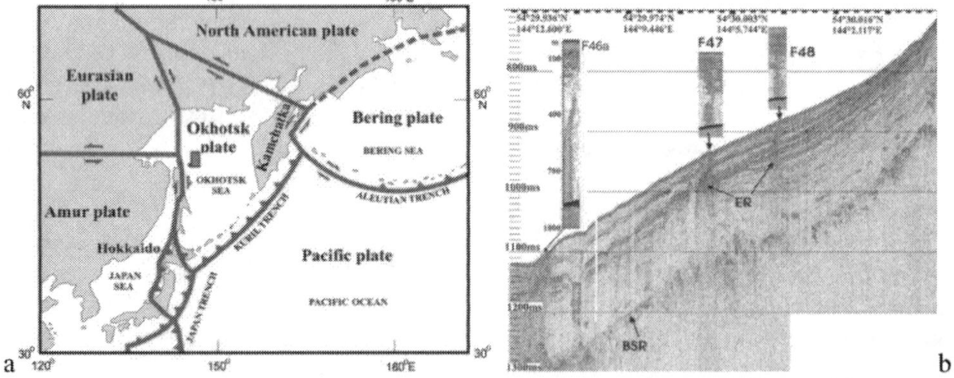

Fig. 3. a) Scheme to locate Okhotsk plate between fellow plates: Amur, Europe-Asian, North-American, Bering, Pacific. Lines between plates are seismic active contact between its and zones different faults. b) Profile with methane fluxes from sediment to water (F46a, F47, F48). BSR – (bottom simulating reflector), gas hydrate is locate in sediment between surface and BSR, ER – different chimneys (gas tubes)

Methane distribution in the Okhotsk Sea was studied since 1984 [7, 8, 9]. The first purpose was to search oil and gas deposit to use gas (especially hydrocarbon) as indicator of oil and gas structures. In period this investigation the first methane flux (1988) and gas hydrate (1991) was found [3, 8]. After it many international expedition were provided on the Sakhalin shelf and slope of the Okhotsk Sea to search methane fluxes and gas hydrates [6, 10, 14].

Every year a new methane bubbles fluxes (flares) from sediment to water column were found. Methane concentration in bottom water column reaches 1000-10000 nl/l and in sediment are 10-100 ml/l on mostly of areas with methane fluxes.(figures 4a and 4b).

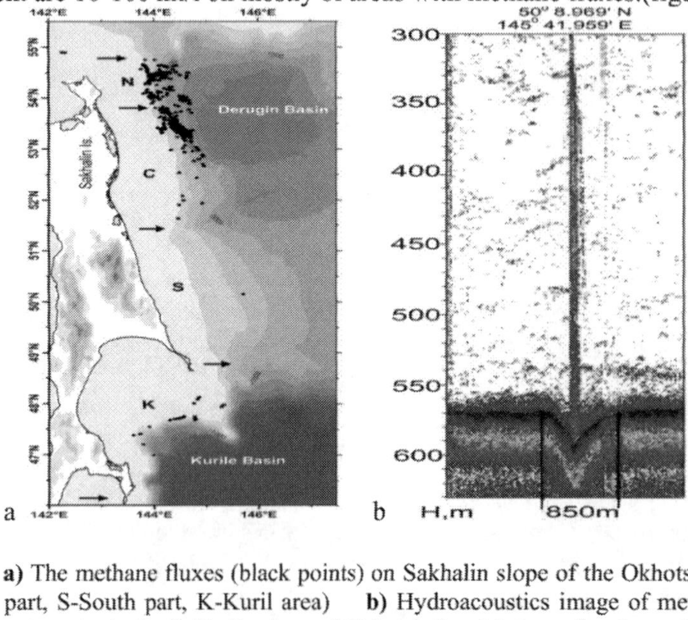

Fig. 4. a) The methane fluxes (black points) on Sakhalin slope of the Okhotsk Sea (N-North part, C-Center part, S-South part, K-Kuril area) **b)** Hydroacoustics image of methane babbles flux from sediment to water in the Sakhalin slope of Okhotsk Sea. Methane flux located in the point of area S

Monitoring investigation show that numbers of methane fluxes were increased during in the period of seismic-tectonic activity in the Okhotsk Sea. It starts from 1988 and continue now (figures 5a and 5b). It regularity supports episodes of earthquakes – 1995-Neftegorsk, 2001-Uglegorsk, 2003-Khokkaido, 2007-Nevel'sk, Fukusima (2011) and other [2, 10].

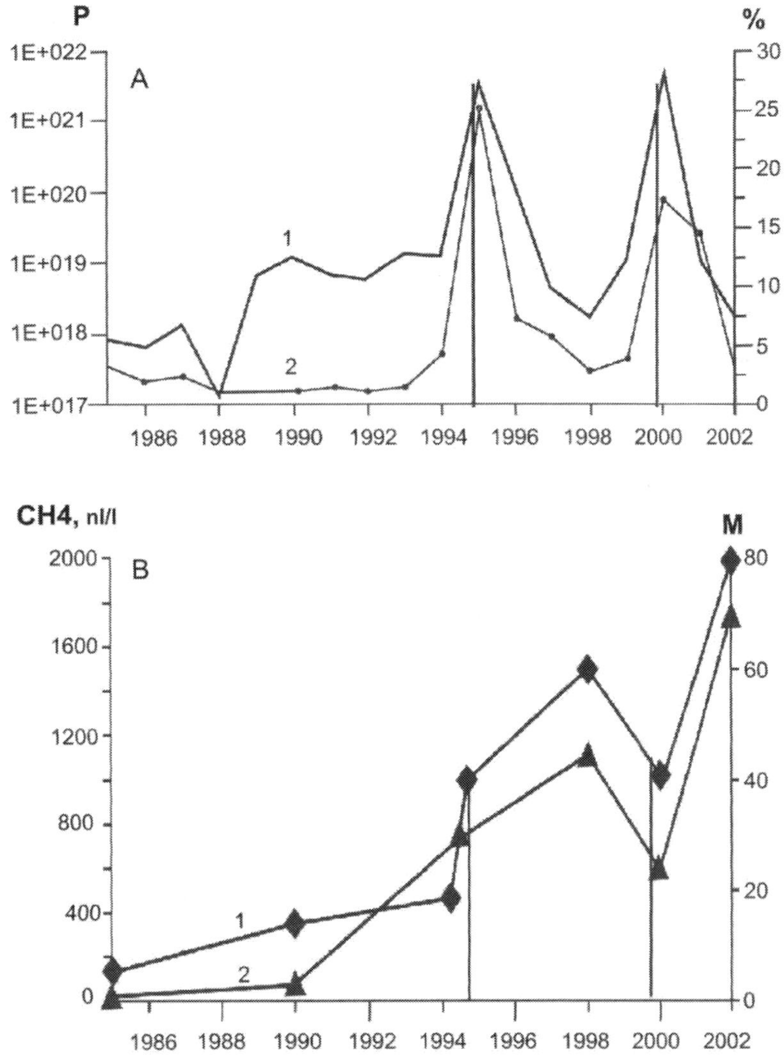

Fig. 5. a) Line 1 show energy (P, in 1Erg + 0.17 and more) of earthquake in different years (from 1986 to 2002). Line 2 is number of earthquakes magnitude M>3. b) Line 1 show methane concentration in bottom water in the various years, Line 2 is number of methane fluxes every year.

When methane bubbles come up from oil-gas deeper layers to the seafloor gas hydrate is formed in near surface layers sediment in gas hydrate stability zone. Below this zone of gas hydrate-bearing layers of sediment (BSR), free gas accumulates. If zone fault activities pass way increase and free gas (mostly methane) come up in surface sediment and water. Methane reaches the seafloor and in cool temperature and enough pressure to create gas hydrate in the stable condition of gas hydrate forming. A second (new) gas hydrate layers form near seafloor sediment usually with cm-scaled layer or fragments in the water depth deeper 400 m.

In period seismic activity methane bubbles come up to the sea floor from the oil-gas-bearing sediment to use zone fault and to form gas hydrate. In period seismic stability methane fluxes absent and gas hydrate could not form. In this period mud sediment accumulates. Its processes are repeated in seismic activity (figure 6a).

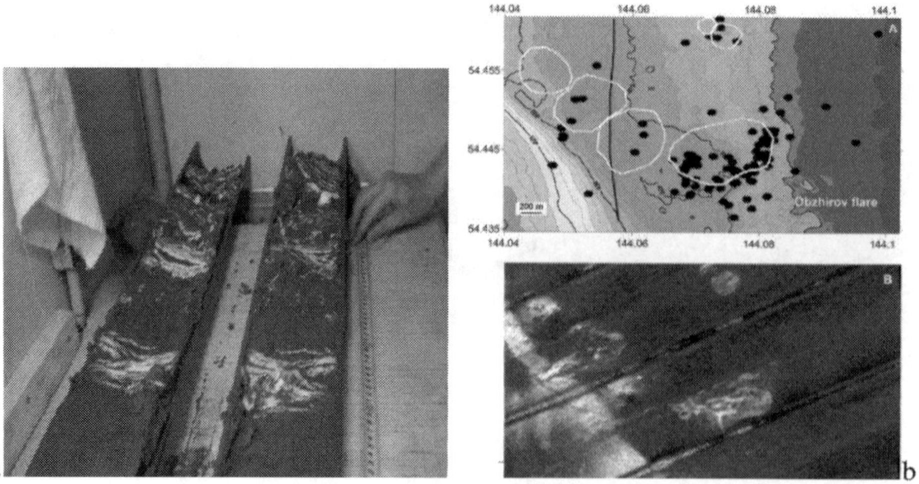

a b

Fig. 6 a) Grey color is layers with silt sediment. White color is layers with gas hydrate. b) (top part) Black points are methane fluxes in the area of the Okhotsk Sea near Obzhirov flare. (low part) White some circular areas are side sonar scanner image on surface sediment where gas (mostly methane) fluxes come up from sediment to water and to destroy sediment surface and forming hillocks and pits.

Sea floor in the area with methane flux is destroyed. Holes and hillocks like mud volcanoes with diameter of 100-800 m (figure 6b, top and low) are formed in this area. In sea floor of methane flux area a new assemblage, many shell and other benthos, crab and fish usually appeared. In near surface sediment layers carbonate fragment and concretions are formed.

4 Conclusion

So, methane fluxes and high methane concentration in the bottom water are considerable indicators to search gas hydrates in marine sediments. The increasing quantities of methane flares and methane concentrations in water column since 1988 are connected with seismic and tectonic activity of the faults in the Okhotsk Sea. In the other side, in the area with methane fluxes the morphology of sea floor is changing (hills and holes appear). The chimney structure is forming in the sediments with raising up gas. In this area sediments shells, carbonate concretions and other mineral association appear.

Geological, geophysical, gas geochemical, hydro-acoustic, morpho-structure surveys are very important indicators to search gas hydrates and to understand regularity forming and destroying of gas hydrates [10, 11, 12, 14, 16]. Thus, integrated complex of investigations with international cooperation allow us to discover methane fluxes, gas hydrates and to understand formation/dissociation gas hydrates in the Okhotsk Sea. This researches help to find much geological regularities around gas hydrate in the Okhotsk Sea as well as to examine relationships between methane fluxes, gas hydrates and oil and gas deposits.

Methane of thermogenic origin forms gas hydrates. It is important to understand because that feature influences on searching gas hydrate method, and on examine of methane volume in gas hydrate field and on mining methane from gas hydrates [13]. In general gas hydrate reserves methane and become a good cap to save all hydrocarbons. In geological period with seismic activity methane and other gases go up to the surface through fault zones and form gas hydrate in stability conditions. In period changing geological conditions sediment accumulation occurs and gas hydrate is buried in deeper zone. When conditions become unstable gas hydrates transform to methane. One part of methane forms gas and oil deposits and other part goes up to form a new gas hydrate. This argument helps to work out hydrocarbons from gas hydrates.

References

1. Akihiro Hachikubo, Alexey Krylovjjj, Hirotoshi Sakagami, Hirotsugu Minami, Yutaka Nunokawa, Hitoshi Shoji, Tatiana Matveeva, Young K. Jin, Anatoly Obzhirov. Geo-Marine Letters, **30:3-4** (2010)
2. Baranov B.V., Jin Y.K., Shoji H., Obzhirov A., Dozorova K.A., Salomatin A., Gladysh V. (2008) Gas Hydrate System of the Sakhalin Slope: Geophysical approach. (Scientific Report of the Sakhalin Slope Gas Hydrate Project 2007. KOPRI)
3. Ginsburg G.D., Soloviev V.A., Cranston R.E., Lorenson T.D., Kvenvolden K.A. Geo-Marine Letters, **13** (1993)
4. Kharakhinov V.V. *Oil and gas geology of the Sakhalin region* (Scientific World, Moscow, 2010) (in Russian)
5. Lein A.Yu., Ivanov M.V. *Biogeochemical cycle of methane in ocean* (Nauka, Moscow, 2009) (in Russian)
6. Hirotsugu Minami, Kazuya Tatsumi, Akihiro Hachikubo, Satoshi Yamashita, Hirotoshi Sakagami, Nobuo Takahashi, Hitoshi Shoji, Young Keun Jin, Anatoly Obzhirov, Nataliya Nikolaeva, Alexander Derkachev. Geo-Marine Letters. **32:5** (2012)
7. Obzhirov A.I., Kazansky B.A., Melnichenko Yu.I. Pacific Geology, **2** (1989) (in Russian).
8. Obzhirov A.I. *Gas geochemical fields in bottom water of seas and oceans* (Science Publ., Moscow, 1993) (in Russian).
9. Obzhirov A.I. et .al. *Methane monitoring in the Sea of Okhotsk* (Dalnauka, Vladivostok, 2002) (in Russian).
10. Obzhirov A., Shakirov R., Salyuk A., Suess E., Biebow N., Salomatin A. Geo-Marine Letter, **24:3** (2004).
11. A.Obzhirov, V.Yusupov, A.Salomatin. *Gas-Hydroacoustic complex to estimate seismic-tectonic activity.* (Patent. RU № 78333 U1, 2008. Bull. No 32)
12. Obzhirov A.I. *Method to search hydrocarbon deposit* (Patent RU 2359290 C1. 2009. Bull. No 17)
13. Obzhirov A.I. Tagiltsev A.A. *Technological complex to mining methane from gas hydrate in the open Sea* (Patent. RU 2386015 C1. 2010. Bull. No 10)
14. Operation Report of Sakhalin Slope Gas Hydrate (SAKHALIN) Project 2011, R/V Akademik M.A.Lavrentiev Cruise 56 (2012) New Energy Resources Research Center, Kitami Institute of Technology, Kitami, H. Shoji, Y.K. Jin, A.Obzhirov and B. Baranov, 140 p.
15. Young-Gyun Kim, Sang-Mook Lee, Young Keun Jin, Boris Baranov, Anatoly Obzhirov, Alexander Salomatin, Hitoshi Shoji. Marine and Petroleum Geology, **45** (2013)
16. Yusupov V., Salomatin A., Obzhirov A. *Gas-Geophysical complex to search underwater gas hydrate.* (Patent. RU № 70377 U1, 2008. Bull. No 2)

Radon transport model into a porous ground layer of finite capacity

Roman Parovik[1,2,*]

[1] *Institute of Cosmophysical Research and Radio Wave Propagation of the Far Eastern Branch of Russian Academy of Science, 684034, Kamchatskiy kray, Paratunka, Mirnaya str. 7, Russia.*
[2] *Vitus Bering Kamchatka State University, 683032, Kamchatskiy kray, Petropavlovsk-Kamchatskiy, Pogranichaya str. 4, Russia.*

Abstract. The model of radon transfer is considered in a porous ground layer of finite power. With the help of the Laplace integral transformation, a numerical solution of this model is obtained which is based on the construction of a generalized quadrature formula of the highest degree of accuracy for the transition to the original - the function of solving this problem. The calculated curves are constructed and investigated depending on the diffusion and advection coefficients.The work was a mathematical model that describes the effect of the sliding attachment (stick-slip), taking into account hereditarity. This model can be regarded as a mechanical model of earthquake preparation. For such a model was proposed explicit finite- difference scheme, on which were built the waveform and phase trajectories hereditarity effect of stick-slip.

1 Introduction

Mathematical modeling of radon (^{222}Rn) transfer processes is an actual problem due to various applications. One of such applications is the study of precursors of strong earthquakes with the aim of their possible prediction [1, 2]. In these papers, it is shown that radon is an indicator of the stress-strain state of the geomedium and its responses are observed into the field of subground radon in the form of various anomalies.

Mathematical modeling of radon transfer processes is considered within the framework of the emanation method, which is described in detail in [3, 4]. As a rule, the main contribution to the transport mechanism of Rn is made by the diffusion and advection processes, however, the migration capacity of Rn is influenced by meteorological conditions, groundwater discharge, effusion, geological parameters and properties of the geomedium, etc. [3, 5]. Therefore, the mathematical model of the transfer of Rn in the geo-environment is adopted with some assumptions.

In this paper, we will assume that the non-stationary transfer of Rn occurs in a homogeneous layer of porous ground only due to the diffusion-advection mechanism, and we also assume that the influence of meteorological conditions on the transport process Rn is insignificant and can be neglected. Consider the following problems.

* e-mail: parovik@ikir.ru

2 Statement of the problem and methods of solution

The first we consider the problem in the case when the nonstationary transfer of Rn occurs in a homogeneous layer of porous ground with infinite power.

Problem 1. There is an emanating homogeneous layer of ground of infinite power with free emanation Rn. It is required to find the diffusion-advective distribution of radon volume activity (VA Rn) $A(z,t)$ in time t and spatial coordinate z [3].

Mathematical model of this problem:

$$\frac{\partial A(z,t)}{\partial t} = D\frac{\partial^2 A(z,t)}{\partial z^2} + \upsilon\frac{\partial A(z,t)}{\partial z} - \lambda\left(A(z,t) - A_\infty\right), \tag{1}$$

$$1)\ z = 0, A = 0,\ 2)\ z \to \infty, A = A_\infty, 3)\ t = 0, A = A_\infty, \tag{2}$$

where D - diffusion coefficient Rn into the ground, m^2/s; υ - Rn advection speed into the ground, m/s; λ - decay constant Rn, 1/c; $A_\infty = {}_{em}A_{Ra}\rho_s(1-\eta)$ - volume activity of radon, which is in radioactive equilibrium with radium (^{226}Ra) to a certain depth;; K_{em} - emanation Coefficient Rn, in relative units; A_{Ra} - Specific activity ^{226}Ra, Bq/kg; ρ_s -density of solid ground particles, kg/m^3; η – porosity, in relative units.

Using the integral Laplace transform and the initial condition 3, problem (1) - (2) can be written for the image $F(z,p)$ as follows:

$$D\frac{d^2F(z,p)}{dz^2} + \upsilon\frac{dF(z,p)}{dz} - (p+\lambda)F(z,p) - \frac{A_\infty(p+\lambda)}{p} = 0, \tag{3}$$

$$1)\ z = 0, F = 0,\ 2)\ z \to \infty, F = \frac{A_\infty}{p}. \tag{4}$$

The solution of equation (1) can be written:

$$F(z,p) = C_1\exp\left(z\left[-\frac{\upsilon}{2D} + \sqrt{\frac{\lambda+p}{D} + \frac{\upsilon^2}{4D^2}}\right]\right) + C_2\exp\left(-z\left[\frac{\upsilon}{2D} + \sqrt{\frac{\lambda+p}{D} + \frac{\upsilon^2}{4D^2}}\right]\right) + \frac{A_\infty}{p} \tag{5}$$

where C_1, C_2 – constants of integration, which are determined from the conditions (4). Taking account this condition, solution (5) has the form:

$$F(z,p) = \frac{A_\infty}{p}\left(1 - \exp\left(-z\left[\frac{\upsilon}{2D} + \sqrt{\frac{\lambda+p}{D} + \frac{\upsilon^2}{4D^2}}\right]\right)\right), \tag{6}$$

Turning to the original for (6), using the reference [6], we obtain:

$$A(z,t) = A_\infty\left(1 - \frac{1}{2}\left(e^{z^-}erfc(u^-) + e^{z^+}erfc(u^+)\right)\right),$$

$$z^- = z\left(\frac{\upsilon}{2D} - \sqrt{\frac{\lambda}{D} + \left(\frac{\upsilon}{2D}\right)^2}\right), z^+ = z\left(\frac{\upsilon}{2D} + \sqrt{\frac{\lambda}{D} + \left(\frac{\upsilon}{2D}\right)^2}\right), \tag{7}$$

$$u^- = \frac{z}{2\sqrt{Dt}} - \sqrt{\left(\frac{\upsilon^2}{4D} + \lambda\right)t}, u^+ = \frac{z}{2\sqrt{Dt}} + \sqrt{\left(\frac{\upsilon^2}{4D} + \lambda\right)t},$$

$$erfc(z) = 1 - erf(z), erf(z) = \frac{2}{\sqrt{\pi}}\int_0^z e^{-z^2}dz.$$

Solution (7) is given and studied in [7], but for a two-layer ground-atmosphere model. The case $\upsilon = 0$ is given in [3].

Noted, that in the case of transfer of Rn in a layer of infinite power, due to the second condition (4), the problem of image reversal is simplified. Let us consider a somewhat different problem.

Problem 2. There is an emanating homogeneous layer of ground with a thickness l of free emanation. It is required to find the diffusion-advective distribution of radon volume activity $A(z,t)$ in time and space.

$$\frac{\partial A(z,t)}{\partial t} = D\frac{\partial^2 A(z,t)}{\partial z^2} + \upsilon\frac{\partial A(z,t)}{\partial z} - \lambda\left(A(z,t) - A_\infty\right), 0 < z < l. \tag{8}$$

$$1)\ z = 0, A = 0,\ 2)\ z = l, A = A_\infty, 3)\ t = 0, A = A_\infty. \tag{9}$$

or into the image $F(z,p)$:

$$D\frac{d^2F(z,p)}{dz^2} + \upsilon\frac{dF(z,p)}{dz} - (p+\lambda)F(z,p) - \frac{A_\infty(p+\lambda)}{p} = 0, \tag{10}$$

$$1)\ z = 0, F = 0,\ 2)\ z = l, F = \frac{A_\infty}{p}. \tag{11}$$

The solution of equation (10) coincides with the solution of (3), and, taking into account conditions (11), it can be written as:

$$F(z,p) = \frac{A_\infty}{p}\left(1 + \frac{\exp\left(-\frac{z\upsilon}{2D}\right)\sinh\left((z-l)\sqrt{\frac{\lambda+p}{D} + \frac{\upsilon^2}{4D^2}}\right)}{\sinh\left(l\sqrt{\frac{\lambda+p}{D} + \frac{\upsilon^2}{4D^2}}\right)}\right), \tag{12}$$

$\sinh(z) = \frac{\exp(z) - \exp(-z)}{2}$ - hyperbolic sine.

We note the solution for the image (12) in the limit $l \to \infty$tends to the solution for the image (6). For the image (12), there is no tabular transition to the original, even if it can be reduced to tabular form, the original will be expressed in terms of integrals from the theta-function [6], which makes it difficult to visualize the solution.

For the inversion (12), we use the numerical method for constructing the generalized quadrature formula of the highest degree of accuracy (GQFHDA), proposed in [8]. The GQFHDA method can be written as follows for the inversion formula:

$$f(t) = \frac{1}{2\pi i}\int_{c-i\infty}^{c+i\infty} F(p)e^{pt}dp = \frac{t^{s-1}}{2\pi i}\int_{c-i\infty}^{c+i\infty} p^{-s}\varphi_s(p/t)e^p dp, c > 0,$$

a quadrature formula of the form is constructed:

$$\frac{t^{s-1}}{2\pi i}\int_{c-i\infty}^{c+i\infty} p^{-s}\varphi_s(p)e^p dp \approx \sum_{k=1}^{n} A_k\varphi_s(p_k), \tag{13}$$

A_k, p_k – coefficients and nodes of the quadrature formula, $\varphi_s(p) = p^s F(p)$. Next, we introduce a positive number a. Region of regularity of the function $\varphi_s(p)$ we choose any pair wise distinct points

p_k, $k = 1, ..., n$ and we require that the quadrature formula (13) is exact for the function $\varphi_s(p) = p^{-am}, m = 0, 1, ..., n-1$, which is equivalent to the fulfillment of the relations:

$$\sum_{k=1}^{n} A_k p_k^{-am} = \frac{1}{\Gamma(s+am)}, m = 0, 1, ..., n-1. \quad (14)$$

Note that the numbers p_k^{-a} are the roots of the polynomial $\omega_n(x) = \prod_{k=1}^{n}\left(x - p_k^{-a}\right)$ which are uniquely determined by virtue of orthogonality:

$$\frac{t^{s-1}}{2\pi i}\int_{c-i\infty}^{c+i\infty} p^{-s}\varphi_n(p^{-a})\,p^{-am}dp = 0, m = 0, 1, ..., n-1. \quad (15)$$

For example, if the polynomial $\omega_n(x)$ is written as: $\omega_n(x) = x^n + b_1 x^{n-1} + ... + b_n$,

Then the orthogonality condition with respect to the coefficients is equivalent to the system of equations:

$$\frac{b_1}{\Gamma(s+(n+k-1)a)} + ... + \frac{b_n}{\Gamma(s+ka)} = -\frac{1}{\Gamma(s+(n+k)a)}, k = 0, 1, ..., n-1. \quad (16)$$

Noted that the polynomial $\omega_n(x)$ exists and is uniquely determined. Algorithm for building GQFHDA contains the following stages:

1. the calculation of the coefficients of the equations systems (16) with respect to unknowns b_k and the solution of this system;
2. finding the nodes p_k of the quadrature formula from equation (15): $\omega_n\left(p_k^{-a}\right) = p_k^{-an} + b_1 p_k^{-a(n-1)} + ... + b_n = 0$;
3. finding the coefficients A_k of the quadrature formula from the conditions (14). In [8] a theorem is proved whose result is an estimate for the error $\varepsilon_n(t^a)$ of the GQFHDA for any $t > 0$:

$$|\varepsilon_n(t^a)| \le M\sigma_n(t^a, \rho), M = \left(\frac{1}{2\pi i}\int_0^{2\pi}\left|P_{as}(1+\rho\exp(i\theta))\right|^2 d\theta\right)^{1/2}, P_{as} = \varphi_s\left(t^{-1/a}\right), \quad (17)$$

$$\sigma_n(t^a, \rho) = \left(\sum_{m=2n}^{\infty}\left|\sum_{j=2n}^{m}\binom{m}{j}\frac{(-1)^j}{\Gamma(s+aj)} - \sum_{k=1}^{n}A_k\left(1 - p_k^{-a}\right)^m\right|^2\left(\frac{t^a}{\rho}\right)^{2m}\right)^{1/2}, t^a < \rho \le t^a + r.$$

r - radius of a circle .

3 Results of modeling

Consider the solution of problem 2 depending on the different values of the diffusion D and advection v coefficients for the following values of the parameters: $A_\infty = 1000\ Bq/m^3$; $\lambda = 2.1 \cdot 10^{-6}c^{-1}$; $l = 100cm$.

For the algorithm GQFHDA: $a = 0.5, s = -1, m = 100, n = 20$.

DOI: 10.1051/e3sconf/20172003004

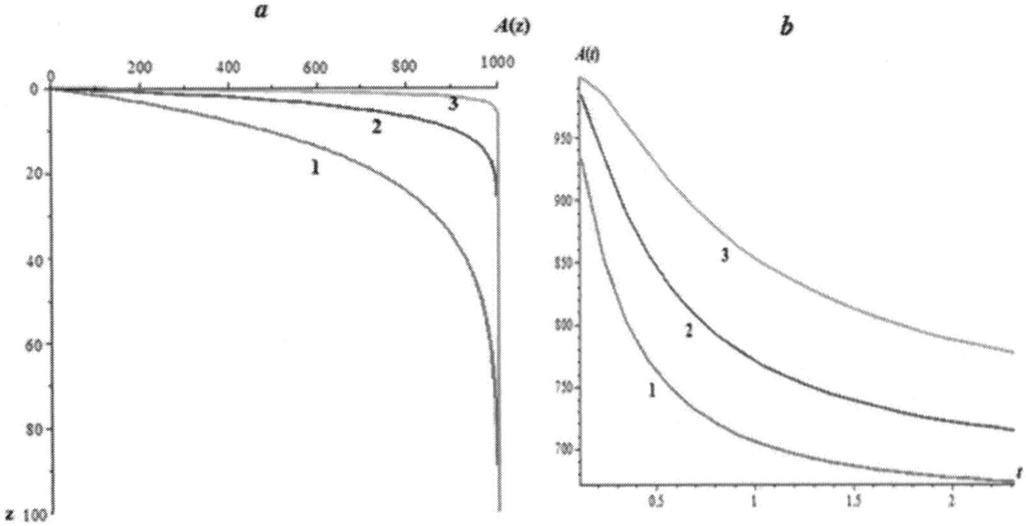

Figure 1. Calculated curves of the distribution of Rn into the ground in depth: ***a***) $z \in [0, 100]$ at a fixed time moment $t = 10^7$ c :curve 1 - $\upsilon = 10^{-6}$ cm/s, $D = 5 \cdot 10^{-4}$ cm^2/s; curve 2 - $\upsilon = 5 \cdot 10^{-5}$ cm/s, cm^2/s; curve 3 - cm/s, - cm^2/s; ***b***) z = 5cm: curve 1 - cm/s, $D = 5 \cdot 10^{-4}$cm^2/s; curve 2 - cm/s, $D = 2.5 \cdot 10^{-4}$cm^2/s; Curve 3 - cm/s, - cm^2/s

In Fig. 1a shows the calculated curves of the distribution of Rn into the ground in depth , depending on the values of the diffusion and advection coefficients for a fixed time c. It is seen that as the value of the advection coefficient increases, the migration capacity of Rn increases, (curve 3). This is manifested by a rapid exit to the background level A_∞ at shallow depths .

In Fig. 1b shows the calculated curves of Rn distributions into the ground as a function of time $t \in [0, 2.31]$ at the point $z = 5$ cm. It is seen that as the value of the advection coefficient increases, the concentration of Rn at a point decreases more rapidly (curve 3), i.e. faster goes to stationary mode.

These results show that taking into account the advective component is important in modeling the migration of Rn [9]. The diffusion and advection coefficients can be estimated using experimental data [10] or tables [3].

Noted that the error of the GQFHDA method by formula (17) was, which indicates an increased accuracy of the quadrature formula (13).

4 Conclusion

In [11, 12], using georadar studies, it was established that the registration points have a layered structure, with two or even three layers clearly identified. Each layer has its own physico-geological parameters. Therefore, problem 2 can be generalized to the case of a two-layer or three-layer geomedium. Of course, the consideration of the properties of the geomedium, namely, the in homogeneity, which can be expressed, for example, by fractal properties [13] or a piecewise-constant anisotropic layered structure with inclusions, deserves close attention [14]. Such problems will be solved by numerical methods as in [15].

Acknowledgments

The paper on the order of RFFI grant No. 16-05-00162 "Presentation of strong earthquakes into the field of subground radon in the Petropavlovsk-Kamchatka geodynamic test range".

References

[1] V.P. Rudakov, J. Geokhim. **9** (1986)
[2] P.P. Firstov , V.P. Rudakov, J. Vulk. i seysm. **6** (2002)
[3] G.F. Novikov *Radiometricheskaya razvedka* (Leningrad, Nedra, 1989)
[4] R.I. Parovik *Matematicheskiye modeli klassicheskoy teorii emanatsionnogo metoda* (Petropavlovsk-Kamchatskiy, KamGU im. Vitusa Beringa, 2014)
[5] V.S. Yakovleva, P.M. Nagorsky, A.G. Kondratyeva, N.V. Mishina *The influence of meteorological parameters and other factors on soil radon dynamics. Innovative Technologies in Engineering VII International Scientific Practical Conference* (Conference Proceedings. National Research Tomsk Polytechnic University. 2016.)
[6] V.A. Ditkin, A.P. *Prudnikov Spravochnik po operatsionnomu ischisleniyu* (Moskva, Vysshaya shkola, 1965).
[7] R.I. Parovik Vestnik KRAUNTS. Fiz.-mat. nauki **1**, 1 (2010)
[8] T.A. Matveyeva *Nekotoryye metody obrashcheniya preobrazovaniya Laplasa i ikh prilozheniya* (Diss. kand. fiz.-mat. nauk. Sankt-Peterburg. 2003).
[9] P.P. Firstov, R.I. Parovik, V.S. Yakovleva, O.P. Malysheva *Svyaz' skorosti advektsii i plotnosti potoka radona s sil'nymi zemletryaseniyami yuzhnoy Kamchatki 2000-2008 gg. Solnechno-zemnyye svyazi i predvestniki zemletryaseniy* (Dal'nevostochnoye otdeleniye, Institut kosmofizicheskikh issledovaniy i rasprostraneniya radiovoln. 2010).
[10] V.S. Yakovleva Vestnik KRAUNTS. Fiz.-mat. nauki. **8**, 1 (2014)
[11] P.P. Firstov, R.I.Parovik, R.R. Akbashev, Ye.O. Makarov *Stroyeniye verkhney chasti geologicheskikh razrezov v punktakh monitoringa pochvennogo radona na petropavlovsk-kamchatskom geodinamicheskom poligone. Vulkanizm i svyazannyye s nim protsessy* (Materialy regional'noy nauchnoy konferentsii, posvyashchonnoy Dnyu vulkanologa. 2014).
[12] P.P. Firstov, Ye.O. Makarov, S.Yu. Paskar', R.I. Parovik ANRI 4 (**87**) (2016)
[13] R.I. Parovik, B.M. Shevtsov J. Math. Mod. and Comp. Sim. **2**, 2 (2010)
[14] A.R. Nafikova , V.N. Krizskiy *Matematicheskoye modelirovaniye protsessov perenosa radona v kusochno-postoyannykh anizotropnykh sloistykh sredakh s vklyucheniyami* (Sterlitamak: BashGU, 2016)
[15] V.S. Yakovleva, R.I. Parovik Nukleonika **55**, 4 (2010)

On some features of geoacoustic emission signals before earthquakes

Albert Shcherbina[1]

[1]*Institute of Cosmophysical Research and Radio Wave Propagation FEB RAS, Paratunka, Russia*

Abstract. Geoacoustic emission signal amplitudes and arrival directions were analyzed during background periods and earthquake preparation. It was determined that during seismic event activation, geoacoustic emission directivity anisotropy increases with the increase of signal amplitude. Taking this feature into account, geoacoustic emission properties were estimated by the superposition method before 2653 earthquakes which occurred within the period from 01.01.2008 to 01.09.2016. It was determined that depending on the type of the signals under consideration, it is possible to detect geoacoustic emission anomalous behavior at the distances of more than 25% and 65% of the precursor manifestation zone radius suggested by I.P. Dobrovol'skiy.

1 Introduction

The long-term investigations carried out in Kamchatka show that significant increase of geoacoustic emission intensity and its directivity change are observed before strong seismic events. This acoustic effect manifests the most brightly in the kilohertz frequency range and is determined by rock deformations at observation sites at the distances of hundreds of kilometers from the epicenters of preparing earthquakes [1, 2].

Geoacoustic emission is registered in the frequency range up to 11 kHz by a high-frequency combined receiver allowing us to record acoustic pressure and three mutually orthogonal components of its gradient at one space point. The receiver is installed by the bottom of Mikizha Lake that is located in a seismically active zone of Kamchatka peninsula, near Paratunka village. The developed methods for signal processing allow us to distinguish geoacoustic pulses in automatic regime, to determine their amplitude, repetition frequency and the direction to a source [3]. Since 2008 all the collected data are regularly added to a data base.

One of the problems that is interesting to solve is the comparison of the data series on earthquakes taking into account their energy parameters and distances with the data series on geoacoustic field responses in the measurement region. The superposition method gives good results in this problem solution [4].

2 Data analysis and results

To apply the superposition method effectively, we need to form the data series characterizing the geoacoustic field state the most accurately. In the paper mentioned above, the acoustic pressure averaged signal was analyzed before 46 earthquakes. In the result, differentiating features of geoacoustic signal anomalous state were detected. Thus, we received a formalized time series (massive) with the sampling period of one minute in which units indicate the disturbance presence at a definite time moment and the zeros indicate its absence.

The suggested method for series formation requires much of manual operation, consequently, it is difficult to apply it for the processing of a large quantity of data (about several thousands of cases). To solve this problem, we will consider the available information on geoacoustic emission signals.

Geoacoustic pulse information, stored in the data bases, allows us to determine the integral $\Omega(t)$ and differential $D(\alpha,t)$ geoacoustic activity. The first value is the pulse repetition frequency depending on time, the second value is the direction distribution of pulse repetition frequency $d\alpha$ in the horizontal plane. Moreover, there is a possibility to select the signals in the required interval of amplitudes dA.

Fig. 1 illustrates typical examples of directivity azimuthal diagrams. It is clear from the graphs that during seismic event preparation, significant increase of pulse repetition frequency and, that is more important for this work, the increase of the root-mean-square deviation (RMS) of the directivity diagram are observed. On the presented examples, the diagram RMS corresponding to the background period is of the order 5×10^{-5} imp./degr.s., that for the disturbed period is about 2×10^{-4} imp./degr.s.

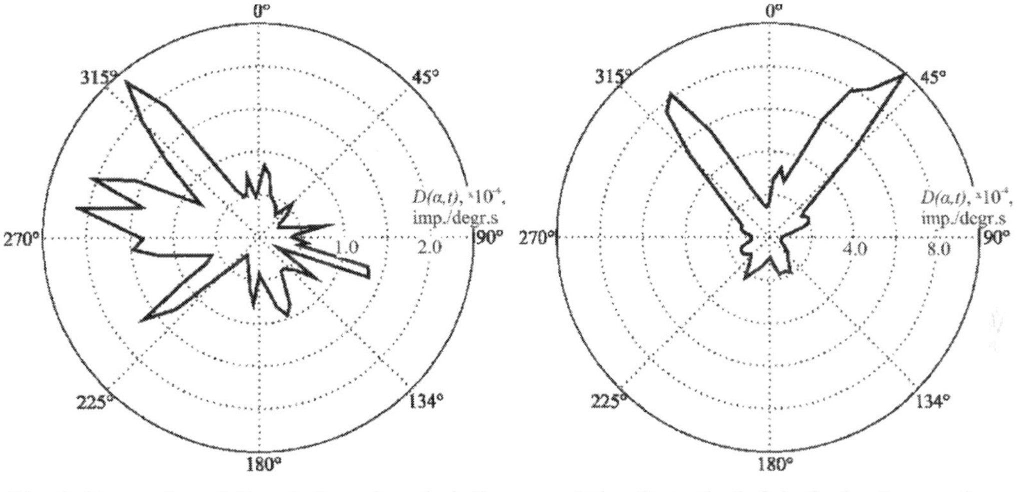

Fig. 1. Examples of directivity azimuthal diagrams $D(\alpha,t)$: to the left is the background period, to the right is the geoacoustic emission registered before the earthquake occurred on 11.07.2017 at 18:09 UT (energy class is 10.3, distance to the epicenter is 79 km).

Analysis of the geoacoustic emission directional properties before 77 seismic events, which occurred within the period from 03.11.2008 to 30.01.2016, showed that geoacoustic signal anisotropy increases as pulse amplitude increases. To demonstrate this, we plot a histogram of geoacoustic pulse amplitude distribution (Fig. 2a). Then we divide the obtained distribution at a conditional point x on the abscissa axis, for example at the value of 10 % from the signal dynamic range (Fig. 2c). The relation of azimuthal distribution diagram RMS for the pulses with the amplitude exceeding this threshold to the signal distribution diagram RMS of less amplitude is determined by the expression

$$\sigma'(x) = \frac{\sigma_h(x)}{\sigma_l(x)}, \quad (1)$$

where:
$\sigma_h(x) = \sigma(D(\alpha,t)_{dA\geq x})$ is the azimuthal distribution diagram RMS for the pulses with amplitude exceeding or equal to the threshold x,
$\sigma_l(x) = \sigma(D(\alpha,t)_{dA<x})$ is the same but for the pulses with the amplitude less than the threshold x,
dA is the geoacoustic pulse amplitude range applied in the calculation.

Having re-counted this expression for all the points x of the signal amplitude dynamic range, we can plot a curve of the relations of diagram RMS of radiation directivity azimuthal distribution $\sigma'(x)$ (Fig. 2b).

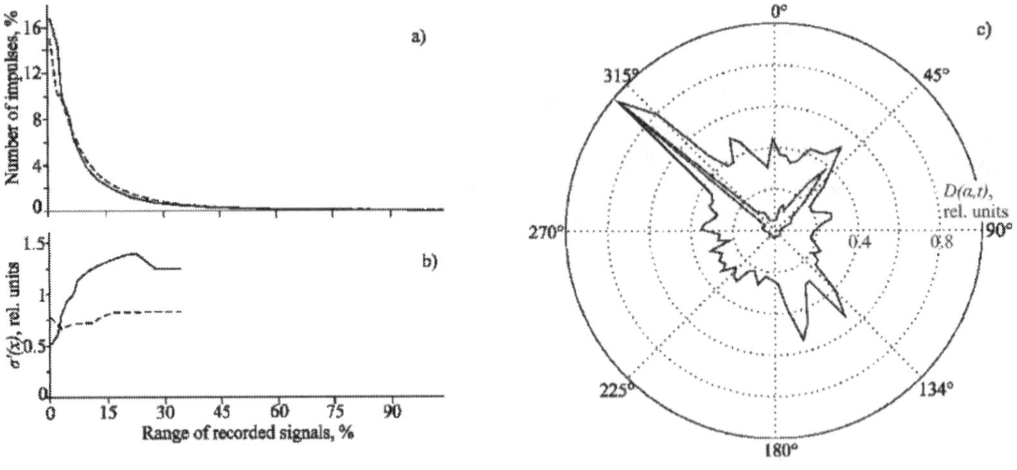

Fig. 2. Histogram of geoacoustic pulse amplitude distribution (a), curve $\sigma'(x)$ the dependance of direcivity azimuthal distribution diagram RMS on signal amplitude (b), examples of diagrams of pulse repetition frequency azimuthal distribution (c). On the graphs (a) and (b), the solid lines indicate the diagrams for the periods of geoacoustic emission disturbances, dashed lines show that for the background periods. On the diagram (c), the solid line indicates the distribution diagram for the pulses with high amplitude, the dashed line is that with small amplitude. The curves (c) are in the relative scale for clarity.

Fig. 2 shows the results of estimation of the graphs $\sigma'(x)$ for the geoacoustic pulses registered within 410 background periods twenty-four hour long and 77 periods with brightly distinguished disturbances of geoacoustic field before seismic events observed from 03.11.2008 to 30.01.2016. The calculation was made for the pulses with the magnitudes from 0 to 35% from the dynamic range of the registered signals. This interval was determined by the fact that the number of geoacoustic pulses exceeding this threshold is about 4 – 5% from the total number. Thus, their directivity azimuthal diagram is not accurate enough.

It is clear from the obtained results that the graph $\sigma'(x)$ for the signals recorded before seismic events grows with the increase of geoacoustic pulse amplitude and becomes more than a unit at the point corresponding to 7% from the dynamic range of the registered signals. Thus, we can state that analysis of function σ_h for the signals exceeding this value is the most informative and reasonable. It is interesting that the graph of $\sigma'(x)$ corresponding to background periods does not have pronounced growth and does not exceed a unit at any point of the range. The obtained result was used to analyze geoacoustic emission properties before earthquakes.

It is logical to suppose that typical changes of geoacoustic emission are observed during the preparation of earthquakes the precursors of which manifest in the area where the signal is registered. The radius of this area can be estimated by the formula suggested by I.P. Dobrovol'skiy [5]:

$$\rho = 10^{0.43M}\ km, \tag{2}$$

where M is the earthquake amplitude.

When the registration system is within the zone of precursor manifestation, the following inequality is fulfilled:

$$\rho/r \geq 1, \tag{3}$$

where r is the distance to an earthquake epicenter.

In the opposite case, when relation (3) is less than a unit, such earthquake precursors will not be recorded by the registering instrumentation.

This assumption was verified by a series of calculations by superposition method. For each group of calculation in a series, the earthquake occurrence times were chosen as the key dates, for which the relation ρ/r was equal or exceeded the threshold e_i where e was measured from 0.3 to 1.9 with a step of 0.1. On the whole, 17 separate calculation groups were derived in which from 2653 to 60 series of geoacoustic data were involved. The key dates were chosen in such a way as to exclude the superposition of close events. Moreover, we did not take into account the earthquakes which occurred at the distances of less than 35 km from the registration site and at the depths of more than 70 km. Basically, it was done to ensure the estimation correctness of relation (3), which can result in high values as the distance from the sensor to an earthquake epicenter decreases.

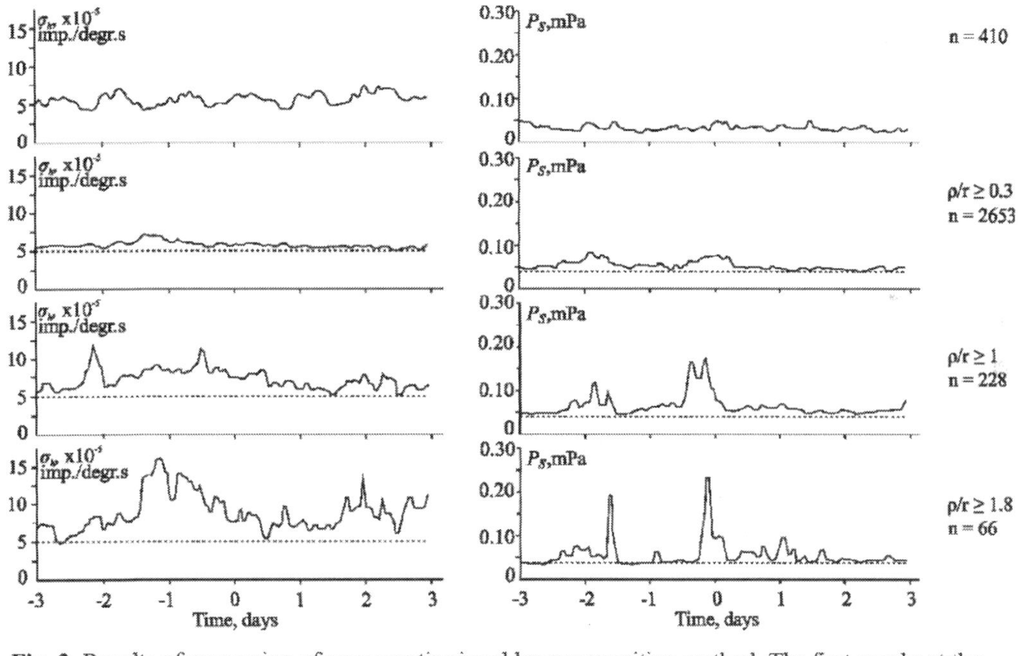

Fig. 3. Results of processing of geoacoustic signal by superposition method. The first graphs at the top correspond to background periods, the rest correspond to the periods of earthquake preparation. Dashed lines indicate the signal levels corresponding to background periods. The letter n indicates the data series number involved in the processing, σ_h are azimuthal distribution diagram RMS, P_S is total 1-second acoustic pressure, ρ is the precursor manifestation zone radius suggested by I.P. Dobrovol'skiy, r is the distance from the registration site to an earthquake epicenter.

For each data series, average graph of $\sigma_h(x)_{x=7\%}$ was estimated as the most informative indicator of anomalous change of geoacoustic emission anisotropy (Fig. 3). Additionally, we determined the total (integral) value of acoustic pressure P_S registered by the corresponding channel of a combined receiver in the frequency range of 2 – 6.5 kHz [1]. The signal was stored in a window of 1 s. For comparison, $\sigma_h(x)_{x=7\%}$ and P_s were calculated for 410 background periods twenty-four hour long.

It is clear from the diagrams that the averaged graph of σ_h for the data series $\rho/r \geq 0.3$ almost does not differ from the curve plotted for the background period. The same refers to the graph of total acoustic pressure P_s. Thus, the earthquakes located at the distances, which exceed the precursor manifestation zone radius ρ (2) by three times, do don affect the geoacoustic field in the region of measurements.

When analyzing the results of processing of the data series $\rho/r \geq 1$, we should note that the graph of σ_h within the period from -2 to +1 day relatively the earthquake time significantly exceed the average value for the background. At the maximal point, the relation is almost double. A similar situation is observed on the graph of total acoustic pressure P_s, but here the background values are exceeded by 3.5 times. Thus, we can determine that significant change of geoacoustic emission properties is observed at the boundary of the precursor manifestation zone suggested by I.P. Dobrovol'skiy.

For systematization and further analysis of the calculation results, graphs of σ_{snr} and P_{snr} were made, similar to the relation signal/noise (Fig. 4)

$$\sigma_{snr} = \frac{max(\sigma_{hs}(x))}{\bar{\sigma}_{hn}(x)}, P_{snr} = \frac{max(P_{ss})}{\bar{P}_{sn}}, \quad (4)$$

where $\max(\sigma_{hs}(x))$ is the maximum value of $\sigma_h(x)_{x=7\%}$ during disturbed periods,
$\bar{\sigma}_{hn}(x)$ is the average value of $\sigma_h(x)_{x=7\%}$ during background periods,
$\max(P_{ss})$ is P_s maximum value during disturbances,
$\bar{P}_{sn}$ is P_s average value during background periods.

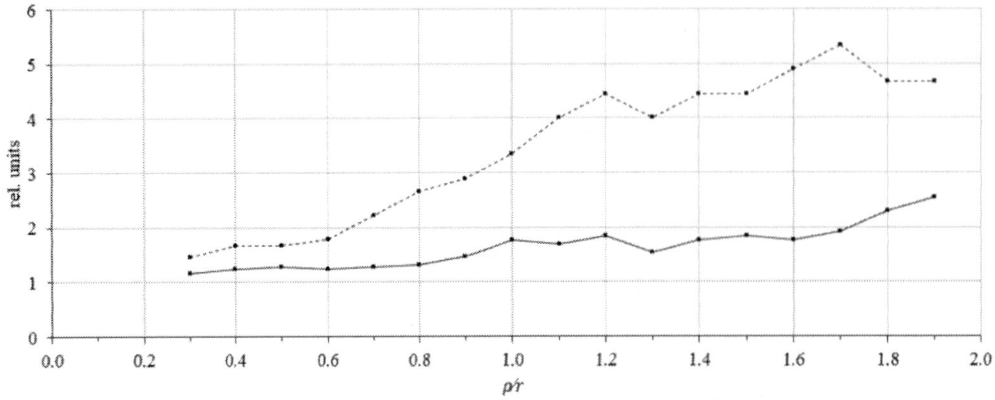

Fig. 4. Relative changes of diagram RMS of geoacoustic emission directivity azimuthal distribution (solid line) and total 1-second acoustic pressure (dashed line) depending on the relation of precursor manifestation zone radius suggested by I.P. Dobrovol'skiy to the distance to an earthquake epicenter (ρ/r). The comparison was made with the background values.

Parts of the graphs close to 1 on the ordinate axis mean that geoacoustic emission anomalous change is not observed before an earthquake. Significant increase of σ_{snr} graph begins from the value $\rho/r \geq 0.8$. Thus, geoacoustic emission anisotropy at the measurement point changes before earthquakes located 25% farther than we can determine by the formula suggested by I.P. Dobrovol'skiy. The total (integral) acoustic pressure P_{snr} grows from the point of 0.6 that corresponds to the precursor manifestation region increased by 65%.

3 Conclusions

The detailed analysis of geoacoustic data recorded in Kamchatka allowed us to determine that geoacoustic emission anisotropy grows as signal amplitude increases. This property was used to determine the effective parameters for signal filtration allowing us to increase the informative value of data statistic processing including the automated techniques. Based on the analysis of more than 2600 earthquakes, occurred from 01.01.2008 to 01.09.2016, by superposition method, the earthquake precursor manifestation zone radii were evaluated.

References

1. Yu.V. Marapulets, B.M. Shevtsov, I.A. Larionov, M.A. Mishchenko, A.O. Shcherbina, and A.A. Solodchuk, Russian Journal of Pacific Geology, **6**, pp. 59-67 (2012).
2. V.A. Gordienko, T.V. Gordienko, N.V. Krasnopistsev, A.V. Kuptsov, I.A. Larionov, Yu.V. Marapulets, A.N. Rutenko, B.M. Shevtsov, Akusticheskij zhurnal, **1**, pp. 97-109 (2008).
3. Yu.V. Marapulets, A.O. Shcherbina, Electronic Journal Technical Acoustics, **14** (2008).
4. M.A. Mishchenko, Bulletin of KRAESC. Phy. & Math. Sci., **1**, pp. 57-65 (2011).
5. I.P. Dobrovol'sky, S.I. Zubkov, V.I. Myachkin, Modeling of earthquake processes. Nauka Publishers, pp. 7-44 (1980).

E3S Web of Conferences **20**, 03006 (2017) DOI: 10.1051/e3sconf/20172003006
Solar-Terrestrial Relations and Physics of Earthquake Precursors

Electromagnetic Manifestation of Earthquakes

Vladimir Uvarov[1,*]

[1]Institute of Cosmophysical Research and Radio Wave Propagation, Far Eastern Branch of the Russian Academy of Sciences (IKIR FEB RAS) 684034, Kamchatskii krai, Elizovo raion, Paratunka, Mirnaya st., 7

Abstract. In a joint analysis of the results of recording the electrical component of the natural electromagnetic field of the Earth and the catalog of earthquakes in Kamchatka in 2013, unipolar pulses of constant amplitude associated with earthquakes were identified, whose activity is closely correlated with the energy of the electromagnetic field. For the explanation, a hypothesis about the cooperative character of these impulses is proposed.

1 Introduction

In this paper the data for year of 2013 regarding seismic activity (represented by the Kamchatka branch of the geophysical service of the Russian Academy of Sciences), and the signal of the vertical electric component of the electromagnetic field of the ELF-VLF range (registered at the Karymshin observation station of the IKIR FEB RAS) we compared.

2 Data analysis

During the visual analysis of the waveforms of the initial data in the time vicinity of the earthquake moment, attention was drawn to the presence of unipolar pulses accompanying earthquakes. The range of these pulses is ten times higher than the background value, while its shape remains unchanged. The pulse duration is about 0.4 milliseconds (Figure 1).

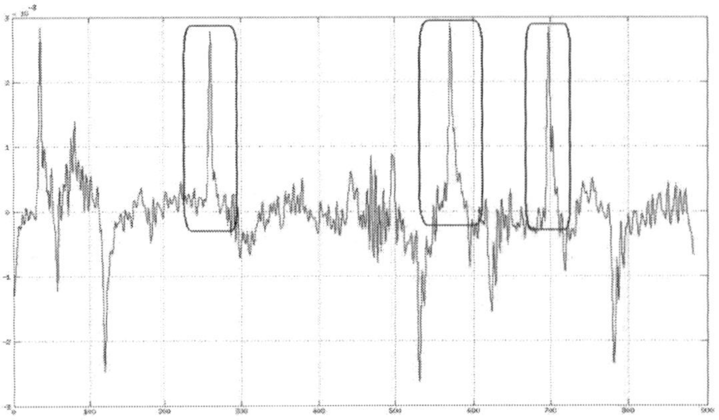

Fig. 1. Unipolar pulses of electromagnetic radiation appearing near the time of the earthquake. Electrical component. Unipolar impulses are singled out.

* Corresponding author: uvarovvn@ikir.ru

For the subsequent analysis of the mentioned phenomenon, a consistent filtering of the original data was carried out using the reference sample obtained by averaging the anomaly value (Fig 2). Stable anomaly of the electric component of the electromagnetic field, which appears near the moment of the main earthquake push. Above is the individual anomaly. Below is the anomaly is averaged by 50 individual implementations.
The Figure 3 shows the results of signal processing using as an example the earthquakes of 2031/02/28 14:00.

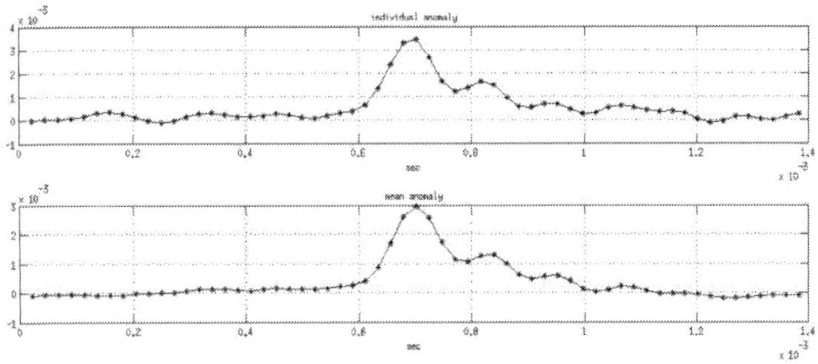

Fig.2. Initial data and processing results. 1 - initial signal. 2 - signal subjected to optimal filtration. 3 - rms value of the original signal. 4 - dispersion of original signal. The moment of the earthquake is marked by the arrow.

3. Pulse properties

Analysis of these graphs shows that
At the time of the earthquake, the activity (the number of pulses per unit time) of unipolar pulses corresponds to the background level

- The activity of unipolar pulses reaches its maximum 2-4 hours before and after the earthquake

- The activity of unipolar pulses is extremely closely correlated with the moments of maximum intensity values

- The amplitudes of different unipolar pulses are the same.

4. Model of deformation process

As is known, the process of deformation of a solid body by its conversion character has two extreme forms - an evolutionary one, at which a continuous change in the body is observed, and a catastrophic one, in which the structure of the body changes abruptly. The process of evolutionary deformation is accompanied by the birth and elimination of structures that impede the evolutionary process of deformation (stoppers) formed from stronger structural heterogeneities by piling up these heterogeneities with the formation of impassable formation. These stoppers can be eliminated in two ways: either as a result of thermal diffusion, or when disrupted by the accumulated stress. If the rate of birth of stoppers is lower than the rate of their annihilation, then evolutionary aseismic development of deformation occurs. The equality of birth and annihilation rates corresponds to the bifurcation region, where intermittency of the evolutionary and catastrophic forms is observed. At a birth rate greater than the rate of annihilation leads to the deformation process occurring with an increasing number of stoppers, an increase in the stress to the threshold of fragile breaking's of the rock, the subsequent disruption of the stoppers, relaxation of the stresses, and a transition to a new equilibrium state. In such case, the process is catastrophic and accompanied by seismic manifestations.

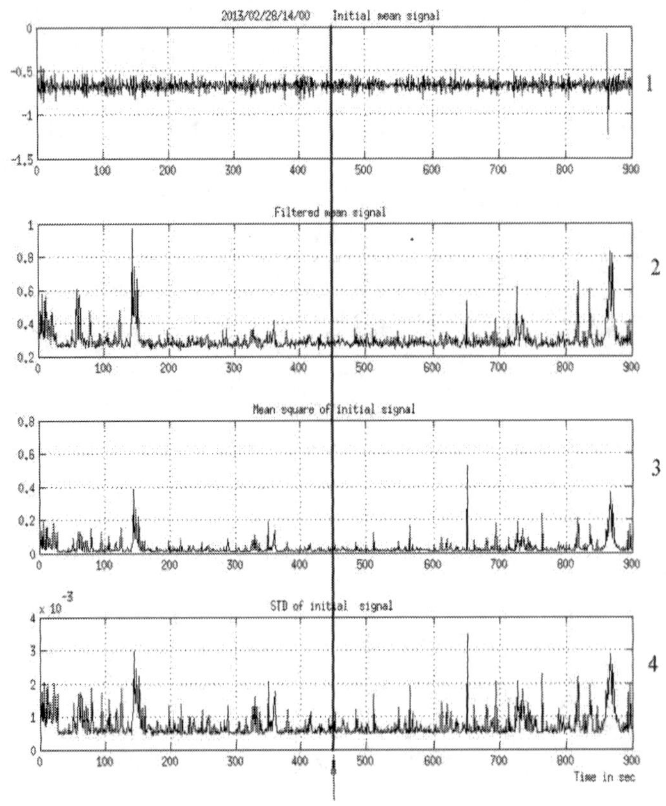

Fig.3. Initial data and processing results. 1 - the initial signal. 2 - signal subjected to optimal filtration. 3 - the rms value of the original signal. 4 - dispersion of original signal. The moment of earthquake is marked by the arrow.

The presence of deformational electromagnetic transformation mechanisms inherent in the rock leads to the existence of electromagnetic satellites of seismic/acoustic perturbations of the earth's crust [1- 3]. Since charges of different signs have different mobility and interact differently with stoppers, during the formation of a stopper electric charges of the same sign are being accumulated, and therefore charges of different signs are being separated. Simultaneously with the accumulation of charge, the environment is polarized to compensate for the excess charge, and a compensating cloud with a core of accumulated charges is being formed. With the subsequent disruption of the stopper, the cloud breaks, during which the total dipole moment increases many times. The rate of growth of the dipole moment is determined by the velocity of the substance motion during the stopper detachment. After the relaxation of mechanical stresses, relaxation of the dipole moment arises.

According to our measurements, the leading edge has a duration of $\Delta t \sim 10^{-4}$ sec, which is determined by the rate of mechanical separation of the charge and the limiting frequency of the transmission of the antenna amplifier (about 10 kHz). At the same time, it is a characteristic time of formation of polarized cracks. If to assume that the rate of formation of such objects is approximately equal to the rate of crack development (according to Griffiths ~ 0.4 speed of sound, and the sound of velocity in the rock about ~ 3000m / s), then the characteristic size of this radiating object (crack) is some $l \approx \Delta t \times 0.4c \approx 0.1$m. That is enough small size.

Trailing edge of the pulse is much longer, at about 3×10^{-4} sec. It is determined by the characteristic relaxation time of the resulting dipole moment and the electrophysical properties of the rock: $\tau = \epsilon \rho$.

6. Discussion and conclusions

The analysis of these facts leads to the following questions:

How to ensure a large amplitude of radiation by a small generating object (and, consequently, a small volume of energy accumulating environment)?

How the nearly constant amplitude of these pulses can be explained?

The only consistent explanation, in our opinion, lies in attraction of the cooperative interaction mechanism of the emitters in the active rock environment (the optical analogue is Dicke's superradiance). Conditions for the cooperative radiation arise when the critical excitation energy density is reached at the wave stress front. In this case, the characteristic volume of the radiation, and, consequently, the amplitude of the emitted pulse is determined by the coherence radius, which is quite a constant value. This radius has the size of a kilometer scale. In the immediate vicinity of the instant of the earthquake (minutes), there are no conditions for the emergence of cooperative radiation.

References

1. Uvarov V., Geophysical Journal, **38,** 6, p.180-185, (2016)

2. Gulielmi A., Physics of the Earth, 3, p. 3-16. (2006)

3. Gulielmi A., Physics of the Earth, 5, p. 65-70. (1994)

AUTHOR INDEX